BARSTOW

MODERN MICROWAVE TECHNOLOGY

MODERN MICROWAVE TECHNOLOGY

Victor F. Veley, Ph.D.

Dean, School of Science and High Technology
Los Angeles Trade-Technical College

PRENTICE-HALL, INC., Englewood Cliffs, New Jersey   07632

Library of Congress Cataloging-in-Publication Data

VELEY, VICTOR F. (*date*)
Modern microwave technology.

Includes index.
1. Microwaves.   2. Microwave devices.   I. Title.
TK7876.V45 1987   621.381'3   86–25361
ISBN 0–13–595414–2

Editorial/production supervision and
interior design: **Kathryn Pavelec**
Cover design: **20/20 Services Inc.**
Manufacturing buyer: **Carol Bystrom**

© 1987 by **Prentice-Hall, Inc.**
A Division of Simon & Schuster
Englewood Cliffs, New Jersey 07632

Printed in the United States of America

10 9 8 7 6 5 4 3 2 1

ISBN 0-13-595414-2 025

PRENTICE-HALL INTERNATIONAL (UK) LIMITED, *London*
PRENTICE-HALL OF AUSTRALIA PTY. LIMITED, *Sydney*
PRENTICE-HALL CANADA INC., *Toronto*
PRENTICE-HALL HISPANOAMERICANA, S.A., *Mexico*
PRENTICE-HALL OF INDIA PRIVATE LIMITED, *New Delhi*
PRENTICE-HALL OF JAPAN, INC., *Tokyo*
PRENTICE-HALL OF SOUTHEAST ASIA PTE. LTD., *Singapore*
EDITORA PRENTICE-HALL DO BRASIL, LTDA., *Rio de Janeiro*

To My Beautiful Wife, Joyce
My Inspiration and Help
for Over Thirty-five Years
and
To Those I Hold Dear
Jackie, Gill, Paula, Phil, Ray,
James, Nicola, Lisa, Peter, Katie, and Paul

CONTENTS

PREFACE

I have written this book primarily for those who intend to study microwave technology at community colleges and vocational trade schools. It is also designed to help technicians who are already working in the microwave field. The purpose is to enable you to *think microwave* rather than in terms of principles that apply only to low-frequency operation. A mastery of such "basics" is essential; only by achieving such a mastery can a technician hope to progress in his or her chosen field.

The book itself is the outcome of many years spent in the instruction of microwave courses as they relate to the training of technicians in the military and for employment with aerospace companies. It is a blend of classical microwave theory and modern developments such as gyrotrons, crossed-field devices, fiber optics, satellite communications, and various solid-state components. I am still teaching at the community college level in the areas of systems, technical mathematics, communications, and microwave technology. In consequence, this book deals thoroughly with the particular problems that will face you as you grapple with microwave fundamentals.

To achieve the book's goals I have used a format which assumes only a basic knowledge of direct- and alternating-current theory. The amount of mathematics has been kept to a minimum and does not exceed the level required for basic alternating current analysis.

The highlights of the book's format are:

1. Every chapter has an introduction that lists the sections into which the material is divided. Each section introduces a limited number of new ideas which are fully explored.
2. The text is written throughout in the international metric system of units.

3. The basic and advanced problems are limited in number (20 per chapter) but have been carefully chosen to avoid redundancy. It would therefore be a good idea for you to attempt all problems.

4. The writing style is "friendly" (I hope) and is aimed at involving you in the discussion of microwave principles.

5. The book illustrates a considerable amount of microwave hardware so that you can be familiar with its appearance.

I hope you will enjoy reading the book and will find your effort to have been worthwhile.

Finally, I would like to acknowledge a considerable debt to my wife, Joyce, for typing the manuscript—without her inspiration and support, the writing of this book would not have been possible. I also wish to thank my secretary, Marta E. Ruiz, for cheerfully and skillfully undertaking the correction of the manuscript.

Victor F. Veley

SURVEY
OF MICROWAVE

1–1 INTRODUCTION

Microwave may be defined as one region of the electromagnetic (EM) wave spectrum. In other words, microwave represents EM energy that can travel through a specialized transmission line such as a waveguide or be propagated through free space. However, the frequency range of the microwave region is open to debate. Some textbooks state that the range begins where the ultra-high-frequency (UHF) TV band finishes—at 890 MHz. Others contend that the start of the microwave band coincides with the beginning (300 MHz) of the UHF band. The most popular view is that microwave extends from 1 gigahertz (1 GHz = 10^9 Hz) to 1 terahertz (1 THz = 10^{12} Hz); this corresponds to a range of wavelengths from 30 cm to 0.3 mm. However, the majority of microwave equipment operates between 1 and 100 GHz, which includes the super-high-frequency (SHF) band of 3 to 30 GHz. Before covering microwave techniques in depth we will briefly summarize some of the features that distinguish microwave from the lower radio-frequency (RF) bands.

Tuned circuits. Knowing that the resonant frequency

$$f_r = \frac{1}{2\pi \sqrt{LC}}$$

we can discover that the values of L and C required for a resonant frequency of 10 GHz are much less than 1 μH and 1 pF, respectively. With these small values it is impossible to use "lumped" circuits since the leads connecting the inductor and the capacitor would probably have much more inductance and capacitance than the total values required. We therefore use distributed circuits which are in the form of resonant cavities (Section 1–3).

Insulators.   All insulators suffer to some degree from dielectric hysteresis loss, which increases as the frequency is raised. In the microwave region this type of loss is too severe to permit us to use such insulators as porcelain, Lucite, Bakelite, waxed paper, and so on. We are therefore required to introduce special insulators, examples of which are polyethylene, polystyrene, and Teflon.

Microwave transmission lines.   At microwave frequencies it is impossible to use a conventional coaxial cable, except for very short lengths. The principal reason for this is the severe dielectric loss, which results in a high degree of attenuation. The main type of practical microwave line is the waveguide, which is basically a hollow metal pipe with a rectangular, circular, or elliptical cross section.

Active devices.   In our discussion on the lighthouse triode (Section 1–5) we will find that conventional tubes are limited to below 2 GHz. As a result of the transit-time effect at higher frequencies, we must employ active devices, which actually use transit time to achieve amplification or oscillation. Such tubes include magnetrons, klystrons, traveling-wave tubes (TWTs), backward-wave oscillators (BWOs), cross-field amplifiers (CFAs), and gyrotrons. There are also solid-state devices which can provide amplification, frequency multiplication, and oscillation.

Our introduction to microwave techniques covers the following topics:

1–2. Electromagnetic Waves in Free Space
1–3. Resonant Cavities
1–4. Comparison of Transmission Lines and Waveguides
1–5. Transit-Time Effect
1–6. Microwave Insulators
1–7. Propagation at Microwave Frequencies
1–8. Advantages of the Microwave Region

1–2 ELECTROMAGNETIC WAVES IN FREE SPACE

An electromagnetic (EM) wave is radiated into space from an antenna system. Such a radiated wave consists of electric and magnetic fields which are in time phase although the two sets of flux lines are 90° apart in space. Consequently, if an EM wave contains electric flux lines which are vertical, the corresponding magnetic flux lines will be horizontal. Both of these sets of flux lines are also at right angles to the direction in which the electromagnetic energy is traveling. In other words, the electric and magnetic fields are transverse to the direction of propagation. Such a

REVIEW
ANTIENNA
BASIC.

wave is of the TEM (transverse electric, transverse magnetic) type, which occurs only in free space and on a conventional transmission line (*not* in a waveguide).

Figure 1–1 illustrates the various features of a TEM wave. The peak conditions of the E and H fields in the path of the wave are shown in Figure 1–1a. Positions X and Z represent adjacent identical conditions and are therefore separated by a distance of one wavelength. For the instantaneous fields at position X to travel to position Z would require a time equal to one period (Figure 1–1c). It follows that the wave travels a distance of one wavelength λ in a time equal to one period T. Since velocity = distance/time, the velocity of the electromagnetic wave in free space,

$$c = \frac{\lambda}{T} = \lambda \times \frac{1}{T} = f\lambda \qquad (1\text{--}1)$$

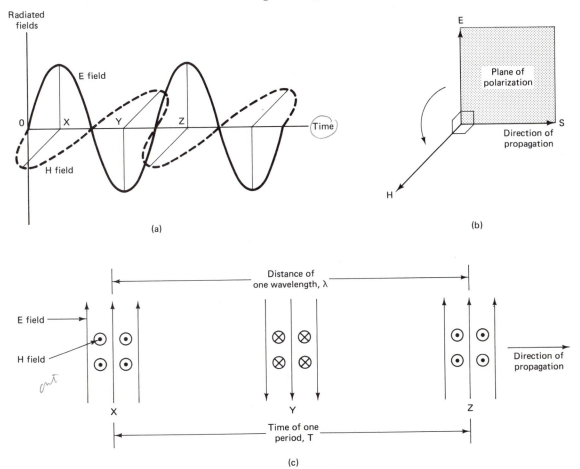

Figure 1–1   (a) Peak conditions for the electric and magnetic fields of a vertically polarized TEM wave; (b) vector diagram of a vertically polarized wave; (c) electric and magnetic fields in the TEM wave.

With the SI system of units, the permeability of free space is $\mu_0 = 4\pi \times 10^{-7}$ H/m, and the permittivity of free space is $\epsilon_0 = 8.85 \times 10^{-12}$ F/m. In 1865, Clerk Maxwell predicted that the velocity of *all* electromagnetic waves in free space would be given by

$$c = \frac{1}{\sqrt{\mu_0 \epsilon_0}} \approx 3 \times 10^8 \text{ m/s} \qquad (1\text{--}2)$$

The value of c is equal to the velocity of light, which is another example of an electromagnetic wave. If the wave is traveling through a medium whose relative permittivity is ϵ_r, the velocity is reduced and equal to $c/\sqrt{\epsilon_r}$.

For a particular medium, the velocity of all electromagnetic waves is a constant. Since this velocity is equal to the product of the frequency and the wavelength, either a wavelength or a frequency scale may be used to distinguish one signal from another.

Frequency Bands

At the Atlantic City Conference of 1947, the following frequency bands were adopted:

Very low frequency (VLF)	3–30 kHz
Low frequency (LF)	30–300 kHz
Medium frequency (MF)	300 kHz–3 MHz
High frequency (HF)	3–30 MHz
Very high frequency (VHF)	30–300 MHz
Ultra high frequency (UHF)	300 MHz–3 GHz
Super high frequency (SHF)	3–30 GHz
Extremely high frequency (EHF)	30–300 GHz

Since the microwave region extends from 1 GHz to 1 THz, it encompasses part of the UHF band and the whole of the SHF and EHF bands. At these frequencies the most convenient unit for the wavelength is the centimeter. The corresponding equation is

$$\lambda = \frac{30}{f} \qquad f = \frac{30}{\lambda} \qquad (1\text{--}3)$$

where

λ = wavelength (cm)
f = frequency (GHz)

The bands shown in Table 1–1 are designated within the microwave region. As an example, the S band is used for radar surface detection systems; at the frequency 3 GHz within this band, the wavelength is 30/3 = 10 cm. By contrast, navigational radar sets generally employ the X band, in which the frequency of 10 GHz corresponds

TABLE 1–1 BANDS IN THE
MICROWAVE
REGION

Letter designation	Frequency range (GHz)
L	1.12–1.7
S	2.6–3.95
C	3.95–5.85
X[a]	8.2–12.4
K	18.0–26.5

[a] The X band is not connected in
any way with x-rays.

to a wavelength of 30/10 = 3 cm. At the upper end of the microwave region the frequency 1 THz (= 1000 GHz) has a wavelength of 30/1000 = 0.03 cm = 0.3 mm.

At the frequency 30 GHz the corresponding wavelength is 1 cm or 10 mm, while at 300 GHz the wavelength is 1 mm. Consequently, the frequency range 30 to 300 GHz is often referred to as the "millimeter waveband."

Beyond microwave lies the infrared region (heat and black light), which extends approximately from 1 to 375 THz. At these frequencies we introduce more convenient smaller units with which to measure the wavelength. These are the micron or micrometer (μm), which is one millionth of a meter, and the angstrom unit (Å), where 1 Å = 10^{-8} cm = 10^{-10} m = 10^{-4} μm. Equation (1–3) is then modified to become

$$\lambda = \frac{300}{f} \qquad f = \frac{300}{\lambda} \qquad (1\text{–}4)$$

where

λ = wavelength (μm)
f = frequency (THz)

Therefore, at the frequency 375 THz the wavelength is 300/375 = 0.8 μm = 8000 Å.

Visible light occupies only a small part of the electromagnetic wave spectrum. From the red end to the violet end the frequency range is only 375 THz (wavelength = 0.8 μm = 8000 Å) to 790 THz (wavelength = 0.38 μm = 3800 Å). As we progress through still higher frequencies (and shorter wavelengths), we pass through the regions of ultraviolet radiation, x-rays, gamma (γ) rays, and finally cosmic rays. The frequencies and wavelengths of these waves are shown in Table 1–2.

Plane of Polarization

Mathematically, the electric field (whose field intensity $\mathscr{E}$ is measured in volts per meter), the magnetic field (whose field intensity, H, is measured in amperes per meter), and the transfer of EM energy in a particular direction represent a right-

TABLE 1-2. ELECTROMAGNETIC WAVE SPECTRUM

Region	Approximate wavelength limits		Approximate frequency limits	
	Maximum	Minimum	Minimum	Maximum
Radio (including microwave)	100,000 m	0.3 mm	3 kHz	1000 GHz
Infrared	0.3 mm	0.8 μm	1000 GHz	375 THz
Visible light	0.8 μm	0.38 μm	375 THz	790 THz
Ultraviolet	0.38 μm	120 Å	790 THz	25,000 THz
X-rays	120 Å	0.067 Å	25,000 THz	45,000,000 THz
Gamma rays	0.067 Å	0.011 Å	45,000,000 THz	270,000,000 THz
Cosmic rays	0.011 Å	Not defined	270,000,000 THZ	Not defined

handed system of vectors. This transfer of EM energy is measured by the Poynting vector S, which represents the amount of radiated energy passing through a unit area (1 square meter) in unit time (1 second). E, H, and S (in that order) form a right-handed system with S as the vector product of E and H; in terms of units, volts/meter (E) $\times$ amperes/meter (H) = volts $\times$ amperes/(meter)^2 = watts/(meter)^2 = joules per square meter per second (S). If the instantaneous E direction is rotated through 90° to lie along the instantaneous H direction, the direction of S is the same as the movement of a right-handed screw which is subjected to the same rotation.

The plane containing E and S is referred to as the plane of polarization (Figure 1–1b). A vertical antenna system will radiate a vertically-polarized wave with a vertical E field and a horizontal H field. Similarly, a horizontally-polarized wave has a horizontal E field and a vertical H field and is associated with a horizontal antenna system. As practical examples, AM broadcast systems have antennas that radiate vertically-polarized waves, while TV broadcast stations use horizontal polarization.

The ratio of $\mathscr{E}$ (volts per meter) to H (amperes per meter) must have the units of ohms. Clerk Maxwell showed that for the EM wave in free space

$$\frac{\mathscr{E}}{H} = \eta_0 \tag{1-5}$$

where

η_0 is the intrinsic impedance of free space (Ω).

From equation (1–2),

$$c = \frac{1}{\sqrt{\mu_0 \epsilon_0}} = 3 \times 10^8 \text{ m/s}$$

or

$$\frac{1}{\sqrt{\epsilon_0}} = \sqrt{\mu_0} \times 3 \times 10^8$$

Combining equations (1-2) and (1-5) gives us

$$\text{intrinsic impedance } \eta_0 = \sqrt{\frac{\mu_0}{\epsilon_0}} = \sqrt{\mu_0} \times \sqrt{\mu_0} \times 3 \times 10^8$$
$$= 4\pi \times 10^{-7} \times 3 \times 10^8$$
$$= 120\pi \approx 377 \ \Omega$$

Summarizing, a radiated electromagnetic wave consists of electric and magnetic fields which are in time phase. The two sets of flux lines are 90° apart in space and each set is at right angles to the direction of propagation. The plane of polarization is defined by the directions of the E field and the Poynting vector S, so that, for example, a vertical antenna system radiates a vertically polarized EM wave in which the electric field is vertical and the magnetic field is horizontal. In free space the ratio of the electric field intensity to the magnetic field intensity is the intrinsic impedence $\eta_0 \approx 377 \ \Omega$ (Section 4-7).

Example 1-1   What is the wavelength in free space of a microwave signal whose frequency is 5.2 GHz?

Solution   The wavelength in free space

$$\lambda = \frac{30}{5.2} = \textbf{5.77 cm} \tag{1-3}$$

Example 1-2   What is the frequency of an infrared signal whose wavelength in free space is 23 μm?

Solution   The frequency

$$f = \frac{300}{23} = \textbf{13.04 THz} \tag{1-4}$$

1-3 RESONANT CAVITIES

In the microwave region it is impossible to employ conventional LC circuits for tuning purposes. For example, at a resonant frequency f_r of 10 GHz, we might attempt to use a capacitor whose lumped capacitance C is 1 pF (this would be about the smallest practical value). To achieve resonance, the corresponding lumped inductance L would be given by

$$L = \frac{0.0253}{f_r^2 C}$$
$$= \frac{0.0253}{(10 \times 10^9)^2 \times 1 \times 10^{-12}}$$
$$= 2.53 \times 10^{-10} \text{ H}$$

Such a low inductance would require a single turn with a diameter of less than 1 mm. Furthermore, the copper wire involved would be less than 1 mm thick and would suffer severely from losses due to skin effect. Clearly, this is not a practical solution, especially since the connecting wires to the coil would probably possess greater inductance.

Since lumped circuitry is impossible at microwave frequencies, we are forced to use some form of "distributed" circuitry. In Section 2–4 we will discover that a quarter-wave section of transmission line (shorted at one end) is electrically equivalent to a parallel LC tank circuit (Figure 1–2a). At the X-band frequency of 10 GHz the wavelength is 3 cm, and therefore the quarter-wave line would appear to be 0.75 cm long before being terminated by a short circuit. In practice, the phase velocity on the line is lower than the free-space velocity, so that the actual length is somewhat less. Irrespective of length considerations, the skin effect on such a line is severe and its Q is correspondingly low.

However, if a second identical quarter-wave line were paralleled with the first at the open end (Figure 1–2b), the total inductance would theoretically be halved while the total capacitance C_T would be doubled. This means that resonant frequency of the parallel combination $(= 1/2\pi \sqrt{L_T C_T})$ is unchanged while the Q $(= R \times \sqrt{C_T/L_T})$ is increased. It follows that more and more parallel quarter-wave lines can be added until we finally create a solid cylindrical "can" (Figure 1–2c). Such a structure has no radiation loss and its RF resistance is extremely low; the result is a high Q value of several thousand. The larger the diameter of the can, the lower is its resonant frequency.

If the cylindrical cavity is excited, the simplest pattern of the magnetic flux is a series of concentric circles with zero intensity at the center and the highest intensities at the curved wall and the ends of the structure. By contrast, the lines of the electric flux are all positioned transverse and parallel to the Z axis of the cylinder, with the maximum intensity at the center and zero at the curved wall. Such a distribution of flux lines is shown in Figure 1–3a and is referred to as the TM_{010} mode. The designation "TM" means that all the magnetic flux lines are entirely transverse (at right angles) to the direction of the cylinder's Z axis; if all the electric flux lines were transverse, the designation would be "TE."

The first subscript (zero) denotes the number of full-wave changes for the electric field around the circumference. The second subscript (one) is the number of half-wave changes for the electric field across the diameter. The third subscript (zero) is the number of half-wave changes along the Z axis. Two other possible field distributions (modes) are illustrated in Figure 1–3b and c and you should verify their designations.

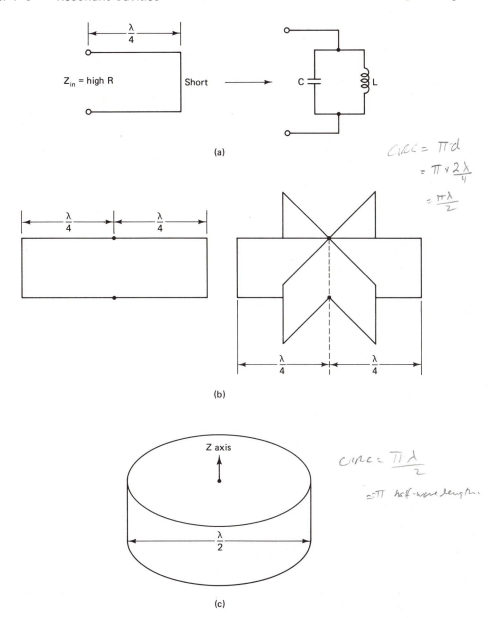

Figure 1–2 Cylindrical "can" as a resonant cavity.

The resonant can or cavity may be mechanically tuned to resonance by a number of methods:

1. The top of the can may be made in the form of an adjustable plunger, which can then alter the operating dimensions of the cavity (Figure 1–4a) and therefore

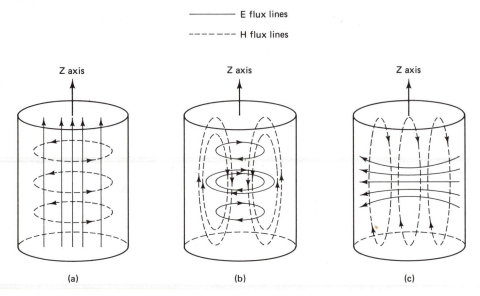

Figure 1–3 Examples of cylindrical cavity modes: (a) TM_{010} mode; (b) TE_{011} mode; (c) TE_{111} mode.

change its resonant frequency; for example, if the distance D is reduced, the cavity is smaller and its resonant frequency is higher. This is the principle behind a common type of absorption microwave frequency meter which is attached directly to the waveguide. A radiating slot or hole is then cut so that, provided that the cavity is at or near resonance, a small amount of RF energy can be absorbed from the waveguide into the cavity. This absorption can be monitored by coupling a detector to the cavity.

One method of coupling is to insert a loop which links with the cavity's magnetic field; the pickup from the loop can then be detected, amplified, and fed to a dc indicator (Figure 1–4b). The plunger is adjusted for a maximum reading in the dc indicator and the frequency can then be read directly off a calibrated scale.

2. To tune some klystrons a metallic screw is inserted into the curved wall of the cavity (Figure 1–4c). Since the screw is not ferromagnetic, the cavity's distributed inductance is lowered and the resonant frequency of the cavity is raised. Alternatively, a paddle can be used in place of the screw. The resonant frequency can be increased by tuning the paddle to a position that is more nearly perpendicular to the direction of the H flux lines.

3. The capacitance at the center of the cavity can be varied by means of a small plate whose position is controlled externally (Figure 1–4d). If the plate is screwed in, the capacitance is increased and the resonant frequency is reduced.

The cylinder is only one of a variety of possible cavity shapes. Instead of a shorted quarter-wave twin lead, we can start with a curved piece of wire (fraction

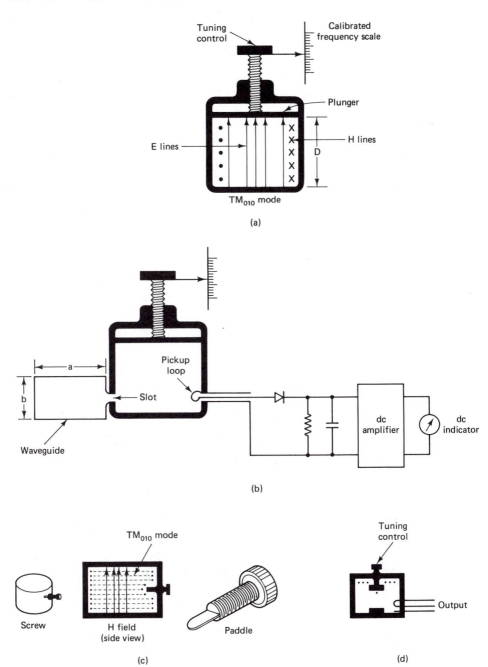

Figure 1–4   Methods of tuning the cylindrical cavity.

of a turn) that will mainly represent the required inductance. The ends of the wire may then run geometrically in parallel to provide most of the capacitance (Figure 1–5a). This shape may then be revolved in a circle to form a surface of revolution; the result is the rhumbatron or reentrant cavity whose electric and magnetic fields are indicated in Figure 1–5b. Figures 1–5c and d illustrate other versions of the reentrant cavity which is commonly used in reflex klystrons.

To tune a reentrant cavity, one or more of the surfaces is made flexible so that the cavity may be mechanically stressed to alter its operating dimensions. An example of this method is shown in Figure 1–5e and is the principle behind the tuning of some reflex klystrons. As the tuner screw is rotated, the distance between the cavity's top and bottom walls is changed; this alters the distributed capacitance

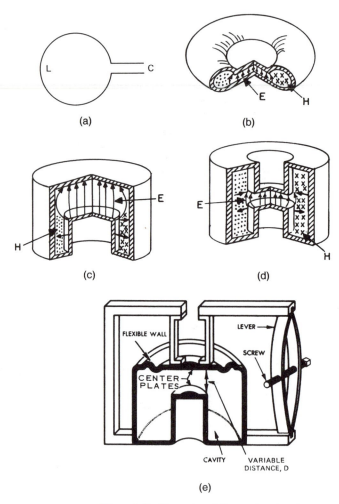

Figure 1–5 Examples of the reentrant cavity.

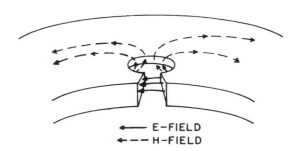

Figure 1–6   "Hole and slot" cavity.

and varies the resonant frequency. For example, if the distance, D, is increased, the distributed capacitance is decreased and the resonant frequency is raised. By this means the frequency of an X-band cavity can be changed by up to 4 GHz. The energy from the cavity may then be passed to a waveguide by means of a loop or a radiating slot.

If the shape of Figure 1–5a is extended into the paper, the result is a "hole and slot" cavity (Figure 1–6) which is cut from a block of solid copper and is commonly used in magnetrons. Another possible cavity shape is the rectangular box whose cross section consists of a narrow dimension b and a wide dimension a which is longer than half a wavelength. This cross section is then extended a distance of half a wavelength along the Z axis to form a closed box (Figure 1–7a). The resulting distributions of the electric and magnetic fields are shown in Figures 1–7b and c. The mode shown is designated as TE_{101}, where the first suffix indicates the number of half-wave variations along the wide dimension a, the second suffix denotes the number of half-wave variations across the narrow dimension b, and the third suffix is the number of half-wave variations in the direction of the Z axis. Such a cavity may be used as an echo box to test the performance of a radar set.

In the various applications of cavity resonators it is necessary both to excite a cavity by introducing energy and to have ways of removing the energy. The most common methods of coupling are the E probe, the H loop, and the slot or aperture.

Figure 1–8a shows the E probe, which represents capacitive coupling and is introduced into the cavity at the position where the electric field is the strongest; this probe can then set up E lines and maintain continuous oscillation. The H loop (Figure 1–8b) is a form of inductive coupling and is positioned where the magnetic field is greatest. The third method is to introduce or remove energy by means of an aperture or slot which is common to the waveguide and the resonator (Figure 1–8c).

In a reentrant cavity (Figure 1–8d) energy may be introduced by a stream of electrons that pass through holes in the center of perforated plates or flat mesh grids. The energy can then be removed by an H loop. This method of exciting a cavity into oscillation is used in the klystron tube.

We have learned that cavity resonators are the microwave equivalent of tuned circuits and can have a variety of shapes, such as the rectangular box, the cylinder,

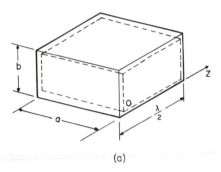

(a)

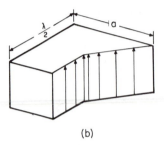 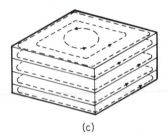

(b)   (c)

Figure 1–7   Rectangular box cavity (TE$_{101}$ mode): (a) closed box; (b) electric field pattern; (c) magnetic field pattern.

the hole and slot, and the rhumbatron; tuning is achieved by altering the dimensions of the cavity. For each shape there are a number of modes which represent the possible distributions of the electric and magnetic fields. Energy may be introduced into or removed from a cavity by such coupling methods as the E probe, the H loop, and the slot.

Example 1–3   Calculate the approximate dimension of a cylindrical cavity that is resonant at 3.8 GHz.

Solution   The diameter is approximately equal to the distance of one half-wavelength in free space:

$$\text{diameter} \approx \frac{1}{2} \times \frac{30}{3.8} \approx \textbf{4.0 cm}$$

1–4 COMPARISON OF TRANSMISSION LINES AND WAVEGUIDES

A twin-lead transmission line (Figures 1–9a and b) may be operated successfully at frequencies well below the microwave region. However, such a line suffers from the following losses:

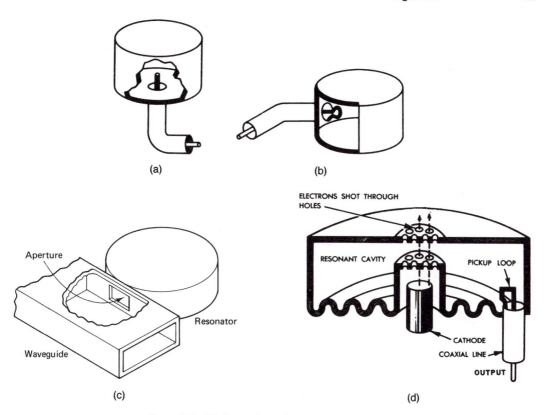

(a)   (b)

(c)   (d)

Figure 1-8 Methods of coupling to and from a resonant cavity.

Copper loss.   This loss is associated with the resistance of the conductors, which will dissipate energy in the form of heat. At microwave frequencies the copper loss is primarily the result of the "skin effect," which restricts the flow of the current to a very small layer at the conductor's surface. The greater the surface area of the conductor, the less are its skin effect and consequent copper loss. A twin-lead line uses a pair of conductors with a small surface area, so that the copper loss is severe. By contrast, the waveguide is a hollow metal pipe with a rectangular, or circular, or elliptical cross section (Figure 1-10). In either case the surface area involved is much larger, so that the copper loss is limited to a low value. In addition, the loss may be further reduced by plating the inside surface of the guide with gold or silver.

Dielectric loss.   This loss is related to the type of insulator between the conductors. All insulating materials suffer to some degree from dielectric hysteresis loss, which causes the insulator to dissipate energy in the form of heat. This effect increases with frequency, and in the microwave region the hysteresis loss for most insulators is extremely severe. However, a waveguide uses a dielectric of dry air or

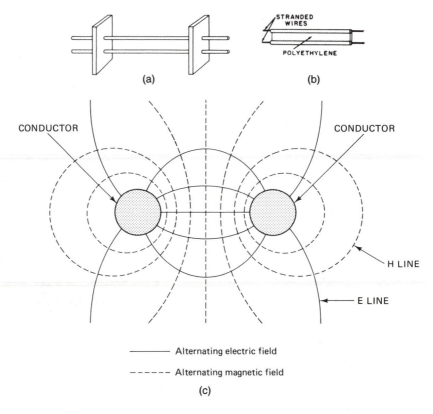

Figure 1–9   Twin lead transmission line: (a) two-wire (air dielectric); (b) two-wire ribbon type; (c) electric and magnetic fields surrounding a two-wire line.

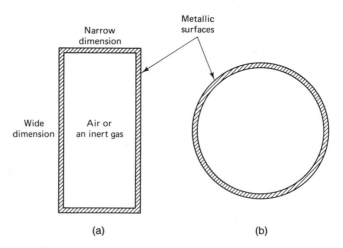

Figure 1–10   (a) Rectangular and (b) circular waveguides.

an inert gas such as nitrogen; for these dielectrics the hysteresis loss is relatively low.

Radiation loss.   The twin-lead line does not confine the electric and magnetic fields in a direction perpendicular to the plane that contains the conductors (Figure 1–9c). The electric and magnetic fields are associated, respectively, with the voltage and the current on the line, and when the separation between the conductors becomes an appreciable fraction of a wavelength, energy will escape from the line by radiation. The higher the frequency, the shorter the wavelength and the greater the radiation loss. By contrast, a waveguide represents a closed metallic surface; the electric and magnetic fields are therefore confined and the radiation loss is negligible.

The coaxial cable (Figure 1–11a) is an improvement over the two-wire line. Since the outer conductor is extended around the inner conductor, the electric and magnetic fields are contained so that there is virtually no radiation loss. The copper loss is reduced because of the outer conductor's larger surface area. However, there remains the dielectric loss, which cannot be eliminated because even if an air dielectric were used, the inner conductor would have to be supported at intervals by insulating spacers. Dielectric hysteresis is the ultimate reason why we must finally turn to waveguides and abandon coaxial cables. At 3 GHz the typical attenuation for a coaxial cable is 0.6 dB/m; with a waveguide, it is only 0.02 dB/m. To look at this situation in practical terms, a 5-m length of coaxial cable would have an attenuation of 5 × 0.6 = 3 dB, which is equivalent to a half-power loss. In other words, if 100 W of RF power were introduced into a 5-m length of coaxial cable, only 50 W would emerge. By contrast, the waveguide would have to be 3/0.02 = 150 m long to produce a half-power loss. At 10 GHz the figures are even more dramatic: 3 dB/m for the coaxial cable but only 0.03 dB/m for the waveguide.

There is a further advantage for the waveguide in terms of power capability. With a coaxial cable the power capability is determined primarily by the spacing between the conductors. If the power being conveyed is excessive, large voltages are established between the conductors and arcing may occur across the dielectric. A circular waveguide can be regarded as a coaxial cable with the inner conductor removed. Consequently, the spacing and the power capability of a circular waveguide are greater than those of a coaxial cable with the same external dimensions (Figure 1–11b).

By now you must be thinking: "Waveguides have all the advantages with regards to attenuation and power capability, so why ever bother with coaxial cables?" The answer, in one word, is "size." We will show in Section 3–2 that in order for an electromagnetic wave to propagate successfully down a rectangular guide, the wide dimension *a* (Figure 1–10a) must exceed one half-wavelength, corresponding to the frequency of the wave. Consequently, at a frequency of 100 MHz, the wavelength is 3 m and therefore *a* must exceed 1.5 m or roughly 5 feet! Such a waveguide would be extremely cumbersome and expensive.

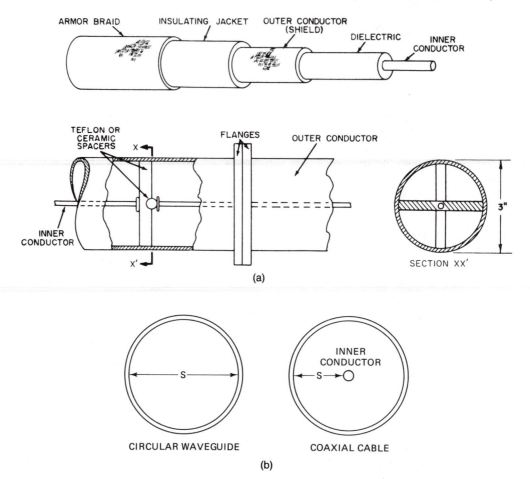

Figure 1–11   (a) Coaxial cable; (b) comparison between the circular waveguide and the coaxial cable. S, Spacing.

If the wide dimension must be greater than a half-wavelength long, it follows that for a particular waveguide there must be a *lower* frequency limit, which is known as the cutoff value, f_c; at this frequency the wide dimension is exactly one half-wavelength long. For example, a WR 284 S-band rectangular guide that operates over 2.6 to 3.95 GHz has an inner wide dimension of 7.214 cm, which corresponds to a cutoff frequency of $30/(2 \times 7.214) = 2.078$ GHz. In the X band of 8.2 to 12.4 GHz, the WR 90 waveguide's inner wide dimension is 2.286 cm and the cutoff frequency is $15/2.286 = 6.557$ GHz (Figure 1–12). For reasons which we shall later see (Section 3–4), the operating frequency range of a rectangular guide is limited between $1.25f_c$ and $1.9f_c$. By contrast, a coaxial cable may be used from dc up to and including the lower part of the microwave region.

Apart from considerations of size and frequency range, a waveguide system is expensive and mechanically rigid. The ideal situation would be one continuous section

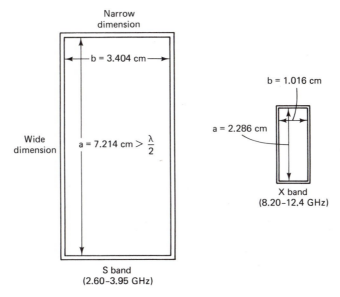

Figure 1–12   Comparison between the sizes of rectangular waveguides.

of waveguide between the transmitting and receiving points. However, this is normally not practical, and it is necessary to join various sections together to form a complete guide. At the joins, care must be taken to prevent discontinuities inside the waveguide and to avoid any leakage of the RF energy.

Summarizing, the waveguide has less attenuation than conventional transmission lines such as the twin wire and the coaxial cable; in addition, the waveguide has greater power capability. However, the disadvantages of size, limited frequency range, lack of flexibility, and expense make a waveguide system impractical at all but micro-wave frequencies.

Example 1–4   A K-band rectangular guide has an inner wide dimension of 1.07 cm. What are its (a) cutoff frequency, and (b) approximate operating frequency range?

Solution   (a) The cutoff frequency

$$f_c = \frac{30}{2 \times 1.07} \approx \textbf{14 GHz}$$

(b) The operating frequency range lies approximately between $1.25f_c =$ **17.5 GHz** and $1.9f_c =$ **26.6 GHz**. This virtually covers the K-band range of 18.0 to 26.5 GHz.

1–5 TRANSIT-TIME EFFECT

The ultimate limitation on conventional tubes, such as the triode, is the transit time taken by an electron to cross from cathode to plate. For example, if an electron is accelerated from rest through a distance d (m) by a potential difference of V volts, the electric field intensity is

$$\mathcal{E} = \frac{V}{d} \text{ V/m}$$

and the force exerted on the electron,

$$F = Q\,\mathcal{E} = ma \tag{1-6}$$

This yields

$$a = \frac{QV}{md} \tag{1-7}$$

where

F = accelerating force (N)
a = electron acceleration (m/s²)
V = accelerating voltage (V)
Q = electron charge (1.602×10^{-19} C)
m = electron mass (9.1096×10^{-31} kg)
d = distance through which the electron is accelerated (m)

The electron's terminal velocity v at the end of the distance d is given by

$$v^2 = 2ad = 2 \times \frac{QV}{md} \times d = \frac{2Q}{m} \times V \tag{1-8}$$

This leads to

$$v = \sqrt{\frac{2Q}{m}} \times \sqrt{V} \tag{1-9}$$

Substitution for

$$Q = 1.602 \times 10^{-19} \text{C}$$

$$m = 9.1096 \times 10^{-31} \text{ kg}$$

yields

$$v = 5.933 \times 10^5 \times \sqrt{V} \text{ m/s} \tag{1-10}$$

Notice that the terminal velocity depends on the accelerating voltage but not on the distance through which the acceleration occurs. If the voltage is 100 V, the terminal velocity is 5.933×10^6 m/s (approximately 2% of the velocity of light) and the average velocity is $5.933 \times 10^6/2 = 2.9665 \times 10^6$ m/s. The time taken for an electron to cross a distance of 1 mm is

$$\frac{10^{-3}}{2.9965 \times 10^6} = \frac{10^{-9}}{2.9665} \text{ s}$$

which is the period corresponding to a frequency of about 3 GHz. Consequently, conventional UHF tubes such as lighthouse triodes are limited to about 2 GHz and below.

Lighthouse Triodes

As the frequency to be amplified increases, the voltage gain that can be achieved through the use of conventional electron tubes decreases until the frequency is reached where such a gain is unity. The reduction of gain is caused by dielectric losses, finite values of lead inductance, interelectrode capacitances, and the effect of transit time.

Dielectric losses have been decreased by tube designs which confine the dielectric material (bases, insulators, envelopes) to positions of the tube where the dielectric stresses are minimized, and by the use of dielectric materials with the lowest possible hysteresis loss.

As the operating frequency is increased, the inductances and capacitances inherent in the tube structure become an increasing portion of the tuned circuit of an amplifier stage. This continues until for all practical purposes, the external tuned circuit disappears or is obliterated by the tube, and no further tuning is possible at this limiting frequency. In addition, as the operating frequency is increased, even the reactances of relatively short leads within the tube become great enough to decrease the size of the actual input signal applied to the tube's electrode.

The transit-time effect is reduced by decreasing the interelectrode spacing in the tube and by increasing the plate voltage. However, decreasing the interelectrode spacing increases the interelectrode capacitances; therefore, electrode dimensions must be decreased in order to maintain a low capacitance. Reducing the size of the electrodes reduces the heat dissipation capabilities of the tube. However, the use of heat sinks will allow the operation of relatively small electrode triode tubes at dissipation levels that permit an appreciable output signal.

The effect of lead inductance (and also skin effect) must be minimized in tubes designed for UHF operation. This is accomplished by the use of large-diameter leads, multiple leads, and planar element construction (the arrangement of cathode, grid, and plate in parallel planes), which allows connections to external circuitry to be made around the periphery of contact disks or over the entire surface of cylinders. Figure 1–13a illustrates a lighthouse triode utilizing planar construction. Figure 1–13b shows the same tube with construction details. The name "lighthouse" is derived from the tube's physical appearance.

In a typical lighthouse triode the separation between cathode and plate is of the order of 1 mm. Its configuration allows the tube to be inserted into tunable coaxial lines which form a part of the external circuitry (Figure 1–13c); the equivalent lumped circuit (Figure 1–13d) is a parallel-fed tuned output–tuned input oscillator which is sometimes found in the microwave system used for radar speed traps. All RF connections are made to the sleeve, disk, and cap that connect to or support the cathode, grid, and plate, respectively. This allows low-inductance leads, which when combined with the close spacing and low interelectrode capacitance make the tube useful at UHF up to approximately 2 GHz. The plate is relatively small, and connection to the heat sink is required to take advantage of the rated plate dissipation.

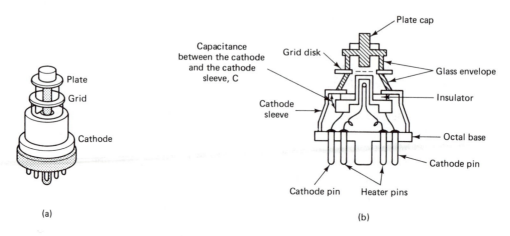

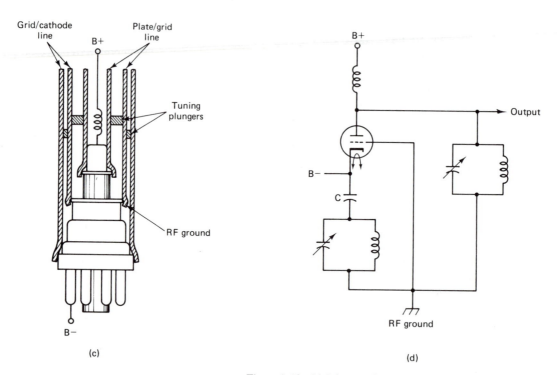

Figure 1–13 Lighthouse triode.

Example 1–5 An electron is accelerated from rest by a potential distance of 10 kV which exists over a distance of 3 cm. Calculate the terminal velocity and compare this velocity with the velocity of light in free space.

Solution The terminal velocity

$$v = 5.933 \times 10^5 \times \sqrt{10{,}000V} = \mathbf{5.933 \times 10^7 \ m/s} \tag{1–10}$$

This is approximately one-fifth of the velocity of EM waves in free space.

1-6 MICROWAVE INSULATORS

If we reverse the polarity of the source voltage applied to an insulator, the electric field intensity $\mathscr{E}$ will also be reversed and this should cause a reversal of the electric flux density D. However, if the polarity of the source voltage is reversed very rapidly, the flux density cannot alter as quickly and will lag behind the change in the electric field intensity. This effect, called dielectric hysteresis, is the electrostatic equivalent of magnetic hysteresis. Dielectric hysteresis causes an energy loss which is particularly severe in microwave circuits, where we must use special insulators such as polyethylene, polystyrene, and Teflon, whose relative permittivities are 2.26, 2.56, and 2.1, respectively. The following is a list of some of the insulators associated with the various frequency bands:

SHF (microwave)	Polyethylene, polystyrene, Teflon, and certain ceramics such as synthetic sapphire and beryllia
UHF	Corning glass and mica
VHF	Pyrex and Micalex
HF	Porcelain, fiber, wax, nylon, and hard rubber
MF	Lucite, slate, Bakelite, and certain oils
LF	Soft rubber
VLF	Plastic insulating tape

1-7 PROPAGATION AT MICROWAVE FREQUENCIES

The word *propagation* is derived from the Latin verb *propagare*, "to travel," and describes the various ways by which a radio wave "travels" from the transmitting antenna to the receiving antenna. Let us start by considering a long half-wave antenna which is fed at one end by a coaxial cable with its outer conductor grounded. The three-dimensional radiation pattern of the wave resembles a large doughnut which is laying on the ground with the antenna as its center (Figure 1–14). Part of the radiated wave moves downward and outward in contact with the ground and is affected by conditions at and below the surface of the earth. This component of the radiation is referred to as the *ground* or *surface wave*.

Higher up the antenna, the radiation is little affected by ground conditions. This portion, called the *space* or *direct wave*, travels in a practically straight line from the transmitting antenna to the receiving antenna (direct or straight-line propagation). Operating in conjunction with the space wave is the ground reflected wave (Figure 1–15), so that the signal may arrive at the receiving antenna through two different paths.

Toward the top of the antenna the radiation moves outward and upward to

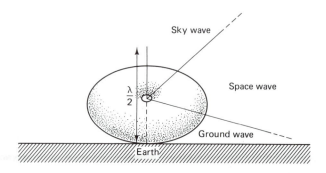

Figure 1–14   Radiation pattern of the half-wavelength antenna.

the ionosphere, which is an ionized layer of gas (primarily hydrogen) which extends from about 35 to 250 miles above the earth's surface. This portion of the radiation, referred to as the *sky* or *indirect wave*, is refracted by the ionosphere back to Earth.

The ground, space, and sky waves are illustrated in Figure 1–16. The ground wave is used primarily for long-distance communication by high-power transmitters at relatively low frequencies; as an example, the signals from commercial AM stations are carried by the ground wave. The space wave operates primarily at very high frequencies for both short-distance and long-distance communications; practical examples are TV and FM broadcasts as well as microwave links and communication systems using satellites. The sky wave, which travels back from the ionosphere, is used for long-distance fixed-service communications and operates in the shortwave band.

Space Waves

As we have already stated, the space wave is the radiation component which travels in a practically straight line from the transmitting antenna to the receiving antenna. This method of propagation is the one most commonly used in modern communications; everyday examples are the various VHF broadcast systems and radar sets which are operated in the microwave bands. The frequencies employed in these systems

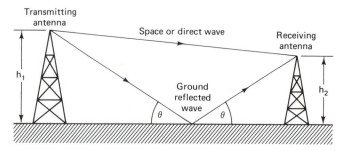

Figure 1–15   Propagation of the space wave and the ground reflected wave.

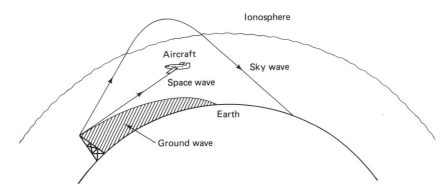

Figure 1–16   Ground, space, and sky waves.

extend from the VHF band upward, and at these high frequencies, the ground-wave range is negligible. Moreover, the radio waves in this part of the EM spectrum cannot *normally* be reflected back to earth by the ionosphere, so that sky-wave propagation is impossible.

If the transmitting and receiving antennas are both located on the ground, the range of the space wave is limited by the curvature of the earth (Figure 1–17a). For example, the "hump" of the Atlantic Ocean between England and America is approximately 200 miles high, and therefore communication by space wave between these two countries can only occur via satellite. Figure 1–17b shows the limiting line-of-sight condition, in which the line joining the tops of the antennas just grazes the earth's surface. The optical range D is given by

$$D = 1.23(\sqrt{h_1} + \sqrt{h_2}) \text{ land miles} \qquad (1\text{–}11)$$

where h_1 and h_2 are the antenna heights (feet).

In practice, the space wave is refracted by the earth's atmosphere, so that 10 to 15% must be added to the optical range. As an example, let us calculate the maximum range when the transmitting and receiving antennas are mounted on board two ships and each antenna is 64 ft above the surface of the sea. The optical range is $1.23 (\sqrt{64} + \sqrt{64}) = 1.23 \times 16 = 19.7$ land miles, and after allowing an additional 10% for atmospheric refraction, the maximum radio range is approximately 22 land miles or 20 sea miles. However, if one of these ships were communicating with an

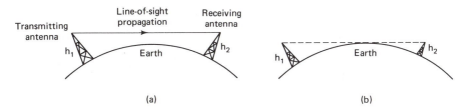

(a)   (b)

Figure 1–17   Space-wave propagation.

aircraft flying at 10,000 ft, the optical range increases to 1.23 ($\sqrt{64} + \sqrt{10,000}$) = 1.23 $\times$ 108 = 133 land miles and the maximum radio range is about 145 miles. For the same reason it is customary to increase the possible range and service area by locating TV and FM antenna systems on mountaintops. In addition to the line-of-sight transmission and the atmospheric refraction, there is a further small increase in the range due to the diffraction around the earth's surface.

Space waves are readily reflected from large metal obstacles so that the received signal will be the result of the space wave traveling directly from the transmitting antenna and all those radio waves that have been reflected from the various obstacles. Since all these waves have taken different paths to the receiving antenna, their signals are not in phase and the result is various forms of interference. One example is TV "ghosting," which appears as a double-image pattern on the receiver screen. Although there is, of course, no absorption of space waves in free space, the situation is different in the actual atmosphere. The two main sources of attenuation are water vapor and oxygen, whose atoms and molecules are caused to vibrate by the EM waves. The effect causes only a minute warming of the atmosphere, but there may be an appreciable attenuation of the EM wave.

Although the attenuation depends on the frequency, it is negligible at frequencies below 12 GHz. The degree of oxygen absorption is also low (0.000015 dB/m) at 0.3 THz but has two strong peaks at 60 and 120 GHz; at these frequencies the attenuations are 0.015 dB/m and 0.0035 dB/m, respectively.

By contrast with the oxygen effect, absorption due to normal water vapor in the atmosphere increases with the frequency, and its attenuation at 0.3 THz is 0.006 dB/m; at 23 and 180 GHz there are two absorption peaks with respective attenuations of 0.00025 and 0.028 dB/m. It follows that the use of microwave frequencies at 23, 60, 120, and 180 GHz should be avoided. However, over the range 23 to 180 GHz, there are frequencies such as 33 and 110 GHz which have relatively low degrees of attenuation; these are referred to as "windows."

If the amount of water vapor increases dramatically, due, for example, to heavy rain, the corresponding attenuation at the higher microwave frequencies may be multiplied by a factor of 10 or more and may become extremely severe.

Tropospheric Ducting

Unusual ranges, well beyond the predictions of line-of-sight transmission, may be caused by abnormal atmospheric conditions a few miles above the earth. These conditions occur in the troposphere, which extends from the earth's surface to a height of approximately 35,000 ft. Under normal conditions the warmest air is found near the surface of the ocean, and the temperature subsequently decreases with height. However, in the tropics a situation can occur where pockets of warm air are trapped between layers of cooler air. The boundaries between the warm air and the cooler air are called temperature inversions; these are capable of refracting space waves which would otherwise continue their line-of-sight propagation. The result is a high tropospheric duct in which the space wave may be trapped and transmitted for hundreds of miles. These high-level ducts typically start at heights of 500 to 1000 ft and extend an additional 500 to 1000 ft into the atmosphere. The signal may then be received by an antenna located in the duct or an antenna on the earth's surface, provided that the lower temperature inversion has disappeared (Figure 1–18a). Under other tropical conditions, the temperature of the air above the surface of the sea will initially increase with the elevation, but at a certain height there will be an inversion and the temperature will subsequently decrease. The result is a surface duct that may extend to an elevation of a few hundred feet. A space wave can be trapped in the surface duct and may be received well beyond the horizon (Figure 1–18b).

Little use can be made of ducts for reliable communications because their occurrence and duration cannot be predicted. Both surface and high-level ducts represent anomalous propagation (sometimes referred to as *anaprop*), in which ranges far in excess of normal are experienced.

Tropospheric Scatter

Tropospheric scatter is a method of propagation that finds an increasing use in modern communications and relies on conditions in the troposphere at an elevation of a few thousand feet. Owing to discontinuities of temperature and humidity in the tropo-

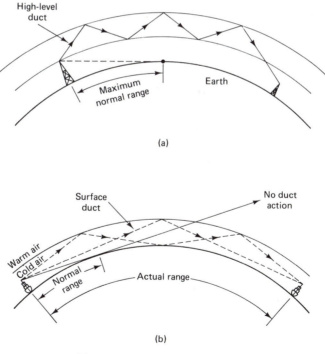

(a)

(b)

Figure 1–18   Tropospheric ducting.

sphere, a scattering region is formed as shown in Figure 1–19. Unlike the creation of ducts, the scattering region always exists and is capable of returning a small fraction of the transmitted power back to earth. An analogy of the process is to consider a stream of water which is directed toward a ceiling with a rough surface. However, most of the transmitted EM energy continues to travel in a straight line and is not received; there is also some scattering in other undesired directions, so that these energy components will also not be intercepted by the receiving system. Typically, the received power is only between 10^{-6} and 10^{-9} times the transmitted power. Consequently, high-power transmitters and extremely sensitive receivers are essential to operate a reliable tropospheric scatter communications system.

The possible frequency range for this propagation method extends from 300 MHz to 10 GHz; however, most commonly used frequencies are 900 MHz, 2 GHz, and 5 GHz, which virtually lie in the microwave region. This wide range is possible because the attenuation rises only slowly as the frequency is increased. The minimum range is about 50 miles and is determined by the size of the scattering volume. The maximum range is 400 to 500 miles and is a geometrical result of the average height of the region from which the energy is scattered.

Apart from the high level of transmitted power, the main disadvantage of tropospheric scatter is the severe fading due to atmospheric changes and the many possible paths by which the energy travels through the scattering volume. The problem of fading may be overcome by some form of diversity reception in which a number of intercepted signals are combined at the receiving system or the strongest of these signals at any instant is selected automatically. The following types of diversity reception are commonly used:

Frequency diversity. The same information is transmitted simultaneously on a number of separate frequencies that suffer different and varying degrees of fading.

Space diversity. A number of receiving antennas are positioned several wave-

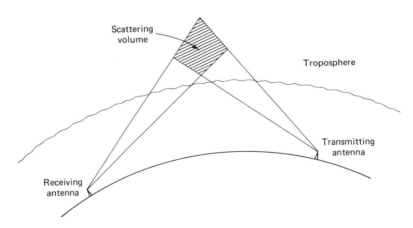

Figure 1–19 Tropospheric scatter.

lengths apart. The strongest signal is selected automatically and fed to the receiver.

Summarizing, except under abnormal conditions, microwave signals are propagated by the simplest method in which the EM wave travels in a practically straight line between the transmitter and the receiver. The primary limitation on the range is due to the curvature of the earth.

Example 1–6   A VHF transmitter is located at the top of a mountain which has a height of 8750 ft. For an aircraft flying at 15,000 ft, what is the optical range of the direct wave?

Solution   The optical range

$$D = 1.23(\sqrt{8750} + \sqrt{15,000})$$
$$\approx \textbf{266 land miles} \tag{1–11}$$

1–8 ADVANTAGES OF THE MICROWAVE REGION

So far we have discussed only the problems associated with the use of the microwave region. We have learned that we need (1) resonant cavities as opposed to conventional tuned circuits, (2) expensive waveguides rather than coaxial cables, (3) special insulators with a low dielectric hysteresis loss, and (4) active devices which are not limited by the transit-time effect. But what of the advantages? These may be summarized as follows:

$$1\% \left(10 \times 10^9 \right) = 10 \times 10^7 = 10^8 = 100 \text{ MHz}$$

Wide Bandwidth

Because microwave frequencies are so high, they can accommodate very wide bandwidths without causing interference problems. This means that a tremendous amount of information can be handled by a single microwave carrier. For example, only low modulating frequencies are associated with telephone calls. Consequently, one system of microwave relay stations can simultaneously transport thousands of telephone conversations across the country. Microwave links are also used extensively for TV transmissions in which frequency modulation is used to impress the information on the carrier. A single link contains from 4 to 10 carriers and the number of channels per carrier lies between 500 and 2500. With the repeater stations about 30 miles apart, more than 20 repeaters will be required between two terminals which are 600 miles apart. A full discussion of microwave links appears in Section 7–8.

Straight-Line Propagation

Most radar systems are of the pulsed type in which a short-duration RF pulse is transmitted by a highly directive antenna system. After being reflected by a target, the echo pulse is received and ultimately appears on the radar display. By timing the interval between the transmission of the pulse and the reception of the echo, the range of the target can be determined. However, this range would be meaningless unless we could assume that the pulse traveled in a virtually straight line to the target and back to the receiver.

High Antenna Directivity

At microwave frequencies it is possible to design highly directive antenna systems of a reasonable size. One of the most common systems employs a paraboloid as a "dish" reflector with a flared waveguide termination placed at the focus (Figure 1–20a). If the paraboloid were infinitely large, the result of the reflection would be to produce a unidirectional beam with no spreading. However, with a paraboloid of practical dimensions there is some degree of spreading and the radiation pattern then contains a major lobe (and a number of minor lobes), as illustrated in Figure 1–20b. The full lobe is, of course, three-dimensional, so that we are showing the pattern in only one plane.

The radiation pattern surrounding an antenna may either represent the field

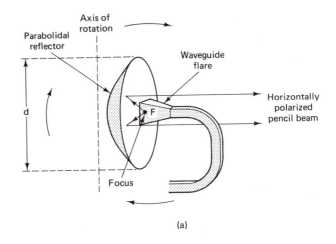

(a)

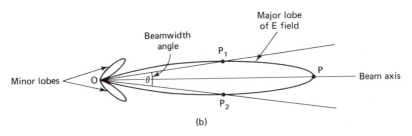

(b)

Figure 1–20   (a) Paraboloidal reflector; (b) radiation pattern of a paraboloidal reflector.

intensity $\mathscr{E}$ (volts per meter) or the power density (watts per square meter) distribution. The two patterns are similar in appearance and we will confine ourselves primarily to the field intensity, which directly measures the strength of the electric field component in the radio wave. Theoretically, the value of $\mathscr{E}$ would be equal to the voltage induced in a conductor 1 m long and positioned parallel to the electric flux lines. Although the basic unit of the electric field intensity is the volt per meter, more practical units are millivolts per meter and microvolts per meter.

The radiation pattern is a plot of the electric field intensity *at a fixed distance* from the (transmitting) antenna versus an angle measured in the particular plane for which the pattern applies. In the case of the half-wave dipole, it is customary to consider two planes. The $\mathscr{E}$ plane contains the antenna itself and the electric flux lines while the magnetic flux lines lie in the H plane, which is at right angles to the antenna at its center point. Combining the results for the two perpendicular planes produces the antenna's complete three-dimensional radiation pattern.

The radiation patterns for a *vertical* dipole are shown in Figures 1–21a and b. The maximum field strength will be associated with the center of the antenna, where the highest current distribution exists; practically, no radiation will occur from the ends where the current is minimal.

The vertical radiation pattern associated with the $\mathscr{E}$ plane consists of two "oval" shapes (not circles) and is shown in Figure 1–21a. In the direction OP the strength of the electric field intensity is at its maximum and is represented by the length OP', while OQ' (which is shorter than OP') is a measure of the field intensity in the direction OQ. For the direction OT, the length intercepted by the radiation plot and consequently the $\mathscr{E}$ value are both zero. Points P_1 and P_2 are those where the electric field intensity is 0.707 times its maximum value (OP'); the beamwidth is then the angle P_1OP_2 between the half-power points and is about 50° for the half-wave dipole.

The vertical dipole radiates equally well in all horizontal directions, so that its pattern in the magnetic plane is a circle (Figure 1–21b). The complete three-dimensional pattern will then resemble a "toroidal" shape or "doughnut" surface with no hole in the center (Figure 1–21c). The same pattern will apply to a vertical dipole used for reception purposes; in any particular direction the length intercepted by the pattern will be a measure of the received signal strength.

Returning to the narrow lobe produced by the antenna with the paraboloid reflector, the beamwidth in the horizontal plane is measured by the angle P_1OP_2 (θ), which is given by

$$\text{beamwidth angle } \theta = \frac{70\lambda}{d} \quad \text{deg} \qquad (1\text{--}12)$$

where

$\lambda = $ wavelength (m)
$d = $ diameter of paraboloid (m)

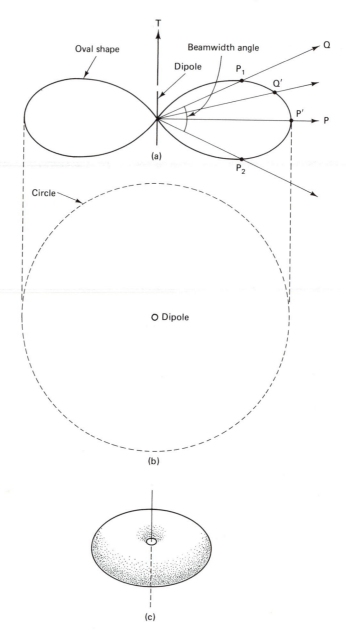

Figure 1–21 Radiation patterns surrounding a vertical dipole: (a) vertical radiation pattern of a vertical dipole; (b) horizontal radiation pattern of a vertical dipole; (c) three-dimensional radiation pattern of a vertical dipole.

In radar systems typical beamwidths are of the order of 1 to 5°. Due to the construction of the paraboloid, the horizontal and vertical beamwidths are not necessarily the same. For a microwave link the normal beamwidth is about 2°.

With such narrow beamwidths the power gains of the antenna systems are extremely high. With respect to a half-wave dipole standard, the power gain G_P is given by

$$G_p = 6\left(\frac{d}{\lambda}\right)^2 \tag{1–13}$$

where d and λ are as defined in equation (1–12). For a paraboloid reflector to be effective, the value of d must be greater than 10λ. Consequently, such reflectors would not be practical in the VHF band due to their excessive size.

Microwave Heating

Compared with low-frequency EM waves, microwave energy is more easily controlled, concentrated, and directed when used for cooking, drying, and physical diathermy. The microwave oven is an obvious example (Section 5–10).

Molecular Resonances

Certain crystal materials possess molecular resonances at microwave frequencies and are used in the manufacture of special diodes. Examples of this effect occur in the Gunn, Read, and IMPATT diodes, all of which are fully discussed in Chapter 6.

Example 1–7   A microwave link operating at 6 GHz uses an antenna with a paraboloid reflector whose mouth diameter is 2.5 m. Calculate the antenna's (a) beamwidth, and (b) power gain.

Solution   (a) The wavelength in free space

$$\lambda = \frac{30}{6} = 5 \text{ cm} \tag{1–3}$$

The beamwidth angle

$$\theta = \frac{70 \times 5 \times 10^{-2}\text{ m}}{2.5\text{ m}} \tag{1–12}$$
$$= \textbf{1.4°}$$

(b) The antenna power gain

$$G_p = 6\left(\frac{2.5}{5 \times 10^{-2}}\right)^2 \tag{1–13}$$
$$= \textbf{15,000}$$

This means that for an effective radiated power of 1 kW along the axis of the main lobe, the equivalent omnidirectional dipole would need to be supplied with a power of 15 MW.

PROBLEMS

BASIC PROBLEMS

1-1. What is the wavelength in free space of a microwave signal whose frequency is 3.46 GHz?

1-2. If the wavelength in free space of a microwave signal is 27 mm, what is its frequency?

1-3. An infrared EM wave has a wavelength in free space of 125 μm. What is its frequency in terahertz?

1-4. A light wave in free space has a wavelength of 6500 Å. What is its corresponding frequency?

1-5. Determine the cutoff frequency for a rectangular waveguide whose inner dimensions are 4.755 cm and 2.215 cm.

1-6. An electron is accelerated from rest through a potential difference of 1 kV. What is the electron's terminal velocity in meters per second?

1-7. A transmitter employing space-wave propagation is located at a height of 1000 ft. The height of the receiving antenna is 25 ft. Calculate the maximum range for straight-line communication.

1-8. An S-band radar set operates at a frequency of 3 GHz. Its antenna system uses a paraboloid reflector with a mouth diameter of 4 m. What is the beamwidth of the major radiation lobe?

1-9. In Problem 1–8, what is the power gain of the antenna system?

1-10. How far does a radar pulse travel in a time of 1.5 μs through free space?

ADVANCED PROBLEMS

1-11. Determine the range of wavelengths associated with the K band.

1-12. What is the maximum frequency that can be propagated through a rectangular waveguide whose inner dimensions are 16.51 cm and 8.255 cm?

1-13. An electron is accelerated from rest over a distance of 1.5 cm. If the accelerating voltage is 125 V, calculate the time interval during which the electron was accelerated.

1-14. A communications system employs a transmitter and a receiver whose effective antenna heights are 100 ft and 64 ft. Using straight-line propagation, could the system operate over a distance of 25 miles?

1-15. An EM wave in free space has an electric field intensity of 500 mV/m. Calculate the value of the wave's magnetic field intensity and the value of its Poynting vector.

1-16. The radiation pattern of an antenna system with a paraboloid reflector has a major lobe whose beamwidth is 1.5°. If the operating frequency is 9.5 GHz, what is the mouth diameter of the paraboloid?

1-17. In Problem 1–16 the power fed to the antenna system is 800 W. Calculate the value of the radiated power in the direction of the major lobe.

1–18. A radar target has a range of 3500 m. What is the total time taken for the radar pulse to travel to the target and back?

1–19. Determine the operating frequency range for a rectangular waveguide whose internal dimensions are 2.85 cm and 1.26 cm.

1–20. An EM wave in free space has a frequency of 65 THz. Express its wavelength in (a) millimeters; (b) micrometers; (c) angstroms.

CHAPTER 2 _____

TRANSMISSION LINES

2–1 INTRODUCTION

Practical considerations require that an antenna be located some distance from its associated transmitter or receiver. A means must therefore be provided to transfer the RF energy between the equipment and the antenna. Basically, an RF transmission line electrically connects an antenna and a transmitter or a receiver. It must do so efficiently with a minimum loss of power or signal strength.

There are four general types of transmission lines: the parallel two-wire line, the twisted pair, the shielded pair, and the concentric (coaxial) line. The use of a particular type of line depends, among other things, on the frequency, the power to be transmitted, and on the type of insulation. However, at microwave frequencies, we normally use specialized transmission lines such as waveguides (Chapter 3).

The two-wire line (Figure 2–1a) consists of two parallel conductors that are maintained a fixed distance apart by means of insulating spacers or spreaders that are placed at suitable intervals. The line has the assets of ease of construction, economy, and efficiency. In practice, such lines used in radio work are generally spaced from 5 to 15 cm apart at frequencies of 14 MHz and below. The maximum spacing for frequencies of 18 MHz and above is 10 cm. To reduce the radiation from the line to a minimum, it is necessary that the wires be separated by only a small fraction of a wavelength. For best results, the separation should be less than one hundredth of a wavelength. At very high frequencies this criterion will limit the amount of RF power that can be conveyed by the line. Consequently, the principal disadvantage of the parallel two-wire line is its relatively high radiation loss. Uniform spacing of a two-wire transmission line may be assured if the wires are embedded in a solid low-loss dielectric such as polyethylene (Figure 2–1b). This so-called ribbon type of line is widely used to connect television receivers to their antennas.

The twisted-pair transmission line is shown in Figure 2–1c. As the name implies, it consists of two insulated wires twisted to form a flexible line without the use of spacers. It is typically used for low-frequency transmission since at the higher frequencies there is excessive loss in the insulation.

The shielded pair (Figure 2–1d) consists of two parallel conductors separated from each other and surrounded by a solid dielectric. The conductors are contained within a copper-braid tubing that acts as a shield. This assembly is covered with a rubber or flexible composition coating to protect the line against moisture.

The principal advantage of the shielded pair is that the two conductors are balanced with respect to ground—that is, the capacitance between each conductor and the ground is uniform along the entire length of the line and the wires are shielded against any pickup from stray fields; this balance is achieved by the grounded shield that surrounds the conductors with a uniform spacing throughout their length.

Coaxial lines, or coaxial cables as they are called, are the most widely used type of RF transmission line. They consist of an outer conductor and an inner conductor held in place exactly at the center of the outer conductor. Several types of coaxial cable have come into wide use for feeding RF power to an antenna system. Figure 2–2 illustrates the construction of flexible and rigid coaxial cables. In both cases one of the conductors is placed inside the other. Since the outer conductor completely shields the inner one, no radiation loss takes place. This is the main advantage of the coaxial cable.

The subject of transmission lines is explored in the following topics:

2–2. Distributed Constants of Transmission Lines

2–3. Matched Lines

2–4. Unmatched Lines

2–5. Artificial Lines

2–6. The Smith Chart and Its Applications

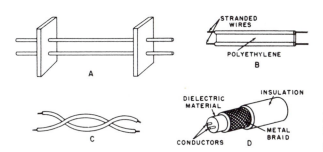

Figure 2–1 Four examples of transmission lines: (a) two-wire (air dielectric); (b) two-wire ribbon type; (c) twisted pair; (d) shielded pair.

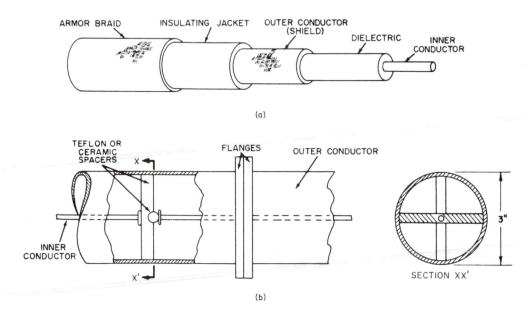

Figure 2–2   (a) Flexible and (b) rigid coaxial cables.

2–2 DISTRIBUTED CONSTANTS OF TRANSMISSION LINES

At low frequencies we are concerned primarily with *lumped* circuitry, where the three electrical properties (resistance, inductance, capacitance) are related to specific components. For example, we regard a resistor as being a lump of resistance and consider all connecting wires to be perfect conductors. By contrast, the resistance R associated with a two-wire transmission line is not concentrated into a lump but is *distributed* along the entire length of the line. The distributed constant of resistance is therefore measured in the basic unit of ohms *per meter* rather than ohms.

Since the straight wires of the parallel line are conducting surfaces separated by an insulator or dielectric, the line will possess the distributed constants of self-inductance L (henrys per meter) and capacitance C (farads per meter). In addition, no insulator is perfect, and consequently there is another distributed constant G (siemens per meter) which is the leakage conductance between the wires. These four distributed constants are illustrated in Figure 2–3; their order of values in a practical line are R (mΩ/m), L (μH/m), C (pF/m), and G (nS/m).

In physical terms the properties of resistance and leakage conductance will relate to the line's power loss in the form of heat dissipation and will therefore govern the degree of attenuation, measured in decibels per meter (dB/m). The self-inductance results in a magnetic field surrounding the wires, while the capacitance means that an electric field exists between the wires (Figure 2–4); these L and C properties determine the line's behavior in relation to the frequency.

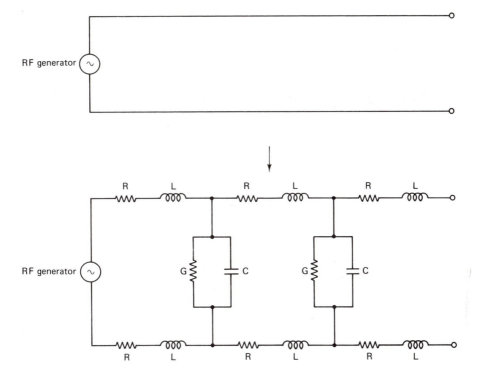

Figure 2-3   Distributed constants of a two-wire line.

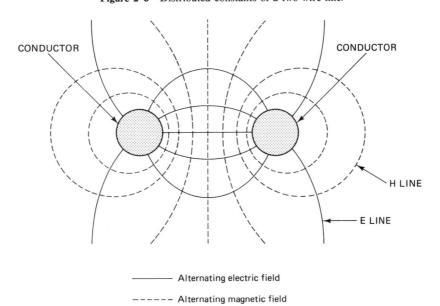

——————— Alternating electric field

- - - - - - - Alternating magnetic field

Figure 2-4   Electric and magnetic fields surrounding a two-wire line.

The two-wire ribbon line that connects an antenna to its TV receiver is often referred to as a 300-Ω line. But what is the meaning of the 300 Ω? Certainly, we cannot find this value with an ohmmeter, so we are led to the conclusion that it appears under working conditions only when the line is being used to convey RF power. In fact, the 300-Ω value refers to the line's surge or characteristic impedance, whose symbol is Z_0. The surge impedance is theoretically defined as the input impedance at the RF generator to an infinite length of the line (Figure 2–5a). In the equivalent circuit of Figure 2–5b, the C and G line constants will complete a path for the current to flow so that an effective current I will be drawn from the RF generator whose effective output voltage is E. Then the input impedance is $Z_{in} = Z_0 = E/I$ ohms.

Referring to Figure 2–6, let us consider the conditions that exist at position X, Y on the infinite line. Since there is still an infinite length to the right of X, Y, the input impedance at this position looking down the line will be equal to the value of Z_0. Consequently, if we remove the section of the line to the right of X, Y and replace it by a resistor whose value in ohms is the same as that of the surge impedance, it will still appear to the generator as if it were connected to an infinite line and the input impedance at the generator will remain equal to Z_0.

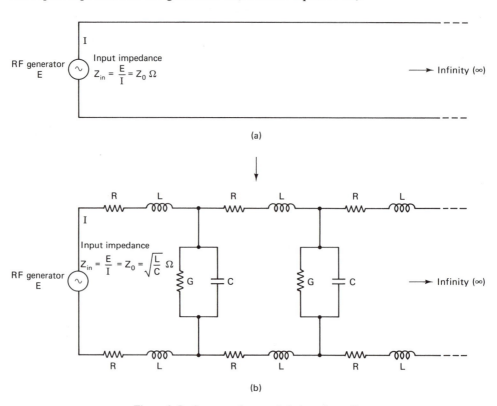

Figure 2–5   Surge or characteristic impedance Z_0.

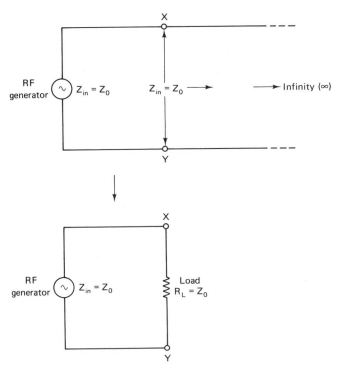

Figure 2–6 Matched line.

When a line is terminated by a resistive load of value Z_0, the line is said to be *matched* to the load. Under matched conditions the line is most efficient in conveying RF power from the generator (e.g., a transmitter) to the load, such as an antenna.

To determine the expression for the surge impedance, we will regard the line as a repeating network consisting of a large number of sections, each with a *series* impedance Z_1 and a *shunt* impedance Z_2 (Figure 2–7a); the terms "series" and "shunt" refer to the positions of Z_1 and Z_2 relative to the RF generator. A particular section is assumed to extend over a short length δl, so that for the series impedance, which contains the distributed resistance R and the distributed inductance L,

$$Z_1 = (R + j\omega L)\,\delta l \qquad (2\text{–}1)$$

where

ω = angular frequency (rad/s)
   $= 2 \times \pi \times$ frequency (Hz)

The shunt impedance consists of the distributed conductance G and the distributed capacitance C, in parallel. Therefore,

$$Z_2 = \frac{1}{(G + j\omega C)\,\delta l} \qquad (2\text{–}2)$$

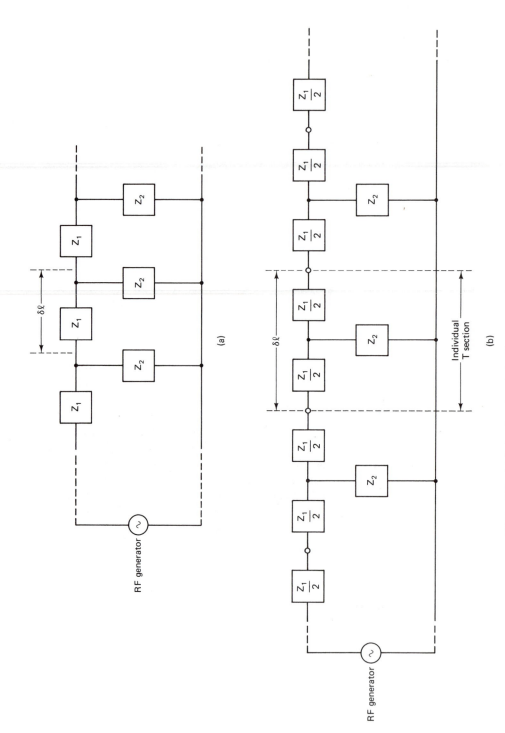

Figure 2-7 Analysis of a transmission line into equivalent tee sections.

To continue our analysis we need to arrange each Z_1, Z_2 combination in a symmetrical formation such as a tee (T) section (Figure 2-7b). It then follows from our previous discussion that if a T section is terminated by a resistive load of value Z_0, the input impedance to the T section must also be equal to Z_0 (Figure 2-8). Consequently,

$$Z_{in} = Z_0 = \frac{Z_1}{2} + \frac{(Z_0 + Z_1/2) \times Z_2}{(Z_0 + Z_1/2) + Z_2}$$

This yields

$$Z_0^2 + \frac{Z_0 Z_1}{2} + Z_0 Z_2 = \frac{Z_0 Z_1}{2} + \frac{Z_1^2}{4} + \frac{Z_1 Z_2}{2} + Z_0 Z_2 + \frac{Z_1 Z_2}{2}$$

$$Z_0^2 = Z_1 Z_2 + \frac{Z_1^2}{4}$$

or

$$Z_0 = \sqrt{Z_1 Z_2 + \frac{Z_1^2}{4}} \qquad (2\text{-}3)$$

Substituting the expressions for Z_1 and Z_2 in equations (2-1) and (2-2) yields

$$Z_0 = \sqrt{\frac{R + j\omega L}{G + j\omega C} + \frac{(R + j\omega L)^2 \, \delta l^2}{4}}$$

If we regard each T section as being infinitesimally short, $\delta l \to 0$ and

$$Z_0 = \sqrt{\frac{R + j\omega L}{G + j\omega C}} \qquad (2\text{-}4)$$

When expressed in polar form

$$Z_0 = \sqrt[4]{\frac{R^2 + \omega^2 L^2}{G^2 + \omega^2 C^2}} \; \left| \; \frac{1}{2}\left(\tan^{-1}\frac{\omega L}{R} - \tan^{-1}\frac{\omega C}{G}\right) \right. \qquad (2\text{-}5)$$

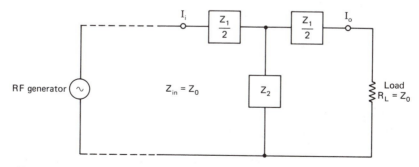

Figure 2-8   Determination of the surge impedance and the propagation constant.

Consider a practical line that is connected to a 300-MHz RF generator. If $R = 50$ mΩ/m, $L = 1.05$ μH/m, $C = 11.7$ pF/m, and $G = 30$ nS/m, then

$$\omega L = 2 \times \pi \times 300 \times 10^6 \times 1.05 \times 10^{-6} = 1979 \ \Omega/m$$

and

$$\omega C = 2 \times \pi \times 300 \times 10^6 \times 11.7 \times 10^{-12} = 22{,}054 \times 10^{-6} \ S/m$$

Since $R = 50 \times 10^{-3}$ Ω/m and $G = 30 \times 10^{-9}$ S/m, $\omega L \gg R$ and $\omega C \gg G$. Therefore,

$$
\begin{aligned}
Z_0 &\rightarrow \sqrt[4]{\frac{\omega^2 L^2}{\omega^2 C^2}} \ \Big/ \frac{1}{2}(90° - 90°) \\[2mm]
&= \sqrt{\frac{L}{C}} \ \Omega \ \underline{/0°} \\[2mm]
&= \sqrt{\frac{1.05 \times 10^{-6}}{11.7 \times 10^{-12}}} \ \underline{/0°} \approx 300 \ \Omega \ \underline{/0°}
\end{aligned}
\tag{2-6}
$$

An RF line therefore behaves resistively and has a surge impedance of $\sqrt{L/C}$ ohms. However, the value of Z_0 must also depend on the line's physical construction; the formulas are as follows:

Two-wire line (Figure 2-9):

$$Z_0 = \frac{2 \times 120\pi}{e \times \sqrt{\epsilon_r}} \ \log_{10} \frac{2S}{d} = \frac{276}{\sqrt{\epsilon_r}} \ \log_{10} \frac{2S}{d} \qquad \Omega \tag{2-7}$$

where

$S =$ spacing between the conductors (m)
$d =$ diameter of each conductor (m)
$\epsilon_r =$ relative permittivity of the insulation
$e = 2.7183$ (the base of the natural logarithms)
$120\pi = 377$ Ω (intrinsic impedance of free space, Section 1-2)

Coaxial cable (Figure 2-10):

$$Z_0 = \frac{120 \ \pi}{e \times \sqrt{\epsilon_r}} \ \log_{10} \frac{D}{d} = \frac{138}{\sqrt{\epsilon_r}} \ \log_{10} \frac{D}{d} \qquad \Omega \tag{2-8}$$

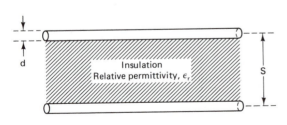

Figure 2-9 Physical factors affecting the value of the surge impedance of a two-wire line.

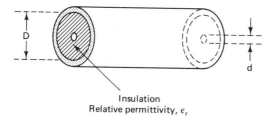

Figure 2–10 Physical factors affecting the value of the surge impedance of a coaxial cable.

Insulation
Relative permittivity, ϵ_r

where

$$D = \text{inner diameter of outer conductor (m)}$$
$$d = \text{outer diameter of inner conductor (m)}$$

Since the surge impedance depends on the ratio $S:d$ (or $D:d$), it follows that a twin lead with thick conductors and a large spacing could have the *same* surge impedance as one with thin conductors and a small spacing. The difference between the two is their power capabilities; the physically larger twin line could carry greater currents and withstand higher voltages between the conductors.

To further our concept of surge impedance, let us consider a repeating network of resistors (Figure 2–11a) with a series resistance R_1 of 10 Ω and a shunt resistance

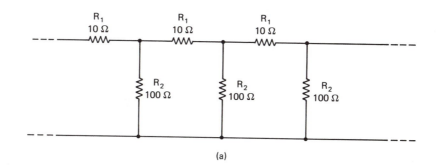

(a)

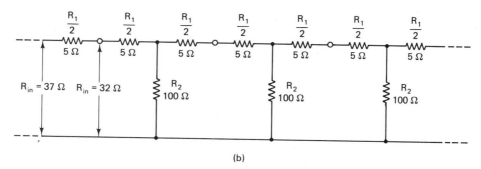

(b)

Figure 2–11 Analysis of a repeating network of resistors.

R_2 of 100 Ω. The input resistance R_{in} of one section is $10 + 100 = 110$ Ω; then if we add more sections, the value of the input resistance changes as follows:

Two sections:

$$R_{in} = 10 + \frac{110 \times 100}{110 + 100} = 62.38 \text{ Ω}$$

Three sections:

$$R_{in} = 10 + \frac{62.38 \times 100}{62.38 + 100} = 48.42 \text{ Ω}$$

Four sections:

$$R_{in} = 10 + \frac{48.42 \times 100}{48.42 + 100} = 42.62 \text{ Ω}$$

Five sections:

$$R_{in} = 10 + \frac{42.62 \times 100}{42.62 + 100} = 39.88 \text{ Ω}$$

Six sections:

$$R_{in} = 10 + \frac{39.88 \times 100}{39.88 + 100} = 38.51 \text{ Ω}$$

The values for seven, eight, nine, and ten sections are respectively 37.80, 37.43, 37.24, and 37.13 Ω. If more and more sections are added, the value of R_{in} tends to approximately 37 Ω. How does this result compare with equation (2–3)?

$$R_{in} = \sqrt{R_1 R_2 + \frac{R_1^2}{4}} = \sqrt{1000 + \frac{100}{4}} = \sqrt{1025} \approx 32 \text{ Ω}$$

Why the difference of $37 - 32 = 5$ Ω in the two results? In Figure 2–11a, each R_1, R_2 combination was an inverted L section, while the equation (2–3) was derived from symmetrical T sections. This distinction is illustrated in Figure 2–11b and we see that the difference in the two values of R_{in} is $R_1/2 = 5$ Ω. This analysis has therefore shown that the input resistance for a large number of identical repeating sections reaches a certain value which virtually does not change however many more sections are added.

Summarizing, a transmission line is used to convey RF power from one position to another. It is an example of distributed circuitry with the four primary constants R, L, G, and C. The surge impedance Z_0 is a secondary constant which is defined as the input impedance to an infinite length of the line; the value of Z_0 depends on the line's physical construction. At radio frequencies a line behaves resistively and has a surge impedance equal to $\sqrt{L/C}$ ohms.

Example 2–1   Each conductor of a two-wire line has a radius of 2.5 mm, and the spacing between the centers of the conductors is 1.4 cm. If an air dielectric is used, what is the value of the surge impedance?

Solution   The line's surge impedance

$$
\begin{aligned}
Z_0 &= \frac{276}{\sqrt{\epsilon_r}} \log_{10} \frac{2S}{d} \\
&= \frac{276}{\sqrt{1}} \log_{10} \frac{2 \times 1.4 \text{ cm}}{2 \times 0.25 \text{ cm}} \\
&= 276 \log_{10} 5.6 \approx \mathbf{200\ \Omega}
\end{aligned}
\tag{2-7}
$$

Example 2–2   The outer conductor of a coaxial cable has an inner diameter of 2.25 cm and the inner conductor has an outer diameter of 1.75 mm. Using polyethylene (relative permittivity, $\epsilon_r = 2.26$) as the insulator between the conductors, what is the value of the surge impedance?

Solution   The line's surge impedance

$$
\begin{aligned}
Z_0 &= \frac{138}{\sqrt{\epsilon_r}} \log_{10} \frac{D}{d} \\
&= \frac{138}{\sqrt{2.26}} \log_{10} \frac{2.25 \text{ cm}}{0.175 \text{ cm}} \\
&\approx \mathbf{100\ \Omega}
\end{aligned}
\tag{2-8}
$$

Example 2–3   An RF transmission line has a distributed inductance of 1.2 μH/m and a distributed capacitance of 13.3 pF/m. Neglecting any losses on the line, what is the value of its surge impedance?

Solution   The surge impedance

$$
\begin{aligned}
Z_0 &= \sqrt{\frac{L}{C}} \\
&= \sqrt{\frac{1.2 \times 10^{-6}}{13.3 \times 10^{-12}}} \\
&= \mathbf{300\ \Omega}
\end{aligned}
\tag{2-6}
$$

2-3 MATCHED LINES

Now let us examine in detail what happens on a matched line. Traveling sine waves of voltage and current start out from the RF generator and move down the line in phase. Due to the small amount of attenuation present on the line, the effective (rms) values of the voltage and current decay slightly, but at all positions $V_{rms}/I_{rms} = Z_0$ (Figure 2–12).

On arrival at the termination the power contained in the voltage and current waves is completely absorbed by the load. If we neglect the losses on the line, the RF power P conveyed down the line is given by

$$
P = V_{rms} \times I_{rms} = I_{rms}^2 \times Z_0 = \frac{V_{rms}^2}{Z_0} \qquad \text{W}
\tag{2-9}
$$

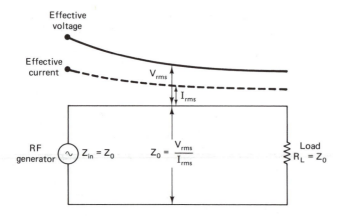

Figure 2–12 Effective voltage and current distribution on a matched transmission line.

These relationships are comparable with those for a resistor; however, it is important to realize that the power is being conveyed down the line and is *not* being dissipated and lost as heat in the surge impedance.

At low frequencies it is assumed that in a series circuit consisting of a source, a two-wire line, and a load, the current is instantaneously the same throughout the circuit. This can only be regarded as true provided that the distances involved in the circuit are small compared with the wavelength of the output from the source. The wavelength, whose letter symbol is the Greek lowercase lambda, λ, is defined as the distance between two consecutive identical states in the path of a wave; for example, the wavelength of a wave in water is the distance between two neighboring crests (or troughs). On a matched transmission line the wavelength is the distance between two adjacent positions where identical voltage (and current) conditions occur instantaneously. Traveling or progressive waves exist on the line because time is involved in propagating RF energy from the source to the load. The wave travels a distance of one wavelength in a time of one period. Since velocity = distance/time,

$$v = \frac{\lambda}{T} = \lambda \times \frac{1}{T} = f \times \lambda \qquad (2\text{–}10)$$

This yields

$$\lambda = \frac{v}{f} \qquad \text{and} \qquad f = \frac{v}{\lambda} \qquad (2\text{–}11)$$

where

v = velocity (m/s)
f = frequency (Hz)
λ = wavelength (m)
T = period (s)

In free space the velocity of an electromagnetic wave (radio wave) is a constant which is approximately equal to 3×10^8 m/s (the speed of light). For example, a 100-MHz radio wave within the FM broadcast band has a wavelength of

$$\lambda = \frac{3 \times 10^8 \text{ m/s}}{100 \times 10^6 \text{ Hz}} = 3 \text{ m}$$

On a transmission line let us consider the generator's output as the reference voltage. At a position A which is a distance of $\lambda/4$ from the generator, the instantaneous voltage will lag by 90° on the reference level while at distances of $\lambda/2$ and $3\lambda/4$ (positions B and C), the voltages will be, respectively, 180° out of phase and 270° lagging (90° leading). A particular phase condition will then travel down at a speed which is called the phase velocity, v_ϕ; this movement of a phase condition is illustrated in Figure 2–13. Consequently, over a distance of 1 m there will be an angular difference which is measured by the phase-shift constant β. The unit of β is the radian per meter, and since there must be a difference of 360° (2π radians) over a distance of λ meters, the phase-shift constant

$$\beta = \frac{2\pi}{\lambda} \qquad \text{rad/m} \tag{2–12}$$

The attenuation α and the phase shift β are both secondary constants which may be obtained from the primary values of R, L, G, and C. The information regarding α and β is contained in the propagation constant, γ. The relationship between α, β, and γ is

$$\text{Propagation constant, } \gamma = \alpha + j\beta \tag{2–13}$$

Therefore, α and β are the "real" and "imaginary" parts of γ, respectively.

Let us consider a T section which again extends over the short distance δl and is terminated by a resistive load whose value is equal to that of the surge impedance Z_0 (Figure 2–8). On the basis of exponential decay, an rms current I_i which enters the T section emerges as a current I_o given by

$$I_o = I_i e^{-\gamma \delta l} \tag{2–14}$$

or

$$e^{\gamma \delta l} = \frac{I_i}{I_o} \tag{2–15}$$

Using the current-division rule yields

$$I_o = I_i \times \frac{Z_2}{Z_2 + Z_0 + \dfrac{Z_1}{2}} \tag{2–16}$$

or

$$\frac{I_i}{I_o} = 1 + \frac{Z_0}{Z_2} + \frac{Z_1}{2Z_2} \tag{2–17}$$

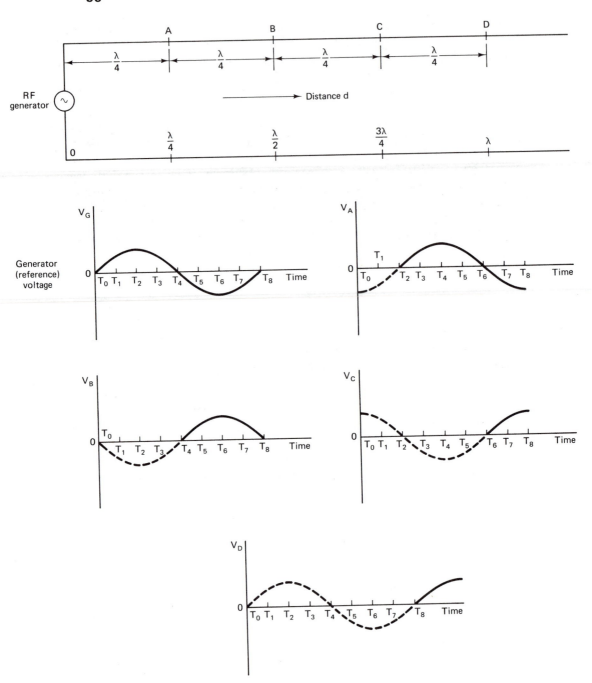

Figure 2–13 Progressive voltage wave traveling down a matched transmission line.

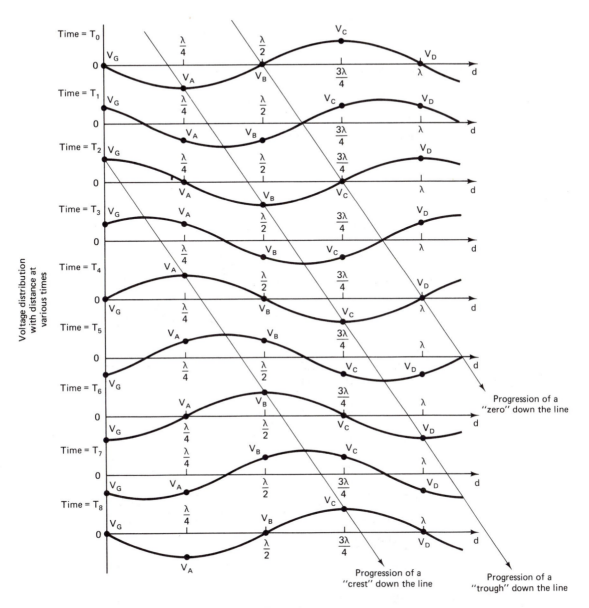

Figure 2–13 (*Cont.*)

Substituting

$$Z_0 = \sqrt{\frac{R + j\omega L}{G + j\omega C}}, \quad Z_2 = \frac{1}{(G + j\omega C)\delta l} \text{ and } Z_1 = (R + j\omega L)\,\delta l$$

from equations (2–4), (2–2), and (2–1),

$$\frac{I_i}{I_o} = 1 + \frac{\sqrt{(R+j\omega L)(G+j\omega C)}}{1!}\delta\ell + \frac{(R+j\omega L)(G+j\omega C)}{2!}\delta\ell^2 \qquad (2\text{--}18)$$

Expanding $e^{\gamma\delta\ell}$ in terms of the exponential series, we have

$$e^{\gamma\delta\ell} = 1 + \frac{\gamma\delta\ell}{1!} + \frac{\gamma^2\delta\ell^2}{2!} + \cdots \qquad (2\text{--}19)$$

Comparing equations (2–18) and (2–19) and neglecting terms involving $\delta\ell^3$, $\delta\ell^4$, and so on, gives us

$$\gamma = \sqrt{(R+j\omega L)(G+j\omega C)} \qquad (2\text{--}20)$$

Therefore

$$\gamma = \alpha + j\beta = \sqrt{(R+j\omega L)(G+j\omega C)}$$

It may be shown that

$$\text{Attentuation constant, } \alpha = 8.686 \times \left(\frac{R}{2}\sqrt{\frac{C}{L}} + \frac{G}{2}\sqrt{\frac{L}{C}}\right)$$
$$= 8.686 \times \left(\frac{R}{2Z_0} + \frac{GZ_0}{2}\right) \quad \text{dB/m} \qquad (2\text{--}21)$$

since $Z_0 = \sqrt{L/C}$ ohms.

In our example of a practical line, $R = 50$ mΩ/m, $G = 30$ nS/m, and $Z_0 = 300\ \Omega\ \underline{/0°}$. The value of the attenuation constant is

$$\alpha = 8.686 \times \left(\frac{50 \times 10^{-3}}{2 \times 300} + \frac{30 \times 10^{-9} \times 300}{2}\right)$$
$$= 8.686 \times (8.333 \times 10^{-5} + 0.45 \times 10^{-5})$$
$$= 8.686 \times 8.783 \times 10^{-5}$$
$$= 7.63 \times 10^{-4} \text{ dB/m}$$

If the attenuation and therefore the losses on the line are neglected,

$$\text{Phase shift constant, } \beta = \sqrt{\omega L \times \omega C} = \omega\sqrt{LC} \quad \text{rad/m} \qquad (2\text{--}22)$$

With $L = 1.05\ \mu$H/m, $C = 11.7$ pF/m, and $f = 300$ MHz,

$$\beta = 2 \times \pi \times 300 \times 10^6 \times 1.05 \times 10^{-6} \times 11.7 \times 10^{-12}$$
$$= 6.61 \text{ rad/m}$$

Since the phase velocity $v_\phi = f \times \lambda$ and $\beta = 2\pi/\lambda$,

$$v_\phi = f \times \frac{2\pi}{\beta}$$
$$= \frac{\omega}{\beta} \qquad (2\text{--}23)$$
$$= \frac{1}{\sqrt{LC}} m/s$$

Since $L = 1.05 \ \mu H/m$ and $C = 11.7 pF/m$,

$$v_\phi = \frac{1}{\sqrt{1.05 \times 10^{-6} \times 11.7 \times 10^{-12}}}$$
$$= 285 \times 10^6 \text{ m/s}$$

The phase velocity is the speed at which the voltage and current waves, as well as the electric and magnetic fields, apparently move down the line.

The equations for the traveling waves of voltage and current on a loss-free line are

$$v = E \sin\left(\omega t - \frac{2\pi d}{\lambda}\right)$$
$$= E \sin\left(2\pi ft - \beta d\right) \qquad (2\text{–}24)$$

and

$$i = I \sin\left(\omega t - \frac{2\pi d}{\lambda}\right)$$
$$= I \sin\left(2\pi ft - \beta d\right) \qquad (2\text{–}25)$$
$$= \frac{E}{Z_0} \sin\left(2\pi ft - \beta d\right)$$

where

$v =$ instantaneous value of the voltage wave (V)
$i =$ instantaneous value of the current wave (A)
$E =$ peak value of the voltage wave (V)
$I =$ peak value of the current wave (A)
$\omega =$ angular frequency (rad/s)
$f =$ frequency (Hz)
$\beta =$ phase-shift constant (rad/m)
$d =$ distance from the generator (m)
$t =$ time (s)
$Z_0 =$ surge impedance (Ω)

The value of v_ϕ is, in fact, always less than the velocity of light, c. The value of v_ϕ/c is called the velocity factor, δ. The value of δ varies from 0.66 for certain types of coaxial cable to 0.975 for an air-insulated two-wire line. It follows that since $v_\phi = f \times \lambda$, the wavelength on the line is shorter than the wavelength in free space.

The features of a matched line may be summarized as follows:

1. Traveling waves of voltage and current move down the line in phase and their power is completely absorbed by the load.
2. The ratio of the effective voltage to the effective current is constant over the entire line and is equal to the surge impedance Z_0.

3. The input impedance at the generator is equal to the surge impedance and is independent of the line's length.

4. The power losses on the line are subdivided into
   (a) Radiation and induction losses, which are a problem with parallel wire lines.
   (b) The dielectric hysteresis loss, which increases with frequency and depends on the type of insulator used. At microwave frequencies of a few gigahertz the dielectric loss is the ultimate reason for abandoning coaxial lines and using waveguides instead.
   (c) The copper loss, which is associated with the conductor's resistance. At high frequencies this loss is increased by the skin effect, which confines most of the electron flow to the surface (skin) of the conductor and therefore reduces the available cross-sectional area. The larger the surface area of the conductors, the less is this type of loss.

Example 2–4   The primary constants of a matched transmission line are $R = 45$ $m\Omega/m$, $L = 1.43$ $\mu H/m$, $C = 8.89$ pF/m, and $G = 12nS/m$. If the frequency of the generator feeding the line is 300 MHz, calculate the line's (a) characteristic impedance, (b) attenuation constant, (c) phase-shift constant, (d) phase velocity, and (e) wavelength.

Solution   (a) The characteristic or surge impedance

$$Z_0 = \sqrt{\frac{L}{C}}$$
$$= \sqrt{\frac{1.43 \times 10^{-6}}{8.89 \times 10^{-12}}} \approx 400\ \Omega \tag{2-6}$$

(b) The attenuation constant

$$\alpha = 8.686 \times \left(\frac{R}{2Z_0} + \frac{GZ_0}{2}\right)$$
$$= 8.686 \times \left(\frac{45 \times 10^{-3}}{2 \times 400} + \frac{12 \times 10^{-9} \times 400}{2}\right) \tag{2-21}$$
$$= 8.686 \times (563 \times 10^{-6} + 2.4 \times 10^{-6})$$
$$= 8.686 \times 565.4 \times 10^{-6}$$
$$\approx 4.9 \times 10^{-3}\ dB/m$$

(c) The phase-shift constant

$$\beta = \omega\sqrt{LC}$$
$$= 2 \times \pi \times 300 \times 10^6 \times \sqrt{1.43 \times 10^{-6} \times 8.89 \times 10^{-12}} \tag{2-22}$$
$$= 6.73\ rad/m$$

(d) The phase velocity

$$v_\phi = \frac{1}{\sqrt{LC}}$$
$$= \frac{1}{\sqrt{1.43 \times 10^{-6} \times 8.89 \times 10^{-12}}} \tag{2-23}$$
$$\approx 280 \times 10^6\ m/s$$

(e) The wavelength

$$\lambda = \frac{v_\phi}{f}$$
$$= \frac{280 \times 10^6}{300 \times 10^6}$$
$$= 0.93 \text{ m}$$

(2-11)

Note that this compares with a wavelength of 1 m in free space.

2-4 UNMATCHED LINES

If an RF line is terminated by a resistive load that is not equal to the surge impedance, the generator will still send voltage and current waves down the line in phase, but their power will be only partially absorbed by the load. A certain fraction of the voltage and current waves that arrive (are incident) at the load will be reflected back toward the generator. At any position on the line the instantaneous voltage (or current) will be the resultant of the incident and reflected voltage (or current) waves; these combine to produce so-called standing waves. In the extreme cases of open- and short-circuited lines, no power can be absorbed by the "load" and total reflection will occur.

Let us consider the production of standing waves on the last wavelength of an open-circuited line (Figure 2-14). The arrow $\rightarrow$ is used to refer to an incident wave, while $\leftarrow$ indicates a reflected wave. All distances are measured from the line's open-circuited end, where a voltage may exist but where the current at all times must be zero. The voltage which is incident at the open circuit cannot abruptly change its polarity as it turns back toward the generator; consequently, there is an "in phase" reflection in which the incident traveling wave is regarded as extended beyond the open-circuited end and then "folded back." The incident and reflected waves are added together to produce the voltage standing wave at an arbitrarily chosen time T. As the incident voltage wave moves down the line toward the open circuit, we can derive the changes in the instantaneous standing wave for each subsequent interval of one-eighth of the period. The instantaneous wave is therefore illustrated in Figure 2-14 for the times, T, $T + \frac{1}{8f}$, $T + \frac{1}{4f}$, . . . , $T + \frac{7}{8f}$. Finally, all instantaneous voltage standing waves are collected together in one representation (Figure 2-15). At distances of 0, $\lambda/2$, and λ from the open-circuit termination, there are *fixed* positions of maximum voltage variation with time; such variations are called voltage antinodes or loops. However, at the distances of $\lambda/4$ and $3\lambda/4$, there are *stationary* positions where the voltage is at all times zero; these are called voltage nodes or nulls. The stationary points that occur with the standing wave are in contrast with the traveling wave, where, for example, a zero-voltage condition would move down the line at a speed equal to the phase velocity.

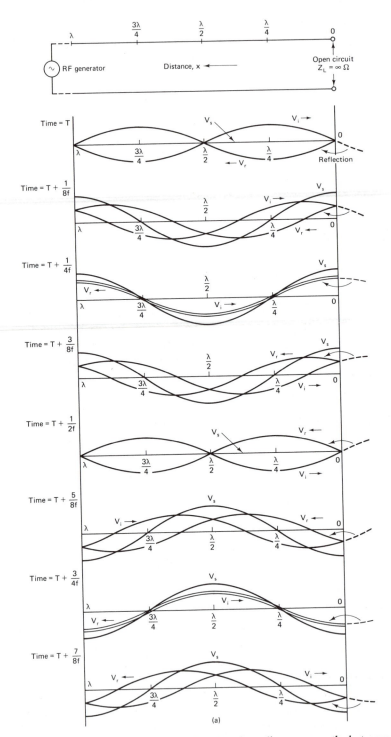

Figure 2–14 Instantaneous incident, reflected, and standing waves on the last wavelength of an open-circuited transmission line: (a) voltage distribution; (b) current distribution.

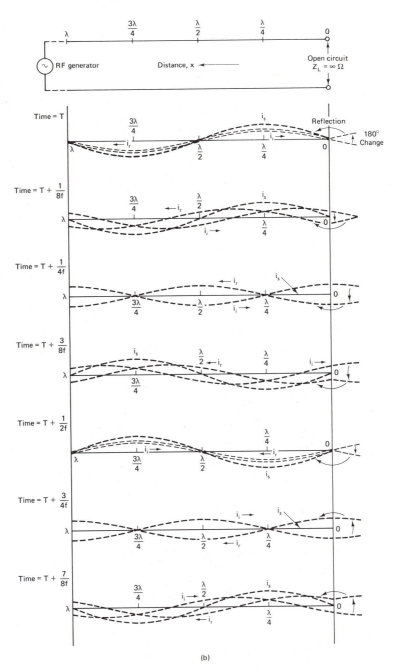

(b)

Figure 2–14 (*Cont.*)

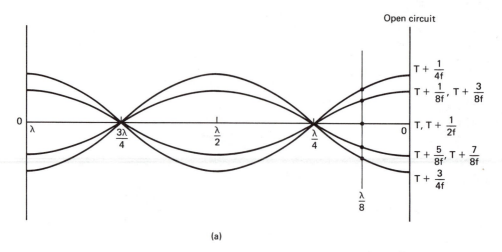

(a)

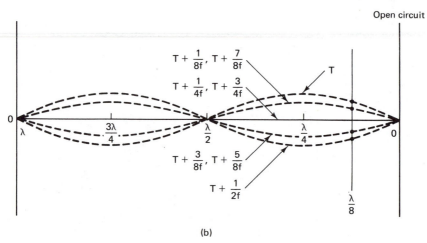

(b)

Figure 2–15 Instantaneous standing waves on an open-circuited transmission line: (a) voltage; (b) current.

It is important to realize that a standing wave represents a sinusoidal distribution with both time and distance. The incident traveling wave can be represented by

$$v_i = E \sin\left(2\pi ft + \frac{2\pi x}{\lambda}\right) \qquad (2\text{-}26)$$

where

v_i = instantaneous value of the incident voltage wave (V)
E = peak value of the incident voltage wave (V)
f = frequency (Hz)
x = distance measured from the open-circuit termination (m)
λ = wavelength existing on the line (m)

The equation of the reflected wave is

$$v_r = E \sin\left(2\pi ft - \frac{2\pi x}{\lambda}\right) \tag{2-27}$$

where v_r is the instantaneous value of the reflected voltage wave (V). The instantaneous standing-wave voltage

$$v_s = v_i + v_r = E\left[\sin\left(2\pi ft + \frac{2\pi x}{\lambda}\right) + \sin\left(2\pi ft - \frac{2\pi x}{\lambda}\right)\right] \tag{2-28}$$

$$= 2E \sin 2\pi ft \cos \frac{2\pi x}{\lambda}$$

sinusoidal distribution   sinusoidal distribution
with time   with distance

It is customary to represent the standing wave in terms of its effective rms value, V_s, which is expressed by $\sqrt{2}\,E\,\cos\,(2\pi x/\lambda)$ and is illustrated in Figure 2-16. Notice that we consider only the magnitude of this expression and that no positive or negative sign is involved.

Now let us examine the current (dashed line) standing-wave distribution on an open-circuited line. At the open-circuit termination the current must reverse its direction, and therefore we have an out-of-phase reflection in which the wave is extended beyond the open circuit, then phase shifted by 180° and afterward folded back to provide the reflected wave. Incident and reflected waves are then combined

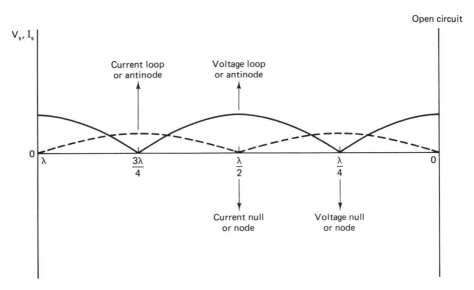

Figure 2-16 Effective voltage (solid line) and current (dashed line) standing-wave distributions on the last wavelength of an open-circuited transmission line.

to produce the current standing waves at times T, $T + \dfrac{1}{8f}$, $T + \dfrac{1}{4f}$, . . . , $T + \dfrac{7}{8f}$ (Figure 2–14b). Finally, all eight instantaneous current standing waves are collected together in one presentation (Figure 2–15b). The incident current wave which is in phase with the incident voltage is presented by

$$
\begin{aligned}
i_i &= I \sin\left(2\pi ft + \frac{2\pi x}{\lambda}\right) \\
&= \frac{E}{Z_0} \sin\left(2\pi ft + \frac{2\pi x}{\lambda}\right)
\end{aligned}
\tag{2-29}
$$

where

i_i = instantaneous value of the incident current wave (A)
I = peak value of the incident current wave (A)

The equation of the reflected current is

$$
\begin{aligned}
i_r &= -I \sin\left(2\pi ft - \frac{2\pi x}{\lambda}\right) \\
&= -\frac{E}{Z_0} \sin\left(2\pi ft - \frac{2\pi x}{\lambda}\right)
\end{aligned}
\tag{2-30}
$$

where the minus sign preceding I indicates the out-of-phase reflection, and i_r is the instantaneous value of the reflected current wave (A).

The instantaneous standing-wave current

$$
\begin{aligned}
i_s = i_i + i_r &= I\left[\sin\left(2\pi ft + \frac{2\pi x}{\lambda}\right) - \sin\left(2\pi ft - \frac{2\pi x}{\lambda}\right)\right] \\
&= \frac{2E}{Z_0} \cos 2\pi ft \, \sin \frac{2\pi x}{\lambda}
\end{aligned}
\tag{2-31}
$$

Notice that because of the "sin $2\pi ft$" and "cos $2\pi ft$" terms, v_s and i_s are 90° out of phase in terms of time. The rms value of the current standing wave, I_s, has a magnitude of $(\sqrt{2}E/Z_0) \sin (2\pi x/\lambda)$, which is illustrated in Figure 2–16. Current nulls exist at distances of zero, $\lambda/2$, and λ from the open-circuit termination, while current antinodes occur at distances of $\lambda/4$ and $3\lambda/4$.

The magnitude of the impedance Z at any position is given by

$$
\begin{aligned}
Z = \frac{V_s}{I_s} &= \frac{\sqrt{2}E \cos (2\pi x/\lambda)}{(\sqrt{2}E/Z_0) \sin (2\pi x/\lambda)} \\
&= Z_0 \cot \frac{2\pi x}{\lambda}
\end{aligned}
\tag{2-32}
$$

Since v_s and v_i are 90° out of phase, the nature of this impedance must be reactive. But at what positions on the line is the impedance capacitive and at what positions is it inductive? Let us start by considering the impedance at the position which is at a distance $\lambda/8$ from the open-circuit termination (Figure 2–14). From this position the values of the instantaneous voltage and current standing waves are plotted versus a time scale that contains T, $T + \dfrac{1}{4f}$, $T + \dfrac{1}{2f}$, $T + \dfrac{3}{4f}$, $T + \dfrac{1}{f}$ (Figure 2–17). It is clear that i_s leads v_s by 90°, so that the impedance is capacitive. Physically, over the last eighth of a wavelength, the voltage distribution, starting from a maximum at the open-circuited end, is greater than the current distribution. The electric field associated with the distributed capacitance will dominate the magnetic field produced by the distributed inductance. If $x = \lambda/8$, $\cot(2\pi x/\lambda) = \cot(\pi/4) = \cot 45° = 1$ and therefore the input impedance to a $\lambda/8$ line is a capacitive reactance whose value is equal to Z_0. A similar analysis for the open-circuited $3\lambda/8$ line will show that i_s lags v_s and that the input impedance is an inductive reactance of value Z_0. At the $\lambda/4$ position there is a voltage null and a current antinode; theoretically, the impedance is zero, but on a practical line the impedance would be equivalent to a low value of resistance. The conditions are reversed for the $\lambda/2$ position, where there is a voltage null and a current antinode; the impedance is theoretically infinite but in practice is equivalent to a high value of resistance.

The variation of impedance along the open-circuited line is shown in Figure 2–18. One is struck with the similarity to the behavior of series and parallel LC circuits. Remember that the impedance response of the series circuit changes from capacitive through resistive (low value) to inductive; by contrast, the parallel circuit has an impedance response which varies from inductive through resistive (high value) to capacitive. These equivalent LC circuits have been included in Figure 2–18. Since the line exhibits resonant properties it may be referred to as "tuned"; by contrast, the matched line is called "flat," "untuned," or "nonresonant."

A short-circuited line (Figure 2–19) also represents a complete mismatch, so that total reflection will occur at the termination. However, the nature of the reflection is different; the incident voltage wave will undergo an out-of-phase reflection while the current wave will be reflected in phase. Since the voltage and the current conditions are interchanged when compared with the open-circuited line, the impedance at the $\lambda/8$ position is an inductive reactance of value Z_0. At the $\lambda/4$ position there is a voltage antinode and a current node, so that the impedance is equivalent to a parallel LC circuit (for frequencies in the ultra-high-frequency (UHF) band, lumped circuitry may be impossible and a quarter-wave section of line shorted at one end is then used as a tank circuit). The impedance will be a capacitive reactance equal in value to Z_0 at the $3\lambda/8$ position, while at the $\lambda/2$ position the impedance is the same as that of a series LC circuit. Summarizing, there is a $\lambda/4$ shift between the impedance conditions on the open- and short-circuited lines. But which one is generally preferred? Because of "end" capacitance effects, it is impossible to achieve infinite resistance at the termination. However, if a bar of low resistance is placed across a line's conductors, the result is a good approximation to a short circuit.

We now examine the effect of terminating a lossless line with a resistive load which is not equal to the surge impedance. The amount of reflection will be reduced

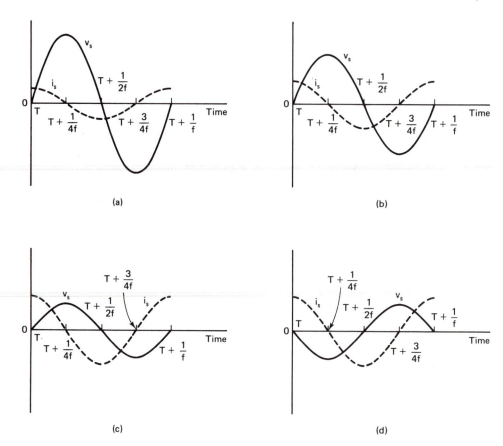

(a)

(b)

(c)

(d)

Figure 2–17 Values and nature of impedance at various positions on the last wave-length of an open-circuited transmission line: (a) distance of $\lambda/16$ from open-circuited end, impedance is a high value of capacitive reactance; (b) distance of $\lambda/8$ from open-circuited end, impedance is a capacitive reactance whose value is equal to the line's surge impedance Z_0; (c) distance of $3\lambda/16$ from open-circuited end, impedance is a low value of capacitive reactance; (d) distance of $5\lambda/16$ from open-circuited end, impedance is a low value of inductive reactance; (e) distance of $3\lambda/8$ from open-circuited end, impedance is an inductive reactance whose value is equal to the line's surge impedance Z_0; (f) distance of $7\lambda/16$ from open-circuited end, impedance is a high value of inductive reactance; (g) distance of $9\lambda/16$ from open-circuited end, impedance is a high value of capacitive reactance; (h) distance of $5\lambda/8$ from open-circuited end, impedance is a capacitive reactance whose value is equal to the line's surge impedance Z_0; (i) distance of $11\lambda/16$ from open-circuited end, impedance is a low value of capacitive reactance; (j) distance of $13\lambda/16$ from open-circuited end, impedance is a low value of inductive reactance; (k) distance of $7\lambda/8$ from open-circuited end, impedance is an inductive reactance equal in value to the line's surge impedance Z_0; (l) distance of $15\lambda/16$ from open-circuited end, impedance is a high value of inductive reactance.

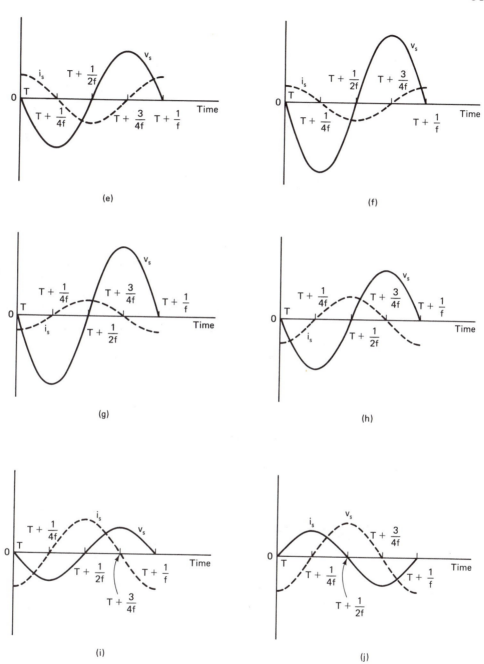

(e)

(f)

(g)

(h)

(i)

(j)

Figure 2–17 *(Cont.)*

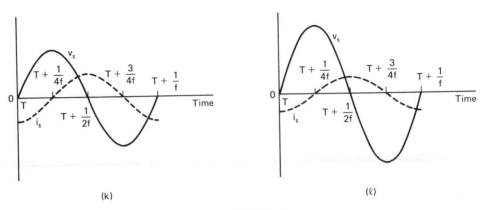

(k)   (ℓ)

Figure 2–17 (*Cont.*)

so that the nodes and antinodes will be less pronounced. Neglecting any losses on the line, the effective voltage and current distribution for the three possible cases of a resistive load, $R_L < Z_0$, $R_L = Z_0$, and $R_L > Z_0$, are shown in Figure 2–20.

The fraction of the incident voltage and current reflected at the load is called the reflection coefficient, and has a symbol the Greek lowercase letter rho, ρ. If $R_L > Z_0$, the effective standing wave voltage at the load is

$$V_L = V_i + V_r = V_i + \rho V_i = V_i(1 + \rho) \qquad (2\text{--}33)$$

where

$$V_i = \text{effective value of the incident voltage wave}$$
$$V_r = \text{effective value of the reflected voltage wave}$$

The effective standing-wave current through the load is

$$I_L = I_i + I_r = I_i - \rho I_i = I_i(1 - \rho) \qquad (2\text{--}34)$$

The negative sign indicates the out-of-phase reflection of the current wave compared with the in-phase reflection of the voltage wave. Then

$$\text{Load resistance, } R_L = \frac{V_L}{I_L} = \frac{V_i(1 + \rho)}{I_i(1 - \rho)} = \frac{Z_0(1 + \rho)}{1 - \rho} \qquad (2\text{--}35)$$

This yields

$$\rho = \frac{R_L - Z_0}{R_L + Z_0} \qquad (2\text{--}36)$$

If $R_L = \infty \ \Omega$ (open-circuited line), $\rho = +1$.
If $R_L = Z_0 \ \Omega$ (matched line), $\rho = 0$.
If $R_L = 0 \ \Omega$ (short-circuited line), $\rho = -1 = 1 \ \underline{/180°}$.

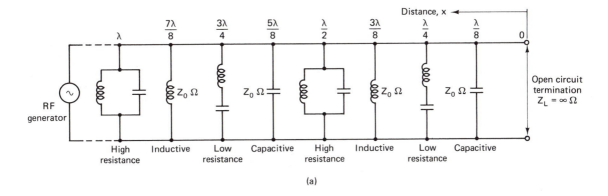

(a)

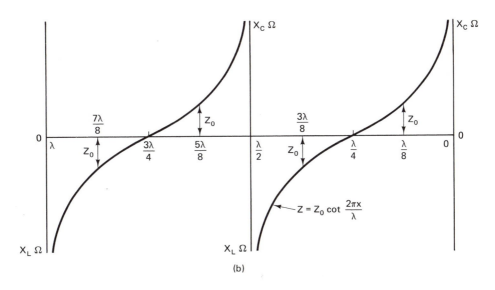

(b)

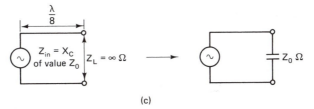

(c)

Figure 2–18   Variation of impedance along an open-circuited transmission line: (a) examples of impedance variation; (b) graphical representation of impedance variation; (c) open-circuited line, $\lambda/8$ long (or $5\lambda/8$, $9\lambda/8$, $13\lambda/8$, etc., long); (d) open-circuited line, $\lambda/4$ long (or any *odd* multiple of a quarter-wavelength); (e) open-circuited line, $3\lambda/8$ long (or $7\lambda/8$, $11\lambda/8$, $15\lambda/8$, etc., long); (f) open-circuited line, $\lambda/2$ long (or any *even* multiple of a quarter-wavelength). *Note*: There is no impedance transformation over a distance of one half-wavelength or any multiple of that distance.

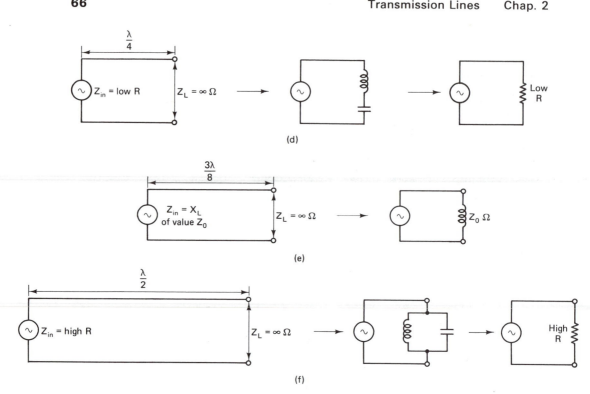

Figure 2–18 (*Cont.*)

The negative sign indicates that when $R_L < Z_0$, the current wave is reflected in phase but the voltage reflection is out of phase. These results are true only for purely resistive loads.

If the load is a general impedance possessing both resistance and reactance, ρ is not just a number but is a complex quantity with a magnitude P which ranges in value between 0 and 1 and a phase angle θ whose value may lie between $+180°$ and $-180°$.

Since the reflected voltage is P times the incident voltage and the reflected current is P times the incident current, it follows that the reflected power, P_r, is P^2 times the incident power, P_i. Then the power absorbed by the load, P_L, is

$$P_L = P_i - P_r = (1 - P^2)P_i = \frac{1 - P^2}{P^2} P_r \qquad (2\text{–}37)$$

Voltage Standing-Wave Ratio

Referring to Figure 2–20, we see that the effective value of a voltage antinode is E_{max} and the effective value of an adjacent voltage node is E_{min}. The degree of standing waves is measured by the voltage standing-wave ratio (VSWR or, more simply, SWR), whose letter symbol is S. The VSWR is defined as the ratio E_{max} :

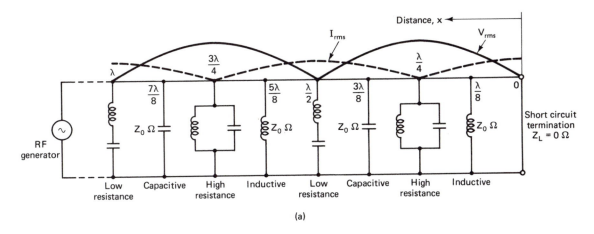

(a)

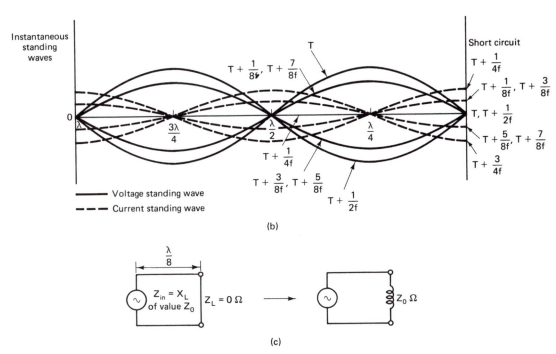

(b)

(c)

Figure 2–19 Standing-wave distribution and impedance variation along the last wavelength of a short-circuited transmission line: (a) examples of impedance variation; (b) instantaneous voltage and current standing waves; (c) short-circuited line, $\lambda/8$ long (or $5\lambda/8$, $9\lambda/8$, $13\lambda/8$, etc., long); (d) short-circuited line, $\lambda/4$ long (or any *odd* multiple of a quarter-wavelength); (e) short-circuited line, $3\lambda/8$ long (or $7\lambda/8$, $11\lambda/8$, $15\lambda/8$, etc., long); (f) short-circuited line, $\lambda/2$ long (or any *even* multiple of a quarter-wavelength). *Note*: There is no impedance transformation over a distance of one half-wavelength or any multiple of that distance. (g) On a line $\lambda/2$ long, the input impedance is equal to the load impedance.

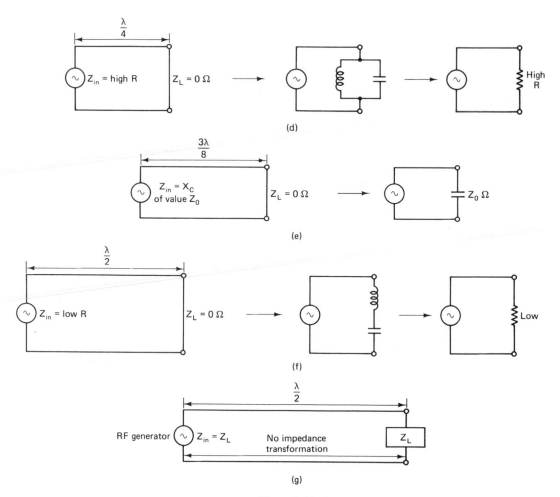

Figure 2–19 (*Cont.*)

$E_{\min}$, which is also equal in magnitude to $I_{\max} : I_{\min}$. Since $E_{\max}$ is the result of an in-phase condition between the incident and reflected voltages, while $E_{\min}$ is the result of a 180° out-of-phase situation,

$$E_{\max} = V_i + V_r = V_i(1 + P)$$
$$E_{\min} = V_i - V_r = V_r(1 - P)$$

Then

$$\text{VSWR, } S = \frac{E_{\max}}{E_{\min}} = \frac{1 + P}{1 - P} \qquad (2\text{–}38)$$

If $P = 0$ (matched line), $S = 1$.

If $P = 1$ (open- or short-circuited line), $S = \infty$.

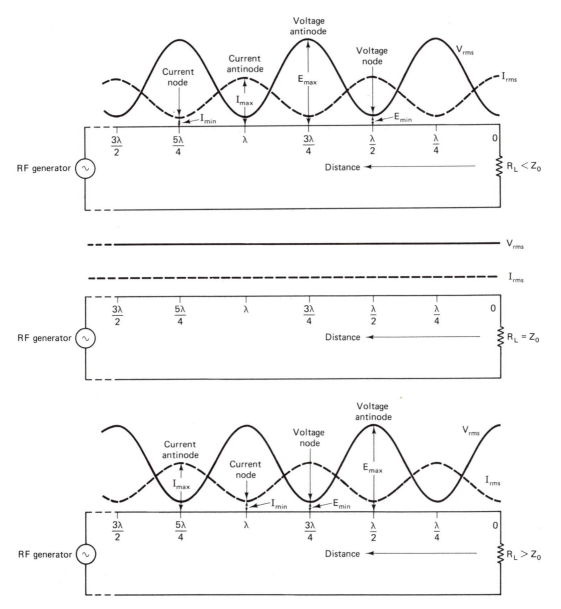

Figure 2–20   Effective voltage and current distributions on matched and mismatched lines.

The VSWR is just a number whose value will range from 1 to ∞. On a practical matched system a value of S that is less than 1.2 is normally regarded as acceptable. It is sometimes preferable to measure the VSWR in decibels, in which case VSWR = $20 \log_{10} S$ decibels.

Equation (2–38) yields

$$P = \frac{S-1}{S+1} \qquad (2\text{–}39)$$

If a resistive load R_L is greater than Z_0, the magnitude of the reflection coefficient is given by

$$P = \frac{R_L - Z_0}{R_L + Z_0} \qquad (2\text{–}40)$$

Combining equations (2–38) and (2–40) gives us

$$S = \frac{1+P}{1-P} = \frac{R_L}{Z_0} \qquad (2\text{–}41)$$

If R_L is less than Z_0,

$$P = \frac{Z_0 - R_L}{Z_0 + R_L}$$

and

$$S = \frac{Z_0}{R_L} \qquad (2\text{–}42)$$

At the positions where the voltage antinodes and the current nodes coincide, the impedance of the line is a maximum and is resistive. The maximum impedance

$$Z_{max} = \frac{E_{max}}{I_{min}} = \frac{E_i(1+P)}{I_i(1-P)} = Z_0\left(\frac{1+P}{1-P}\right) = SZ_0 \qquad (2\text{–}43)$$

At the voltage-node position, there is a minimum resistive impedance which is given by

$$Z_{min} = \frac{E_{min}}{I_{max}} = \frac{E_i(1-P)}{I_i(1+P)} = Z_0\left(\frac{1-P}{1+P}\right) = \frac{Z_0}{S} \qquad (2\text{–}44)$$

As an example, an RF power of 100 W is being conveyed down a 50-Ω loss-free line which is terminated by a 40-Ω resistive load. The following may be calculated:

$$\text{VSWR, } S = \frac{Z_0}{R_L} = \frac{50}{40} = 1.25$$

$$\text{Reflection coefficient, } P = \frac{S-1}{S+1}$$

$$= \frac{0.25}{2.25}$$

$$= 0.11$$

$$\text{Reflected power, } P_r = P^2 \times P_i = (0.11)^2 \times 100$$
$$= 1.2 \text{ W}$$

$$\text{Load power, } P_L = P_i - P_r = 100 - 1.2 = 98.8 \text{ W}$$

Summarizing, the presence of standing waves on a practical transmission line has the following disadvantages:

1. The incident power reaching the termination is not fully absorbed by the load. The difference between the incident power and the load power is the power reflected back toward the generator.

2. The voltage antinodes may break down the insulation (dielectric) between the conductors and cause arc-over.

3. Since power losses are proportional to the square of the voltage and to the square of the current, the attenuation on the practical line increases due to the presence of standing waves.

4. The input impedance at the generator is a totally unknown quantity that varies with the length of the line and the frequency of the generator.

By contrast with these disadvantages, resonant lines have a number of useful applications. To review, when a quarter-wave line is shorted at one end and is excited to resonance by the circuit frequency applied at the other end, standing waves of voltage and current appear on the line. At the short circuit the voltage is zero and the current is a maximum. At the input end the current is nearly zero and the voltage is at its peak. The input impedance to the resonant quarter-wave line is extremely high, and therefore this section behaves as an insulator. Figure 2–21 shows a quarter-wave section of line acting as a standoff insulator (support stub) for a two-wire transmission line. At the particular frequency that makes the section a

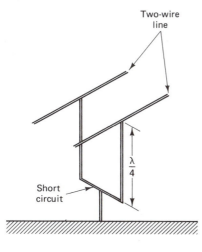

Figure 2–21 Quarter-wave support stub (metallic insulator).

quarter-wavelength line, the stub acts as a highly efficient insulator. However, at other frequencies the stub is no longer resonant and will behave as an inductor or a capacitor; such behavior causes a mismatch and standing waves will appear on the two-wire line. At the second harmonic frequency the stub is $\lambda/2$ long and since there is no impedance transformation over the distance of one half-wavelength, the stub will place a short circuit across the main line. No power at the second harmonic frequency can then be transferred down the line. In this way a section of the line that is $\lambda/4$ long at a fundamental frequency and is shorted at one end will have the effect of filtering out all even harmonics.

The quarter-wave section may also be used to match two nonresonant lines with different surge impedances (Figure 2–22). When a $\lambda/4$ line of surge impedance Z_0 is terminated by a load Z_L the input impedance Z_{in} is given by

$$Z_{in} = \frac{Z_0^2}{Z_L} \tag{2–42}$$

Then

$$Z_0 = \sqrt{Z_{in} \times Z_L} \tag{2–43}$$

This formula can be used to derive some results that we have already obtained. For example,

If $Z_L = \infty\ \Omega$ (an open circuit), $Z_{in} = 0\ \Omega$ (a short circuit).
If $Z_L = Z_0\ \Omega$ (matched conditions), $Z_{in} = Z_0\ \Omega$.
If $Z_L = 0\ \Omega$ (a short circuit), $Z_{in} = \infty\ \Omega$ (an open circuit).

In our example we need to match 70-Ω and 300-Ω nonresonant lines. The $\lambda/4$ section would need to have a surge impedance of $\sqrt{70 \times 300} = 145\ \Omega$. The effective load on the 300-Ω line as seen from the generator is $145^2/70 = 300\ \Omega$, while the 70-Ω line is effectively connected to a $145^2/300 = 70$-Ω termination. Both nonresonant lines are therefore matched and the standing waves exist only on the $\lambda/4$ section.

Example 2–5   A transmission line whose surge impedance is 50 Ω has an air dielectric and is terminated by a resistive load of 75 Ω. If a 150-MHz generator delivers 100 W of incident power to the load and the line is assumed to be loss-free, calculate the

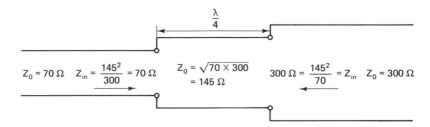

Figure 2–22   Principle of the quarter-wave impedance transformer.

values of the (a) VSWR, (b) reflection coefficient, (c) reflected power, (d) load power, (e) maximum and minimum impedances on the line, (f) maximum and minimum effective voltages and currents on the line, and (g) distance between two adjacent voltage nulls.

Solution   (a) The VSWR

$$S = \frac{R_L}{Z_0} = \frac{75}{50} = \mathbf{1.5} \tag{2–41}$$

(b) The reflection coefficient

$$\begin{aligned} P &= \frac{S-1}{S+1} \\ &= \frac{0.5}{2.5} \\ &= \mathbf{0.2} \end{aligned} \tag{2–39}$$

(c) The reflected power

$$\begin{aligned} P_r &= P^2 \times P_i \\ &= (0.2)^2 \times 100 \\ &= \mathbf{4\ W} \end{aligned} \tag{2–37}$$

(d) The load power

$$\begin{aligned} P_L &= P_i - P_r \\ &= 100 - 4 \\ &= \mathbf{96\ W} \end{aligned} \tag{2–37}$$

(e) The maximum impedance

$$\begin{aligned} Z_{max} &= S Z_0 \\ &= 1.5 \times 50 \\ &= \mathbf{75\ \Omega} \end{aligned} \tag{2–35}$$

The minimum impedance

$$\begin{aligned} Z_{min} &= \frac{Z_0}{S} \\ &= \frac{50}{1.5} \\ &= \mathbf{33.33\ \Omega} \end{aligned} \tag{2–44}$$

(f) The incident effective voltage

$$\begin{aligned} V_i &= \sqrt{100\ W \times 50\ \Omega} \\ &= 70.7\ V \end{aligned}$$

The maximum effective voltage

$$\begin{aligned} E_{max} &= V_i(1 + P) \\ &= 70.7 \times 1.2 \\ &= \mathbf{84.84\ V} \end{aligned} \tag{2–38}$$

The minimum effective voltage

$$E_{min} = V_i(1 - P)$$
$$= 70.7 \times 0.8$$
$$= \mathbf{56.56 \ V}$$

The maximum effective current

$$I_{max} = \frac{E_{min}}{Z_{min}} \tag{2-44}$$

$$= \frac{56.56 \ V}{33.33 \ \Omega}$$
$$= \mathbf{1.70 \ A}$$

The minimum effective current

$$I_{min} = \frac{E_{max}}{Z_{max}}$$
$$= \frac{84.84 \ V}{75 \ \Omega} \tag{2-43}$$
$$= \mathbf{1.13 \ A}$$

(g) With an air dielectric, the wavelength on the line, λ, is approximately $300/150 = 2$m. Consequently, the distance between two adjacent nulls is $\lambda/2 = \mathbf{1 \ m.}$

Example 2-6   A transmission line with an air dielectric has a surge impedance of 50 Ω and is terminated by a load consisting of a 33-Ω resistance in series with a 21-Ω capacitive reactance. Calculate the values of the (a) reflection coefficient, and (b) voltage standing-wave ratio.

Solution   (a) The load impedance

$$Z_L = 33 - j21 \ \Omega$$

The reflection coefficient

$$\rho = \frac{Z_L - Z_0}{Z_L + Z_0}$$
$$= \frac{(33 - j21) - 50}{(33 - j21) + 50}$$
$$= \frac{-17 - j21}{83 - j21} \tag{2-40}$$
$$= \frac{27 \ \underline{/-129°}}{86 \ \underline{/-14°}}$$
$$= \mathbf{0.31 \ \underline{/-115°}}$$

(b) The standing-wave ratio

$$S = \frac{1 + P}{1 - P}$$
$$= \frac{1 + 0.31}{1 - 0.31} \tag{2-41}$$
$$= \frac{1.31}{0.69} = \mathbf{1.90}$$

Example 2–7   A transmission line is terminated by a short circuit. At a certain position, X, on the line there is a 2-dB attenuation between X and the short circuit. Calculate the value of the VSWR at position X.

Solution   The total two-way attenuation $= 2 \times 2 = 4$ dB. The corresponding power ratio

$$P^2 = \log^{-1}\left(\frac{-4}{10}\right)$$
$$= 0.398 \tag{2-37}$$

The reflection coefficient

$$P = \sqrt{0.398} = 0.631$$

The VSWR

$$S = \frac{1+P}{1-P} = \frac{1.631}{0.369} = \mathbf{4.42} \tag{2-41}$$

2–5 ARTIFICIAL LINES

It is possible to use energy considerations to derive some of the results in Sections 2–2 and 2–3. Figure 2–23a represents a lossless open-circuited twin line whose length is l and to which a positive-going step voltage V is applied. The result is to create a step current of magnitude I such that $V/I = Z_0$. The step voltage and the step current move down the line together (Figure 2–23b) by charging the distributed capacitance and creating a magnetic field around the distributed inductance. When the steps reach the open circuit after a time $t = T$ seconds, the current is reduced to zero, so that the magnetic field collapses and induces a voltage which charges to 2V the distributed capacitance at the end of the line. This is the beginning of the returning voltage and current steps, which occur as the result of total reflection from the open-circuited end. The reflected voltage step is also positive going $(+V)$; this is equivalent to the in-phase type of reflection involving sine waves. As this step moves back to the beginning of the line, the distributed capacitance is progressively charged to 2V. However, the reflected current is negative going and equal to $-I$ (similar to the sine-wave out-of-phase reflection), and as this step travels back toward the start, the current on the line is reduced to zero (Figure 2–23c). After a total time of $2T$ seconds, the line voltage everywhere equals 2V but the current is zero throughout (Figure 2–23d).

If the line's total capacitance is C_T, the energy in this capacitance is $\frac{1}{2} C_T (2V)^2 = 2C_T V^2$ joules. This energy is derived from the source that supplied the step voltage V and the step current I for a time of $2T$ seconds. Consequently,

$$V \times I \times 2T = 2C_T V^2 \tag{2-45}$$

Next, consider identical voltage and current steps applied to the same line, which is now short circuited (Figure 2–24a). The initial movement of the steps down the line does not change (Figure 2–24b) and total reflection will again occur at the short

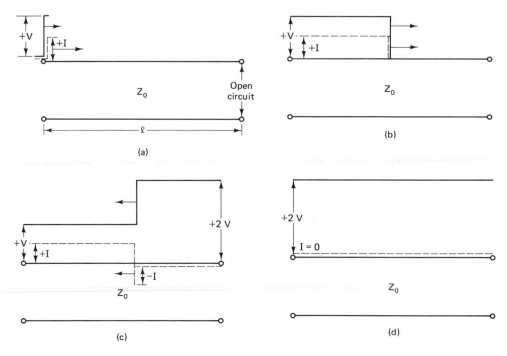

Figure 2–23 Charging of an open-circuited line: (a) $t = 0$; (b) $0 < t < T$; (c) $2T > t > T$; (d) $t = 2T$.

circuit. However, the reflected voltage step will be negative-going $(-V)$, while the reflected current step is positive-going $(+I)$ (Figure 2–24c). After a total time of $2T$ seconds, the line current is everywhere equal to $2I$ but the voltage is zero (Figure 2–24d).

If the line's total inductance is L_T, the energy stored in this inductance is $\frac{1}{2}L_T(2I)^2 = 2L_TI^2$ joules. Again this energy has been supplied from the source that provided the steps, so

$$V \times I \times 2T = 2L_TI^2 \tag{2–46}$$

From equations (2–45) and (2–46),

$$2C_TV^2 = 2L_TI^2$$

or

$$\frac{V}{I} = Z_0 = \sqrt{\frac{L_T}{C_T}} \quad \Omega \tag{2–47}$$

This is the same formula as equation (2–6).

In addition,

$$T = C_T \times \frac{V}{I} = C_T \times \sqrt{\frac{L_T}{C_T}} = \sqrt{L_TC_T} \quad \text{s} \tag{2–48}$$

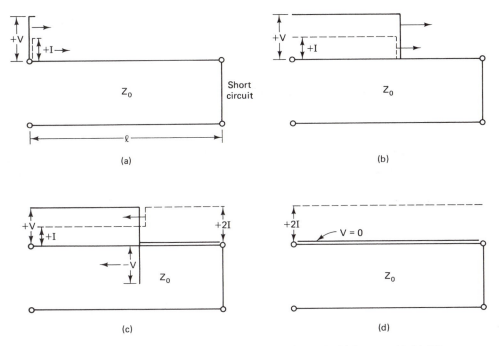

Figure 2–24   Charging of a short-circuited line: (a) $t = 0$; (b) $0 < t < T$; (c) $2T < t < T$; (d) $t = 2T$.

If L and C are, respectively, the line's distributed inductance and capacitance, $L_T = L \times \mathit{l}$ and $C_T = C \times \mathit{l}$. Then

$$T = \sqrt{L \times \mathit{l} \times C \times \mathit{l}} \qquad \text{s}$$

or

$$v = \frac{\mathit{l}}{T} = \frac{\mathit{l}}{\sqrt{LC}} \qquad \text{m/s} \tag{2–49}$$

where v is the velocity of the steps as they move down the line. This formula is the same as the expression for the phase velocity in equation (2–23).

Similar results can be obtained with artificial lines which are often used to form rectangular pulses or to delay pulses. They consist of a repeating network of (lumped) inductors and capacitors (Figure 2–22) which simulate the behavior of a real line and are used when the required length of a real line is far too long to be practical. For example, it might require 270 m of an actual transmission line to delay a pulse by 1 μs. Moreover, the surge impedance of real lines may be inconveniently low.

If an artificial line consists of N sections, each of inductance L and capacitance C, the time T taken for a step voltage to traverse the line once is

$$T = \sqrt{NL \times NC} = N\sqrt{LC} \qquad \text{s} \tag{2–50}$$

If the step voltage traverses the line back and forth (two-way travel), the total time is

$$2T = 2N\sqrt{LC} \tag{2–51}$$

The line's equivalent surge impedance is

$$Z_0 = \sqrt{\frac{L}{C}} \quad \Omega \tag{2–52}$$

The behavior of an artificial line is only comparable with that of an actual transmission line provided that the wavelength is compared with one section. It follows that there is a critical frequency f_c *below* which the operations of the real and artificial lines are approximately the same. The value of this cutoff frequency is

$$f_c = \frac{1}{\pi\sqrt{LC}} \quad \text{Hz} \tag{2–53}$$

A pulse will therefore travel along an artificial line at a definite speed and with no change of shape, provided that the significant harmonic frequencies contained in the pulse are below the cutoff value.

Charging an Open-Ended Artificial Line

Let us consider an open-circuited artificial line that is to be charged from a dc source E whose internal resistance is R_i (Figure 2–25). The manner in which the terminal voltage V ultimately reaches E volts depends on the ratio of R_i to Z_0 (the surge

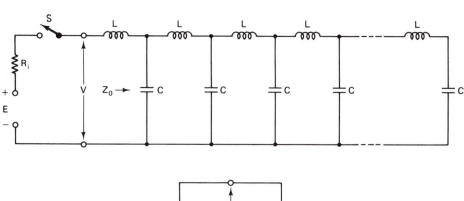

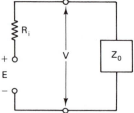

Figure 2–25   Charging an open-ended artificial line.

impedance of the artificial line). There are three possible cases which we have to take into account:

$R_i = Z_0$ **(Figure 2–26a).**   When the switch S is closed, the source voltage will divide equally between R_i and Z_0 so that the initial value of the terminal voltage rises abruptly to $E/2$ volts. This voltage step creates a corresponding current step $I/2$ of value $E/2Z_0$ amperes. The current and voltage steps move down the line together and the capacitors are charged to a voltage of $E/2$. Here an interesting point arises. We are accustomed to saying that the current can flow only in a closed circuit. However, we know that for a capacitor with a perfect dielectric, the electron flow exists in the circuit external to the capacitor but does not flow through the capacitor. Clerk Maxwell resolved this inconsistency by conceiving the idea of a *displacement current* which exists between the plates. A capacitor stores its energy in the form of an electric field whose flux ψ is measured in coulombs (in the SI system). When the capacitor is being charged, the flux is changing at a certain rate, and this rate of change of the flux must be measured in coulombs per second, which is equivalent to amperes (the unit of current). The rate of change of the electric flux is therefore the displacement current, and we can say that the current exists throughout a closed circuit which contains a capacitor.

After a time $T = N\sqrt{LC}$ seconds, both the voltage and current steps have reached the end of the line, which is charged throughout to $E/2$ volts (Figure 2–26b). At the end the current falls to zero and the magnetic field collapses, inducing a voltage which charges the final capacitor to E volts. This is the start of the reflected voltage step, $+E/2$, which is associated with a reflected current step, $-I/2$. As the reflected steps travel back toward the source the line is progressively charged to E volts, while the current is correspondingly reduced to zero (Figure 2–26c). When these steps arrive back at the source, whose internal resistance matches the line's surge impedance, no reflection occurs. The line is totally charged to E volts and the current along the line is zero. The value of the terminal voltage V therefore remained at $E/2$ volts for a total time of $2T = 2N\sqrt{LC}$ seconds and then climbed abruptly to E volts; consequently, only two "steps" were involved in the terminal voltage rising from zero to the value of E (Figure 2–26d).

$R_i > Z_0$ **(Figure 2–27a).**   The artificial line now requires more than two "steps" for the charging process. The higher the value of R_i in relation to Z_0, the smaller is each step and the longer is the time for the terminal voltage V to reach a certain percentage of E.

Let us take a particular case in which $R_i = 2Z_0$. As soon as the switch S is closed, the source voltage will divide between R_i and $2Z_0$ and the terminal voltage, by the voltage division rule, is

$$V = E \times \frac{Z_0}{R_i + Z_0} = E \times \frac{Z_0}{2Z_0 + Z_0}$$

$$= \frac{E}{3} \text{ volts}$$

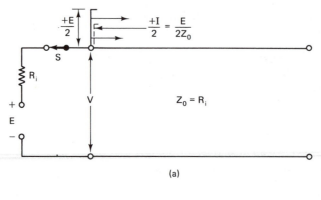

(a)

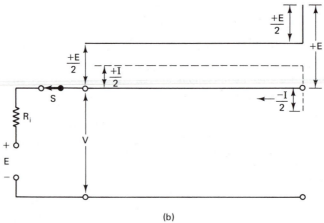

(b)

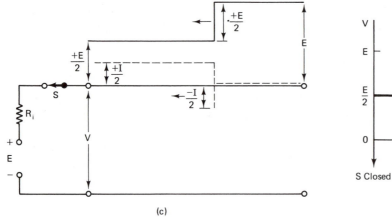

(c)

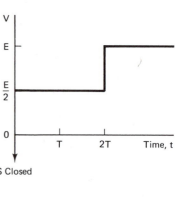

(d)

Figure 2–26   Charging an open-ended artificial line whose surge impedance is matched to the internal resistance of the dc source: (a) $t = 0$; (b) $t = T$; (c) $2T > t > T$; (d) $t > 2T$.

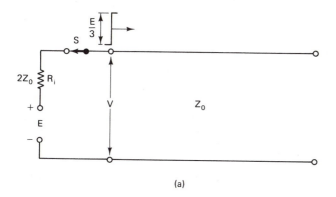

(a)

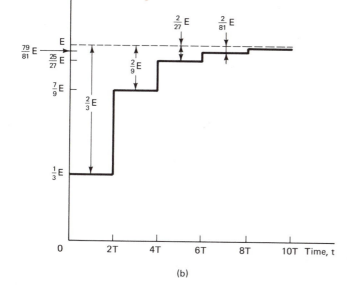

(b)

Figure 2–27 Charging of an artificial line whose surge impedance is less than the charging resistance of the dc source. Time $t = 0$.

Consequently, the initial difference between the values of V and E is $E - \frac{1}{3}E = \frac{2}{3}E$. The $E/3$ step travels down the line and reaches the end after a time $T = N\sqrt{LC}$ seconds. Upon reflection a $+E/3$ step travels back toward the source and progressively charges the line to $2E/3$ volts. After a total time of $2T = 2N\sqrt{LC}$ seconds, the $+E/3$ voltage step arrives back at the source, and since the source is not matched to the line, some reflection will occur. The amount of the step reflected from the source is

$$(E - \tfrac{2}{3}E) \times \frac{Z_0}{Z_0 + 2Z_0} = \frac{E}{9}$$

so that the new value of V is $2E/3 + E/9 = 7E/9$ and the difference between E and V has been reduced to $E - 7E/9 = 2E/9$.

There is another way for us to obtain this result. The reflection coefficient at the source is

$$\frac{2Z_0 - Z_0}{2Z_0 + Z_0} = +\frac{1}{3}$$

[equation (2–36)], so that when the $+E/3$ voltage step arrives back at the source, it is reflected as a $(+E/3) \times (+\frac{1}{3}) = +E/9$ step.

Let us consider a general case in which $R_i = nZ_0$. The initial value of V is

$$\frac{E \times Z_0}{nZ_0 + Z_0} = \frac{E}{n+1}$$

and the difference between E and V is

$$E - \frac{E}{n+1} = \frac{nE}{n+1}$$

The reflection coefficient at the source is

$$\frac{nZ_0 - Z_0}{nZ_0 + Z_0} = \frac{n-1}{n+1}$$

so that after the time $2T$, the value of V is

$$\frac{2 \times E}{n+1} + \frac{E}{n+1} \times \left(\frac{n-1}{n+1}\right) = E\frac{3n+1}{(n+1)^2}$$

The difference between V and E is therefore

$$E - \frac{E(3n+1)}{(n+1)^2} = \frac{E(n^2 - n)}{(n+1)^2} = \frac{nE}{n+1} \times \frac{n-1}{n+1}$$
$$= \rho \times \text{(previous difference between } V \text{ and } E)$$

Therefore, in our example where $R_i = 2Z_0$ and $\rho = +\frac{1}{3}$, the differences between V and E after times $4T$, $6T$, $8T$, and so on, are respectively $2E/27$, $2E/81$, $2E/243$, and so on. Consequently, the value of V after time $8T$ is $241E/243$ and the line is charged to within 1% of the source voltage. The relationship between V and time is illustrated in Figure 2–27b.

$R_i < Z_0$ **(Figure 2–28a).**   Let our example be $R_i = Z_0/2$; consequently, the reflection coefficient at the source is

$$\rho = \frac{(Z_0/2) - Z_0}{(Z_0/2) + Z_0} = -\frac{1}{3}$$

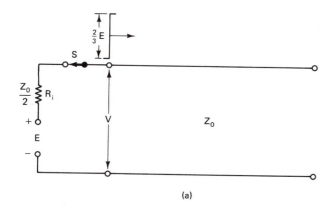

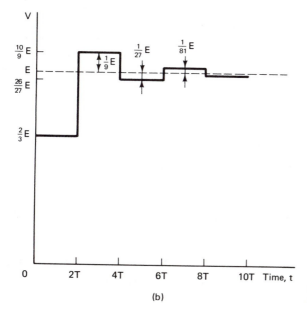

(b)

Figure 2–28 Charging of an open-ended artificial line whose surge impedance is greater than the charging resistance of the dc source. Time $t = 0$.

The initial value of V is

$$\frac{E \times Z_0}{\dfrac{Z_0}{2} + Z_0} = \frac{2}{3} E$$

and the difference between V and E is $E - \frac{2}{3}E = \frac{1}{3}E$. After the time $2T$, the value of V is

$$\tfrac{2}{3}E + \tfrac{2}{3}E + [(+\tfrac{2}{3}E) \times (-\tfrac{1}{3})] = \tfrac{10}{9}E$$

so that V has risen *above* the level of E, and the difference between V and E is $E - \frac{10}{9}E = -\frac{1}{9}E$. The value of V then oscillates about the source voltage as illustrated in Figure 2–28b. The differences between V and E at times of $4T$, $6T$, $8T$, and so on, are, respectively,

$$\left(-\frac{1}{9}E\right)\times\left(-\frac{1}{3}\right)=+\frac{E}{27}, \quad \left(+\frac{E}{27}\right)\times\left(-\frac{1}{3}\right)=-\frac{E}{81}, \quad \left(-\frac{E}{81}\right)\times\left(-\frac{1}{3}\right)=+\frac{E}{243}$$

Therefore, at the time $8T$, $V = E - E/243 = 242E/243$ and the line is charged to within 1% of the source voltage (Figure 2–28b).

Discharging an Artificial Line

After being charged to a certain voltage E, an open-ended artificial line may be discharged through a load resistance R_L. One purpose is to create a rectangular voltage pulse whose magnitude and duration are accurately known. An application of such a pulse occurs in a radar transmitter's modulator unit, where the system's pulse length is determined. In this manner the artificial line acts as a pulse-forming network.

The circuit of the discharge line and its accompanying load resistance are shown in Figure 2–29a. The time taken for the total discharge depends on the relationship between R_L and Z_0 (the line's equivalent surge impedance). As with the charging of the line, there are three possible cases to consider.

$R_L = Z_0$.   When the switch S is closed, the voltage E is divided equally between R_L and Z_0, so that initially the voltage V_L rises abruptly to $E/2$ volts. At the same time the line voltage between points X and Y suddenly falls to $E/2$. This creates a negative voltage step of $-E/2$, which travels down the line and progressively reduces the line voltage to $E/2$ volts. After a time $T = N\sqrt{LC}$ seconds, the step reaches the end of the line and total reflection occurs. Another $-E/2$ step travels back toward the load resistance and reduces the line voltage to zero. After a time $2T = 2N\sqrt{LC}$ seconds has elapsed from the time the switch is closed, the line is completely discharged. The waveform (Figure 2–29b) of V_L versus time then illustrates the rectangular pulse developed across the load R_L.

$R_L > Z_0$ (Figure 2–30a).   The artificial line now requires a longer time in which to discharge. Let us consider one such case where $R_L = 2Z_0$. As soon as the switch S is closed, the voltage V_L rises from zero to

$$E \times \frac{2Z_0}{2Z_0 + Z_0} = \frac{2}{3}E$$

A negative step of $E - \frac{2}{3}E = E/3$ travels down the line and is reflected from the open-circuited end. After a total time $2T = 2N\sqrt{LC}$ seconds, the line is totally discharged to $\frac{1}{3}E$, but since the reflection coefficient at the load resistance is

$$\rho = \frac{2Z_0 - Z_0}{2Z_0 + Z_0} = +\frac{1}{3}$$

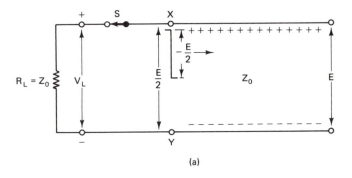

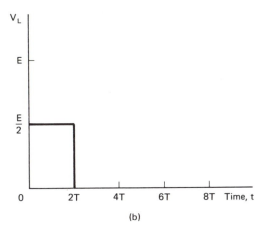

(b)

Figure 2–29   Discharge of artificial line whose surge impedance is matched to the load resistance R_L. Time $t = 0$.

there is a new reflected step whose magnitude is $(-\frac{1}{3}E) \times (+\frac{1}{3}) = -\frac{1}{9}E$. The value of V_L therefore falls to $(\frac{1}{3} - \frac{1}{9})E = \frac{2}{9}E$, which is equal to $\rho \times$ (previous value of V_L). The general analysis for the discharge line is, in fact, comparable with the analysis for the charging process. After times $4T$, $6T$, $8T$, and so on, the values of V_L are respectively $2E/27$, $2E/81$, $2E/243$, and so on, so that after the time $8T$, V_L has fallen to within 1% of its final zero value (Figure 2–30b).

$R_L < Z_0$ **(Figure 2–31a).**   As with the charging process, let our example be $R_L = \dfrac{Z_0}{2}$; consequently, the reflection coefficient at the load resistance is

$$\rho = \frac{(Z_0/2) - Z_0}{(Z_0/2) + Z_0} = -\frac{1}{3}$$

After the switch S is closed, the initial value of V_L is

$$\frac{E \times Z_0/2}{(Z_0/2) + Z_0} = \frac{E}{3}$$

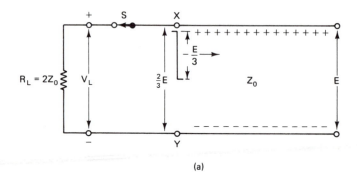

(a)

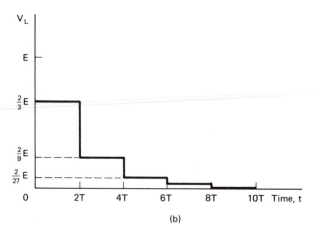

(b)

Figure 2–30 Discharge of artificial line whose surge impedance is less than the value of the load resistance, R_L. Time $t = 0$.

This results in a step of $-\frac{2}{3}E$, which travels down the line and progressively discharges the line to $E/3$ volts. When the $-\frac{2}{3}E$ step is totally reflected at the open-circuited end and travels back toward the load resistance, the line is discharged and then recharged to $-E/3$ volts with the polarity reversed. After a time of $2T = 2N\sqrt{LC}$ seconds, the $-\frac{2}{3}E$ step is reflected as a new step of $(-\frac{2}{3}E) \times (-\frac{1}{3}) = +\frac{2}{9}E$. The magnitude of V_L falls to $-E/3 + 2E/9 = -E/9$ volts, which is equal to $\rho \times$ (previous value of V_L). It follows that V_L oscillates above and below the zero level; after times of $4T$, $6T$, $8T$, and so on, the values of V_L are, respectively,

$$\left(-\frac{E}{9}\right) \times \left(-\frac{1}{3}\right) = +\frac{E}{27}$$

$$\left(+\frac{E}{27}\right) \times \left(-\frac{1}{3}\right) = -\frac{E}{81}$$

$$\left(-\frac{E}{81}\right) \times \left(-\frac{1}{3}\right) = +\frac{E}{243} \text{ volts}$$

and so on (Figure 2–31b).

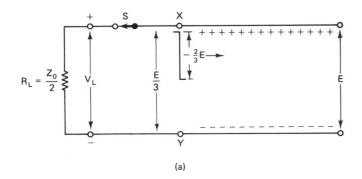

(a)

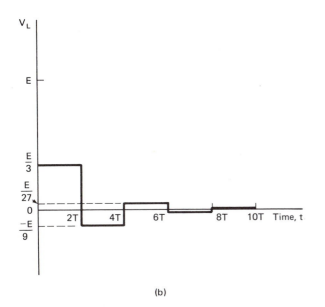

(b)

Figure 2-31 Discharge of artificial line whose surge impedance is greater than the value of the load resistance, R_L. Time $t = 0$.

In practice, the value of the load resistance may fluctuate. To avoid the problems that arise when R_L becomes greater than Z_0, it is common practice to make R_L less than Z_0 and then arrange that the switch is opened after the time $2T$; the waveform of V_L is then the desired rectangular pulse.

In summary, we have learned that an artificial line can replace a true line when the true line's length becomes impractical. Such an artificial line may be used as a pulse-forming network or for delay purposes.

Example 2-8   An artificial delay line consists of four LC sections, each of inductance 5 mH and capacitance 50 pF. Calculate (a) how long does it take a voltage pulse to traverse the line, and (b) what is the line's equivalent surge impedance?

Solution   (a) The delay time

$$T = N \sqrt{LC}$$
$$= 4 \times \sqrt{(5 \times 10^{-3}) \times (50 \times 10^{-12})} \qquad (2\text{-}50)$$
$$= 20 \times 10^{-7} \text{ s} = \mathbf{2 \; \mu s}$$

Notice that practical delay times are of the order of a few microseconds. Relatively long delay times are difficult to create because of the large number of sections required.

(b) The equivalent surge impedance

$$Z_0 = \sqrt{\frac{L}{C}}$$
$$= \sqrt{\frac{5 \times 10^{-3}}{50 \times 10^{-12}}} \qquad (2\text{-}52)$$
$$= 10{,}000 \; \Omega$$
$$= \mathbf{10 \; k\Omega}$$

Example 2–9   An artificial discharge line consists of five *LC* sections, each with an inductance of 100 μH and a capacitance of 100 pF. The line is charged to 1 kV and is then abruptly connected across a 1-kΩ resistive load. What is the waveform that appears across the load?

Solution   The line's surge impedance

$$Z_0 = \sqrt{\frac{L}{C}}$$
$$= \sqrt{\frac{100 \times 10^{-6}}{100 \times 10^{-12}}} \qquad (2\text{-}52)$$
$$= 1000 \; \Omega$$
$$= 1 \; k\Omega$$

The value of the load is equal to the line's surge impedance. Consequently, the waveform across the load is a rectangular pulse with an amplitude of 1 kV/2 = **500 V.** The duration of the pulse is

$$2N \sqrt{LC} \text{ seconds} = 2 \times 5 \times \sqrt{100 \times 10^{-6} \times 100 \times 10^{-12}} \text{ s} \qquad (2\text{-}51)$$
$$= \mathbf{1 \; \mu s}$$

2–6 THE SMITH CHART AND ITS APPLICATIONS

In example 2–6 we were able to determine the conditions that existed at the end of the mismatched line. However, we were in no way able to calculate the value of the impedance existing at any position of the line. Such calculations involve a level of mathematics that is beyond the scope of this book. However, we can obtain reasonably accurate results by the use of the Smith chart, which was originally developed in 1939. In the example of Figure 2–32, the chart refers to a particular surge impedance of 50 Ω.

First, we know that the impedance along the line must vary in a periodic manner since on a lossless line, two identical conditions of voltage (and current) are separated

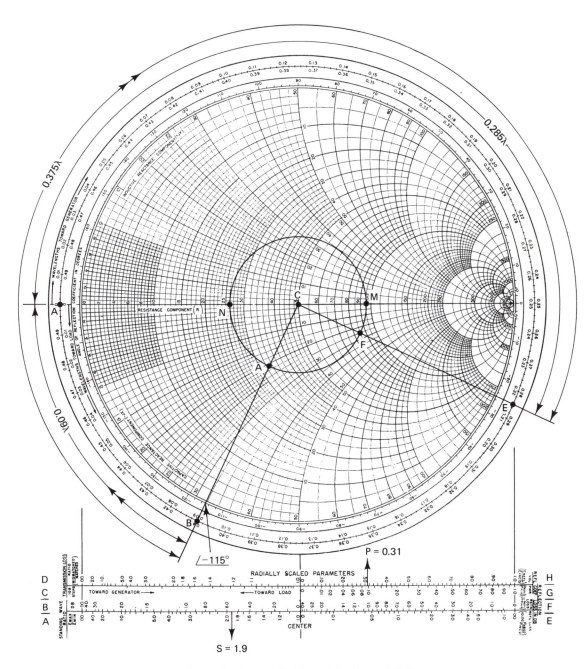

Figure 2–32   50-Ω Smith chart for Example 2–6.

by a half-wavelength. We therefore notice that each of the two outermost circular scales represents a total distance of 0.5λ; in other words, one complete revolution around the chart is equivalent to moving a distance of one half-wavelength along the line. If actual distances are given, we must convert them to equivalent wavelengths by knowing the frequency. For example, if the frequency is 7.5 GHz, the wavelength is 30/7.5 = 4 cm and a distance of 2.8 cm would correspond to 2.8/4.0 = 0.7λ.

The chart itself is composed of a coordinate grid that consists of two sets of circles; examples of these circles appear in Figure 2–32. The first set are complete circles that represent the resistive components of impedance. All these circles touch one another at the common point 0. The other sets of circles are, in fact, partial arcs which also mutually touch at point 0. The arcs to the right and left of the center line, respectively, represent the inductive and capacitive reactance components of the impedance.

The resistance circles and the reactance arcs always intersect at right angles (90°). The point that relates to a particular impedance is found by first locating the two circles which represent the particular resistance and reactance values. The point of intersection of these circles corresponds to the impedance. Of course, if the resistance and reactance values do not exactly correspond to the values of the circles, the impedance point must be found by interpolating between the circles.

To illustrate the use of the Smith chart, let us repeat Example 2–6. By interpolation the 33-Ω resistance circle and the 21-Ω capacitive reactance circle intersect at the impedance point A. We now draw a line through A and the center of the chart C; this line is then extended to the scale that measures the angle of the reflection coefficient. At point B the angle is found to be −115°, which agrees with the mathematical result.

The next step is to draw a circle whose center is C and which passes through point A. This is known as the VSWR circle since its radius is equal to the magnitude of the VSWR. The circle will also enable us to determine the value of the impedance at any point on the line.

To find the magnitude of the reflection coefficient, we measure the distance CA along the radial scale H, which is marked as "Refl. Coef., VOL(E REFL/E INCD)". The value of P is shown to be 0.31. Using the same procedure on the radial scale, A (standing-wave ratio E_{max}/E_{min}), the corresponding value of S is 1.9. It is obvious that these results have been obtained much more quickly and simply with the Smith chart than with the mathematical analysis.

Now let us expand the example by assuming that the load is 21.5 cm from a 7.5-GHz generator. The distance of 21.5 cm is equivalent to 21.5/4 = 5.375λ. To find the input impedance at the generator we must first remember that 5λ represents 10 complete revolutions around the Smith chart; consequently, we only need to consider the remaining 0.375λ. Starting at point B, we must move around the chart in the clockwise direction toward the generator. When point D is reached, we have covered a distance of 0.090λ. Therefore, we still have to travel a further 0.375λ − 0.090λ = 0.265λ, which brings us to point E. The line CE intersects the VSWR circle at point F, which represents the input impedance at the generator. Reading off the coordinates at point F, the value of this input impedance is $85 - j25$ Ω. Points M and N represent, respectively, the maximum and minimum (resistive) impedances of 96 and 26 Ω.

So far it would appear that we need a new chart for each value of the surge impedance. However, if in our example we doubled the load impedance to $2 \times (33 - j21) = 66 - j42$ Ω, the position of this impedance point on a 100-Ω chart would be identical to the point A on the 50-Ω chart. In other words, the position of an impedance point is determined by the ratios of the resistive and reactive component values to the value of the surge impedance. It follows that we can develop a universal Smith chart (Figure 2–33) in which all impedance values are divided by the magnitude of the characteristic impedance before being entered on the chart. This process is called *normalization*; in our examples, the normalized values of $66 - j42$ for $Z_0 = 100$ Ω is $(66 - j42)/100 = 0.66 - j0.42$ (0.66 is the normalized resistance, R/Z_0 and $-j0.42$ is the normalized reactance, $-jX_c/Z_0$). This is the same result as we obtained for $(33 - j21)/50 = 0.66 - j0.42$ with $Z_0 = 50$ Ω. When $0.66 - j0.42$ is entered on the universal chart, its position is the same as that of point A on the 50-Ω chart. However, you must remember that when a solution is finally obtained on the universal chart, the coordinates must be read off and then *denormalized* (multiplied by Z_0) to obtain the actual value of the impedance.

Apart from the fact that the universal chart can be used with any value of surge impedance, it can also be used for admittance coordinates where the full circles represent normalized conductances G/Y_0, with the surge admittance $Y_0 = 1/Z_0$, and the arcs signify the normalized susceptances [$-j(B_L/Y_0)$ to the left of the center line and $+j(B_c/Y_0)$ on the right]. The admittance coordinates will be used in Examples 2–14 and 2–16, which involve the use of stubs.

In our example we made use of the radial scales A and H. The following is the complete list of the radial scales, together with the quantities which these scales measure.

A   Voltage standing-wave ratio, $S = \dfrac{1+P}{1-P}$

B   Voltage standing-wave ratio in dB $= 20 \log S$

C   Transmission-loss scale in dB $= 10 \log P$

D   Loss coefficient (on an unmatched line as compared with the same line

      when matched) $= \dfrac{1+S^2}{2S}$

E   Reflection loss in dB $= -10 \log (1 - P^2)$

F   Return loss in dB $= 20 \log P$

G   Power reflection coefficient $= \dfrac{\text{reflected power}}{\text{incident power}} = P^2$

H   Voltage reflection coefficient $= \dfrac{\text{reflected voltage}}{\text{incident voltage}} = \dfrac{S-1}{S+1}$

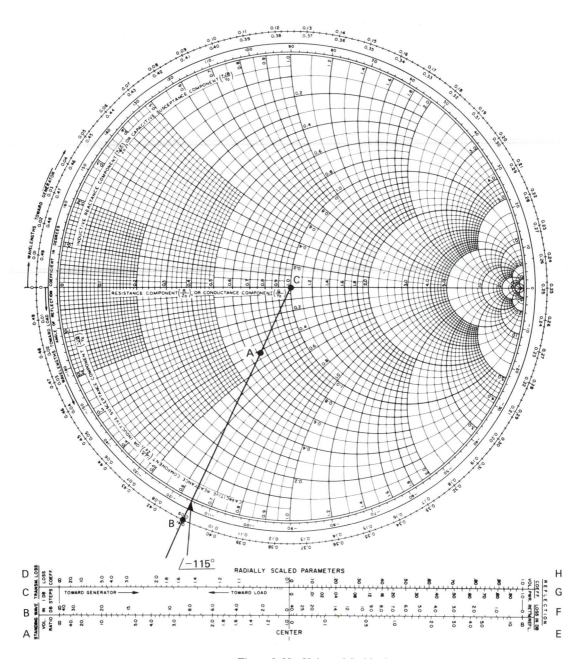

Figure 2–33 Universal Smith chart.

The following examples serve to illustrate the varied uses of the universal Smith chart.

Example 2–10   A 100-Ω transmission line with an air dielectric is terminated by a load of 50 − j80 Ω. Determine the values of the VSWR, the reflection coefficient, and the percentage of the reflected power. If the power incident at the load is 100 mW, calculate the power in the load. If the generator frequency is 3 GHz and the line is 73 cm long, find the input impedance at the generator. What are the values of the maximum and minimum impedances existing on the line?

Solution

Step 1: Normalize the value of the load so that $Z_{LN} = (50 − j80)/100 = 0.5 − j0.8$.

Step 2: Plot Z_{LN} on the universal Smith chart (Figure 2–34) and draw the VSWR circle through Z_{LN}.

Step 3: Mark off a length equal to the circle's radius on the H scale. Then the reflection coefficient $P = \textbf{0.56}$. Following the same procedure on the G scale, the percentage of the reflected power is $P^2 \times 100 = 0.31 \times 100 = \textbf{31\%}$. Then the reflected power is $(31/100) \times 100 = 31$ mW and the power in the load is $100 − 31 = \textbf{69 mW}$.

Notice that the reflected power is $10 \log (31/100) = −5.03$ dB with respect to the incident power while the load power when compared with the incident power is $10 \log (69/100) = −1.61$ dB. By marking off a length equal to the VSWR circle's radius on the E scale (REFL Loss), we can obtain the value of 1.61 dB. Carrying out the same procedure for the F scale (RET'N Loss) gives us our answer of 5.03 dB. By measuring off the same length on the A scale, the value of the VSWR, $S = \textbf{3.6}$.

Step 4: From the center of the chart draw a line through Z_{LN} and extend the line to the peripheral scales. Since all distances on the chart are measured in terms of wavelength, it is necessary to determine the line's wavelength, which is given by

$$\lambda = \frac{c}{f} = \frac{3 \times 10^{10}}{3 \times 10^9} \text{ cm} = 10 \text{ cm}$$

Therefore, 73 cm is equivalent to $73/10 = 7.3\lambda$.

Identical impedance values on a mismatched line repeat every half-wavelength, which is the distance covered by one complete revolution on the Smith chart. A distance of 7.3λ will require 14 complete revolutions (7.0λ) together with an additional rotation of 0.3λ in the clockwise direction toward the generator.

On the inner peripheral wavelength scale, the distance from the Z_{LN} position to the null position, N, is 0.118λ. We must then travel toward the generator a further $0.3\lambda − 0.118\lambda = 0.182\lambda$ on the outermost peripheral scale and arrive at point G.

Step 5: Draw a line from G to the center of the chart. Point Z_{GN} of intersection between this line and the VSWR circle represents the normalized input impedance at the generator; therefore, $Z_{GN} = 1.18 + j1.43$ and the denormalized value of this input impedance is $(1.18 + j1.43) \times 100 = \textbf{118} + j\textbf{143 Ω}$. The minimum

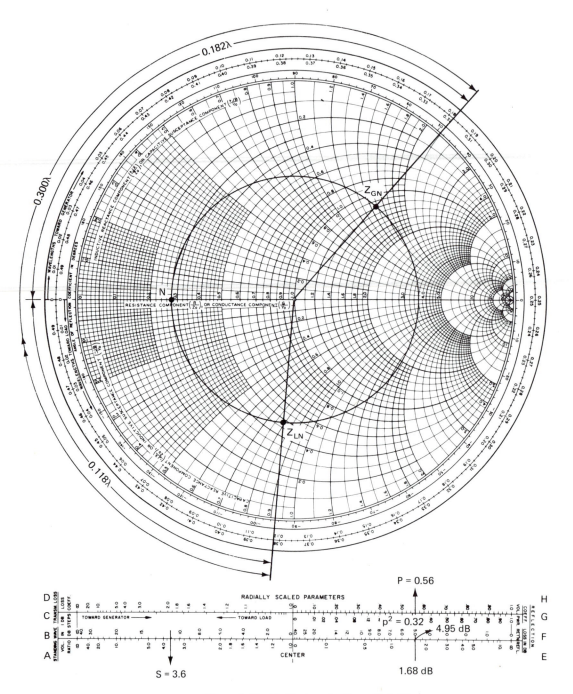

Figure 2–34   Smith chart for Example 2–10.

and maximum impedance values on the line, respectively, occur at points N and M. At point N the normalized value is $0.28 + j0$, so that the minimum impedance is $0.28 \times 100 = \mathbf{28\ \Omega}$ (or Z_0/S). The normalized value at point M is $3.6 + j0$, so that the maximum impedance is $3.6 \times 100 = \mathbf{360\ \Omega}$ (or $S \times Z_0$).

Example 2-11   The VSWR on a 200-Ω transmission line is 2.5. If a null point is located at a position which is 3.38λ from the load, find Z_L, $Z_{\max}$, and $Z_{\min}$, and give the value of the VSWR in decibels.

Solution

Step 1: On the A scale determine the length corresponding to the VSWR of 2.5 and then draw the VSWR circle with this length as its radius (Figure 2-35); using the B scale, the VSWR is read off as **8 dB**.

Step 2: Point N represents the normalized value of the null impedance, $Z_{\min}$; therefore, $Z_{\min} = 0.4 \times 200 = \mathbf{80\ \Omega}$. Similarly, the maximum impedance occurs at the voltage antinode (point M), so that $Z_{\max} = 2.5 \times 200 = \mathbf{500\ \Omega}$.

Step 3: Z_{LN} is found by moving a distance of 3.38λ from the null position in the *counter-clockwise* direction toward the load. The first three wavelengths are equivalent to six complete revolutions on the chart; the additional 0.38λ on the inner wavelength scale brings us to point Z_{LN} on the VSWR circle. The "denormalized" load impedance is then $(0.65 + j0.7) \times 200 = \mathbf{130 + j140\ \Omega}$.

Example 2-12   The VSWR on a 50-Ω line is 1.8 and a null point is located at a certain position on the line. When the load is replaced by a short circuit, the position on the null point shifts a distance of 0.15λ *toward the load*. Find Z_L as well as the magnitude and the phase angle of the reflection coefficient. How far from the load is the first voltage maximum? (Give your answer as a fraction of the wavelength.)

Solution   This example illustrates a common method of determining the value of an unknown load impedance. The position of a voltage null is located with the line terminated by the load. The load is then removed and replaced by a short circuit. The result is a shift in the position of the null (Figure 2-36a). According to the nature of the load (inductive or capacitive), the direction of the shift may be either toward the generator or the load, and the amount of the shift must be 0.25λ or less. If the shift is either zero or exactly 0.25λ, the load is a resistance whose value is respectively less or greater than Z_0 (Figure 2-36b). Knowing the amount and the direction of the null shift, Z_{LN} may be determined by carrying out the following steps.

Step 1: On scale A determine the length corresponding to a VSWR of 1.8 and then draw the VSWR circle with this length as its radius (Figure 2-37). From the null position (point N) travel a distance of 0.15λ on the inner wavelength scale in the counterclockwise direction *toward the load*. Arriving at point L, draw a line from L to the center of the chart. The point of intersection between the line and the VSWR circle is the normalized value of the load impedance. From the impedance coordinates $Z_{LN} = 1 - j0.6$.

Step 2: The "denormalized" load impedance is $Z_L = Z_{LN} \times Z_0 = (1 - j0.6) \times 50 = \mathbf{50 - j30\ \Omega}$.

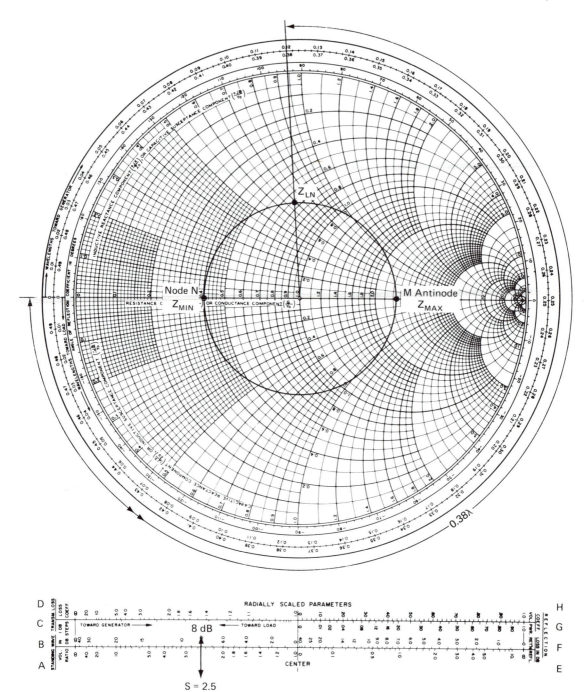

Figure 2–35   Smith chart for Example 2–11.

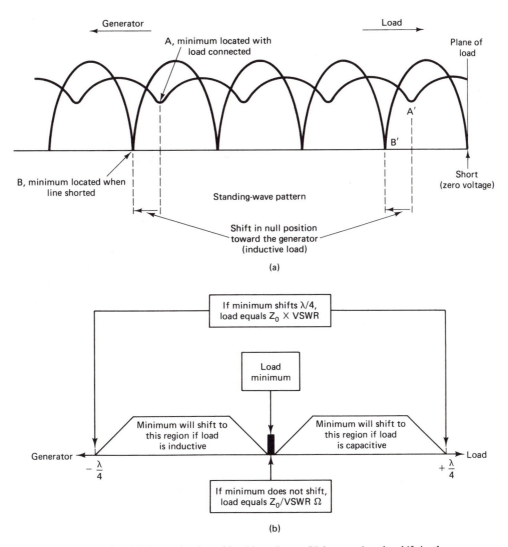

Figure 2-36 (a) Determination of load impedance; (b) interpreting the shift in the null position (Example 2–12).

Step 3: The line passing through L intersects the angle $\theta°$ of the reflection coefficient scale at $-72°$. Marking off the radius of the VSWR circle on the radial scale, G, shows that $P = 0.28$ and therefore the reflection coefficient $\rho = \mathbf{0.28\ \underline{/-72°}}$.

Step 4: The voltage maximum nearest the load exists at point M. The required distance is found by moving from L to M on the wavelength scales in the clockwise direction toward the generator. Moving from L to N on the inner wavelength scale is equivalent to a distance of 0.15λ, and moving from N to M on the outer wavelength scale is a distance of 0.25λ. The total distance is therefore $0.15\lambda + 0.25\lambda = \mathbf{0.4\lambda}$.

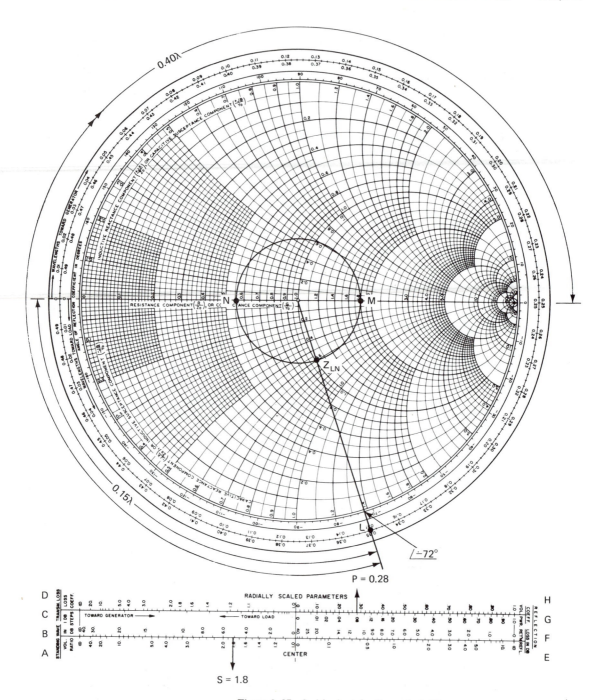

Figure 2-37 Smith chart for Example 2-12.

Example 2–13   A 200-Ω line with an air dielectric is terminated by a load impedance of $150 + j80$ Ω and is excited by a 1-GHz generator. Find the position on the line of a single matching stub, and determine the length of the stub.

Solution   One use of a stub is to match a general load impedance to the surge impedance of the cable that feeds the load. The stub itself may be regarded as a section of transmission line, about one-quarter wavelength long, which is terminated by a movable short (Figure 2–38). The single movable stub is placed across (in parallel with) the twin line feeder, and by sliding the stub along the line, a position is found such that the combination of the stub and the load represents an entirely resistive impedance whose value is equal to the feeder's Z_0. The purpose of the Smith chart (Figure 2–39) therefore, is to determine the lengths L_1 and L_2.

Step 1: The normalized load Z_{LN} is

$$\frac{150 + j80}{200} = 0.75 + j0.4$$

and the wavelength

$$\lambda = \frac{c}{f} = \frac{3 \times 10^{10}}{1 \times 10^9} \text{ cm}$$
$$= 30 \text{ cm}$$

Step 2: Plot Z_{LN} on the Smith chart and draw the corresponding VSWR circle. Since the stub is across or in parallel with the line, an admittance analysis must be used. The normalized load admittance is the reciprocal of the normalized impedance. Therefore, the normalized load admittance

$$Y_{LN} = \frac{1}{Z_{LN}}$$
$$= \frac{1}{0.75 + j0.4}$$
$$= \frac{0.75 - j0.4}{0.75^2 + 0.4^2}$$
$$= 1.04 - j0.55$$

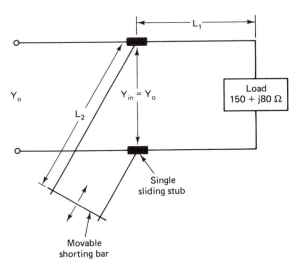

Figure 2–38   Single sliding stub (Example 2–13).

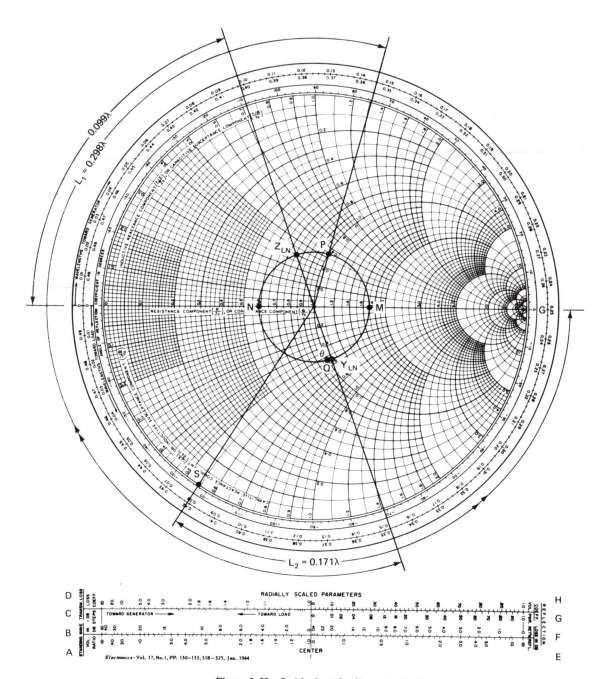

RADIALLY SCALED PARAMETERS

Figure 2–39   Smith chart for Examples 2–13 and 2–14.

When we plot Y_{LN} we find that its position is at the *opposite* end of the diameter passing through Z_{LN}.

Step 3: From the position of Y_{LN}, move around the VSWR circle in the clockwise direction toward the generator until you reach the point of intersection P between the VSWR circle and the circle for which the normalized conductance G/Y_0 = 1. The interval covered in this move is equivalent to the distance $L_1 = 0.152\lambda$ + 0.146λ = 0.298λ or $L_1 = 0.298 \times 30 = 8.94$ cm. At position P the normalized admittance due to the load is $Y_{PN} = 1 + j0.54$, which represents a normalized conductance in parallel with a normalized capacitive susceptance.

Step 4: To achieve a match with the line's surge impedance, the stub must eliminate the normalized capacitive susceptance by contributing an equal amount of normalized inductive susceptance, namely $- j0.54$. This value is therefore entered at point S on the chart. The short at the end of the stub has infinite conductance and is represented by point G. Since this short is the stub's load termination, we must travel from S to G on the inner wavelength scale in a counterclockwise direction toward the load. The length of the stub, L_2, is therefore 0.25λ − 0.079λ = 0.171λ = 0.171 × 30 cm = **5.13 cm.**

Another solution to the problem is represented by point Q. However, the distance L_1 would then be very short (although any multiple of λ/2 could be added to L_1) and the distance L_2 would be much longer.

Example 2–14 In Example 2–13 the line is to be matched to the load by means of a λ/4 transformer (Section 2–4). Find the two possible positions for locating the transformer, and calculate the surge impedance of the transformer in each case.

Solution The λ/4 transformer must be inserted at either of positions M and N, where the line impedance is resistive; only then can the line impedance be matched to the resistive value of the line's surge impedance Z_0. On the outer wavelength scale in Figure 2–39, the position of M is at a distance of 0.25λ − 0.099λ = 0.151λ = 0.151 × 30 cm = **4.57 cm** from the load, while the position of N is a further quarter-wavelength back along the line for a total distance of 0.25λ + 0.151λ = 0.401λ = 0.401 × 30 cm = **12.03 cm.** The impedance on the line at the position M is $Z_{max} = 1.71 \times 200 = 342$ Ω, so that the required surge impedance for the λ/4 transformer is $\sqrt{342 \times 200}$ = **261.5 Ω,** rounded off [equation (2–43)]. At the position of N, $Z_{min} = 0.58 \times 200 =$ 116 Ω and the necessary transformer's surge impedance is $\sqrt{116 \times 200} = $ **150 Ω,** rounded off.

Example 2–15 A load is to be matched to a 100-Ω coaxial line by means of two fixed stubs which are 3λ/8 apart (Figure 2–40a). At the reference plane corresponding to the position of the first stub, the admittance is 0.016 − j0.008 siemens. Find the length of each stub.

Solution The application of a single movable stub is limited to the two-wire line. For a coaxial cable we are frequently required to use two fixed stubs, which are typically 3λ/8 apart. It is then necessary to use the Smith chart (Figure 2–41) to determine the lengths of the stubs to achieve the required impedance match. This problem is solved by means of the following steps.

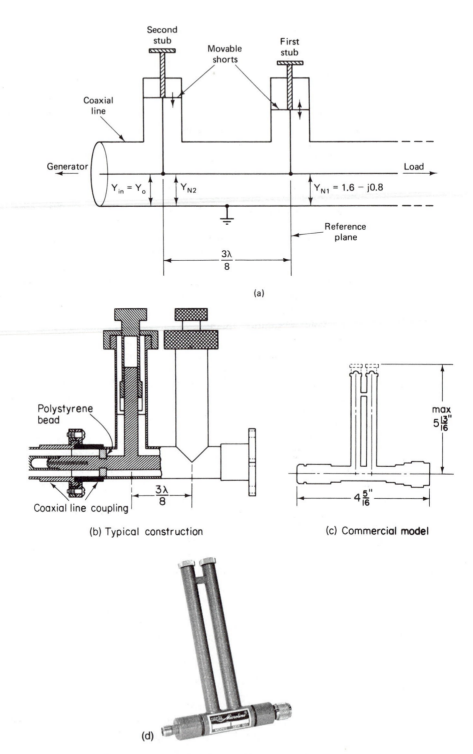

Figure 2–40 Twin-stub impedance matching. Circuit for Example 2–15. (Courtesy of The Narda Microwave Corporation.)

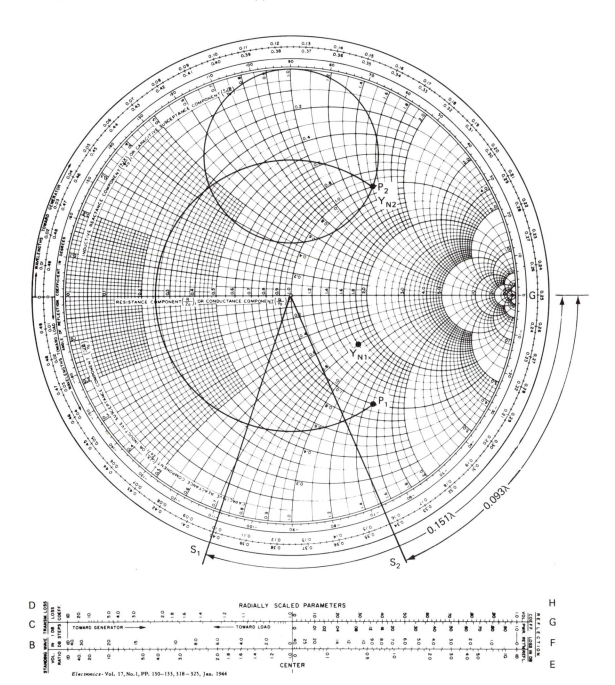

Figure 2–41 Smith chart for Example 2–15.

Step 1: The surge admittance of the line is $1/100 \; \Omega = 0.01$ S. Therefore, the normalized admittance Y_{N1} at the reference plane of the first stub is $(0.016 - j0.008)/0.01 = 1.6 - j0.8$.

Step 2: Enter Y_{N1} on the Smith chart and observe that it lies on the 1.6 normalized conductance circle. Rotate this circle through three-quarters of a revolution in the clockwise direction; this rotation corresponds to the $3\lambda/8$ spacing between the stubs. The center of the circle after the rotation will lie on the line passing through the points $+j1.0$ and $-j1.0$.

Step 3: Determine the point P_2 of intersection between the newly constructed circle and the circle for which $G/Y_0 = 1$. Draw the VSWR circle through P_1 and P_2; then P_1 and P_2 are $3\lambda/8$ apart. Normalized admittance at P_2 is $Y_{N2} = 1 + j1.52$.

Step 4: The effect of the first stub is to change the susceptance $(-j0.8)$ of Y_{N1} at the reference plane to the susceptance $(-j1.52)$ at point P_1. This stub must therefore contribute a susceptance of $-j1.52 - (-j0.8) = -j0.72$, which corresponds to point S_1. The short circuit that terminates the stub is point G. The length of the first stub is therefore $0.25\lambda - 0.095\lambda = \mathbf{0.155\lambda}$.

Step 5: The second stub must change the normalized susceptance from its value $(+j1.52)$ at point P_2 to the center of the chart, where the match to the line's surge impedance occurs, and the normalized susceptance is zero. This stub must therefore contribute a normalized susceptance of $-j1.52$, which is entered at point S_2. The length of the second stub is measured from S_2 to G; this distance is equal to $0.25\lambda - 0.157\lambda = \mathbf{0.093\lambda}$.

Clearly, there is a limitation to the use of the two stubs, which are $3\lambda/8$ apart. If the normalized conductance G/Z_0 is greater than 2, there will be no point of intersection such as point P_2, and no solution can be found. To cover all possibilities, it is necessary to have three stubs, each separated from its neighbor by a distance of $\lambda/4$ (Figure 2–42).

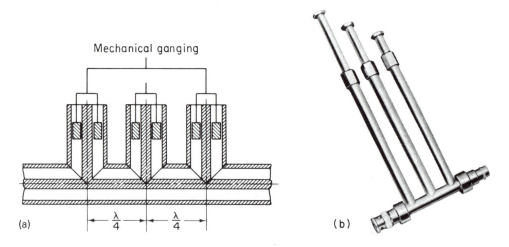

Mechanical ganging

(a)

$\dfrac{\lambda}{4}$ $\dfrac{\lambda}{4}$

(b)

Figure 2–42   Triple-stub impedance matching. (Courtesy of Microlab/FXR.)

Example 2–16   On a slotted line the VSWR is measured and found to be 3.6. When the load is removed and replaced by a short, the position of a null point on the line moves 0.11λ toward the generator and the measured VSWR is 8.6. Find the one-way attenuation in decibels between this null position and the load, the value of the VSWR at the load (with the load connected), and the normalized value of the load.

Solution   When there are appreciable losses associated with the line, the incident voltage and current waves are attenuated as they travel from the generator toward the load. Assuming that the load is mismatched to the line, the reflected voltage and current waves will be further attentuated as they move back from the load toward the generator. The VSWR value will therefore be highest at the load position and will gradually decrease in the direction of the generator. Consequently, as we move from the load toward the generator, we must visualize that the VSWR value will be represented by a decreasing spiral rather than a circle. On the Smith chart (Figure 2–43) the change in the value of the VSWR (scale A) may be used to determine the line's one-way attenuation in decibels (scale C, attenuation in decibel steps).

The method of obtaining the solution is as follows:

Step 1: When the "short" is placed at the end of the line, the VSWR at the position of the short is a maximum and is theoretically infinite; this corresponds to the maximum radial distance on the A scale. However, the VSWR as measured at the null position is only 8.6, and therefore the change from $S = \infty$ to $S = 8.6$ is due to the line's attenuation. After locating the $S = \infty$ and $S = 8.6$ positions on the A scale, mark off the same radial distances on the C scale. The difference between these distances is approximately equivalent to a one-way attenuation of **1.0 dB.**

Step 2: On the A and C scales mark off the radial distances for the VSWR of 3.0. Increase the distance on the C scale by an amount corresponding to the one-way attenuation of 1.0 dB and then apply the same increase in distance to the A scale. The result is a VSWR of approximately **4.5** at the load.

Step 3: Draw the VSWR circle for $S = 4.5$. On the outer wavelength scale move clockwise through a distance of 0.11 λ from the null position N toward the generator. When point P is reached, draw a straight line from P to the center of the chart. The intersection between this line and the VSWR circle reveals the normalized load impedance Z_{LN} to be **0.35 + j0.77.**

Example 2–17   It is desired to achieve a VSWR of 2.5 on an unmatched line by the introduction of attenuation. If the VSWR at the load is 5.4, find the required one-way attenuation in decibels. Determine the value of the loss coefficient for the VSWR of 5.4.

Solution   This problem may be solved by using the radial scales only.

Step 1: Locate the $S = 2.5$ and $S = 5.4$ positions on the A scale and mark off the same radial distances on the C scale. The difference between these distances is equivalent to a one-way attenuation of **2.0 dB** approximately (Figure 2–44).

Step 2: A more accurate estimate of the attenuation may be obtained by using the F scale (return loss in dB), which represents the two-way loss. On this scale $S = 2.5$ and $S = 5.4$ correspond to losses of 7.4 and 3.4 dB, respectively. The one-way attenuation is therefore $(7.4 - 3.4)/2 = $ **2.0 dB.**

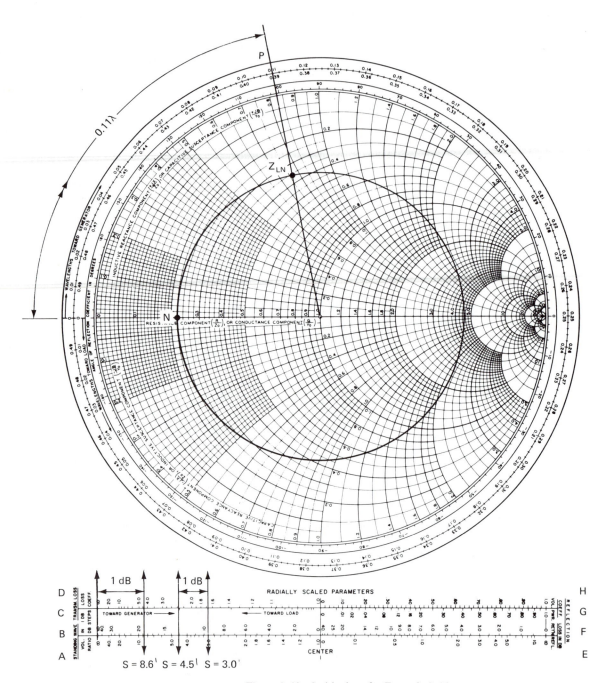

Figure 2–43 Smith chart for Example 2–16.

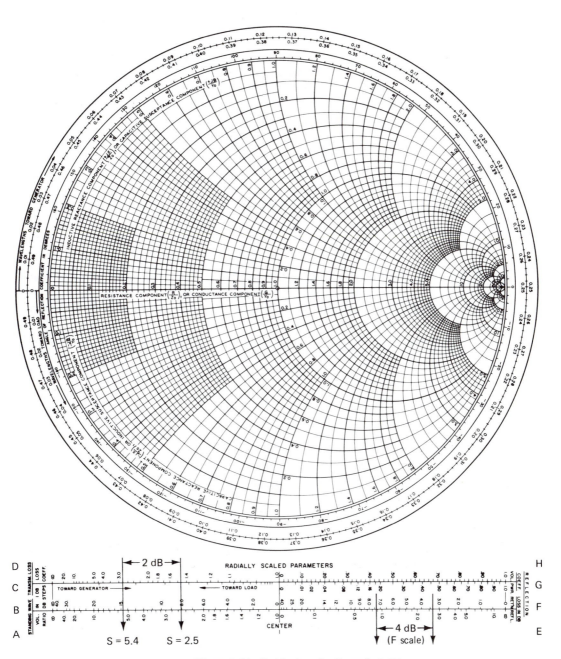

Figure 2–44 Smith chart for Example 2–17.

Step 3: On scale D mark off the radial distance for $S = 5.4$. The corresponding loss coefficient is approximately **2.8**.

We have used the universal Smith chart to solve a variety of examples. Further examples involving the Smith chart appear in the problems. However, with any problem it is important to remember that any change in the frequency will alter the values of the normalized reactance and/or susceptance. This in turn will change the position of the normalized impedance and/or admittance points so that the answers in the solution will be different. The effect of a change in frequency is illustrated in Problem 2–20.

PROBLEMS

BASIC PROBLEMS

2–1. (a) On a lossless RF line the distributed primary constants are $L = 6.5$ μH/m and $C = 8.7$ pF/m. What is the value of the line's surge impedance?

   (b) If the line is matched and the line current is 600 mA (rms), what is the amount of power conveyed down the line?

2–2. A matched RF line (assumed to be lossless) is terminated by a 300-Ω resistive load. If the generator voltage is 150 V (rms), what is the input impedance to the generator, and what is the amount of power delivered to the load?

2–3. A matched RF line has a velocity factor of 0.85 and the generator feeding the line has a frequency of 100 MHz. What is the value of the wavelength on the line?

2–4. A lossless 100-Ω RF line with an air dielectric is terminated by an open circuit.

   (a) If the frequency of the RF generator feeding the line is 150 MHz, how far from the open-circuited end is the first voltage null?

   (b) If the line is 2.25 m in length, what is the input impedance at the generator?

2–5. A lossless 50-Ω RF line with an air dielectric is terminated by a short circuit.

   (a) If the frequency of the RF generator is 75 MHz, what is the distance between two adjacent current nulls?

   (b) If the line is 4.5 m long, what is the input impedance at the generator?

2–6. A lossless 600-Ω line is terminated by a resistive load of 200 Ω. What are the values of the VSWR on the line and the reflection coefficient at the load?

2–7. In Problem 2–6 the RF power incident at the load is 50 W. What are the amounts of the reflected power and the power absorbed by the load?

2–8. (a) A 300-Ω loss-free line is terminated by a load which consists of 210-Ω resistance in series with 450-Ω inductive reactance. Use the universal Smith chart to find the values of the reflection coefficient and the standing-wave ratio, together with the maximum and minimum impedances existing on the line.

   (b) If the incident power at the load is 80 mW, find the values of the reflected power and the power absorbed by the load.

(c) The generator is located 77 cm from the load. If the line is operating at a frequency of 2 GHz, what is the input impedance at the generator?

2–9. On a loss-free line the VSWR is found to be 4.3. If a null is located at a distance of 2.632λ from the load, find the value of the normalized load impedance.

2–10. An artificial line consists of six LC sections, each with an inductance of 250 μH and a capacitance of 200 pF. What is the (one-way) delay time provided by the line, and what is the line's equivalent surge impedance?

ADVANCED PROBLEMS

2–11. Each conductor of a twin-lead transmission line has a radius of 1.6 mm and the spacing between the centers of the conductors is 0.85 cm.

(a) If an air dielectric is used, what is the value of the line's surge impedance?

(b) If the line (assumed to be lossless) is matched and the generator voltage is 125 V (rms), what are the values of the current drawn from the generator and the power delivered to the load?

2–12. The outer conductor of a coaxial cable has an inner diameter of 1.8 cm and the inner conductor has an outer diameter of 2.4 mm.

(a) What is the value of the cable's surge impedance if its dielectric has a relative permittivity of 2.2?

(b) If the line (assumed to be lossless) is matched and 50 W of RF power is delivered to the load, what are the effective values of the line voltage and the line current?

2–13. The primary constants of an RF transmission line are $L = 1.65\ \mu$H/m, $C = 7.75$ pF/m, $R = 55$ mΩ/m, and $G = 12$ nS/m. Calculate the values of the surge impedance and the attenuation constant.

2–14. In Problem 2–13 the frequency of the generator feeding the line is 300 MHz. Find the values of the phase-shift constant, the wavelength on the line, and the phase velocity.

2–15. (a) A transmission line with an air dielectric and a surge impedance of 100 Ω is terminated by a load consisting of 80-Ω resistance in series with 130-Ω capacitive reactance. Calculate the phasor value of the reflection coefficient. What is the magnitude of the VSWR?

(b) If a 100-MHz generator delivers 90 W to the line, which is assumed to be lossless, what are the amounts of the reflected power and the power absorbed by the load? What is the distance between two adjacent voltage nulls on the line?

2–16. An artificial line consists of five LC sections each with an inductance of 200 μH and a capacitance of 200 pF. The line is charged to 500 V and then is abruptly connected across a 250-Ω resistive load. What is the voltage across the load after 4.5 μs has elapsed?

2–17. The VSWR on a 300-Ω line is 2.5 and the position of a null point is first located. When the load is replaced by a short circuit, the position of the null point shifts a distance of 0.12λ toward the generator. Find the value of the load impedance as well as the magnitude and the phase angle of the reflection coefficient. How far from the load is the first voltage antinode? (Give your answer as a fraction of the wavelength.)

2–18. A load is to be matched to a 100-Ω line by means of two fixed stubs which are separated by a *quarter* wavelength. At the reference plane corresponding to the position of the first stub, the *admittance* is $0.008 - j0.014$ S. Find the length of each stub in terms of the wavelength. Discuss any limitation on the use of the two stubs which are $\lambda/4$ apart.

2–19. At a certain null position N on an unmatched line, the SWR is measured and found to be equal to 1.9. When the load is removed and replaced by a short, the position of the short shifts 0.13λ toward the load and the measured VSWR increases to 4.4. Find the one-way attenuation between N and the load, the value of the VSWR at the load (with the load connected), and the normalized value of the load.

2–20. A 50-Ω loss-free line is terminated by a load impedance of $70 - j35$ Ω and is excited by a 600-MHz generator. How far from the load is the position of a single matching stub? Determine the length of the stub. Give your answers in centimeters. If the frequency is reduced to 450 MHz, find the new position of the stub and its new length.

███ CHAPTER 3 ███

RECTANGULAR AND CIRCULAR WAVEGUIDES

3–1 INTRODUCTION

In Section 1–4 we compared the properties of conventional transmission lines with those of waveguides. Owing primarily to dielectric hysteresis, the waveguide has much less attenuation than a coaxial cable when both are operated in the SHF band. However, we also discovered that the disadvantages of size, lack of flexibility, limited frequency range, and expense make a waveguide system impractical at all but microwave frequencies.

Our introduction to waveguides is covered in the following topics:

3–2. Development of the Rectangular Waveguide from the Two-Wire Line

3–3. Development of the Field Patterns in a Rectangular Guide

3–4. Propagation in a Rectangular Waveguide

3–5. Circular Waveguides

3–6. Microstrip and Stripline Connections

3–7. Optical Fibers (Fiber Optics)

3–2 DEVELOPMENT OF THE RECTANGULAR WAVEGUIDE FROM THE TWO-WIRE LINE

A two-wire line with an air dielectric must be mechanically supported at intervals by some form of insulator (Figure 3–1a). However, at high radio frequencies, insulators such as porcelain or plastic have a large dielectric loss, so that they act as a low

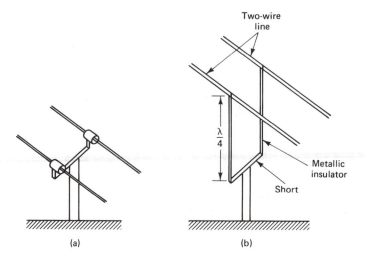

Figure 3–1 Principle of the metallic insulator.

impedance across the line and represent a discontinuity. A superior high-frequency insulator is a quarter-wavelength of RF line which is shorted at one end (Figure 3–1b). In Section 2–4 we discovered that the input impedance to such a line was theoretically infinite, although in practice the input impedance is a very high value of resistance which will have little effect on the two-wire line. This type of metallic insulator is sometimes referred to as a quarter-wave $\lambda/4$ support "stub."

The $\lambda/4$ support stubs can be joined both above and below the twin line. To increase the rigidity of the line, we can add more and more stubs until each makes contact with its neighbor. The result is a hollow rectangular tube (Figure 3–2) with the line itself forming part of the walls. It is clear that the wide dimension, a, must be longer than a half wavelength since if a is less than $\lambda/2$, the stubs would behave

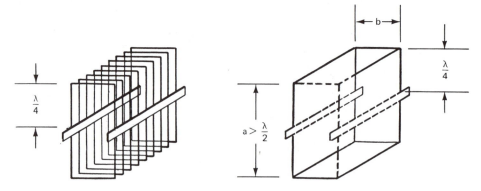

Figure 3–2 Formation of a rectangular waveguide by the addition of quarter-wave support stubs.

inductively and appear as a severe discontinuity across the line. At the cutoff condition, the value of the wide dimension, $a = \lambda_c/2$ or $\lambda_c = 2a$; the cutoff frequency

$$f_c = \frac{c}{\lambda_c} = \frac{c}{2a} \tag{3-1}$$

where

f_c = cutoff frequency (Hz)
λ_c = cutoff wavelength (m)
a = waveguide's wide dimension (m)
$c = 300 \times 10^6$, which is the velocity of electromagnetic waves in free space (m/s)

In practical units, the free-space wavelength

$$\lambda = \frac{30}{f} \text{ cm} \tag{3-2}$$

where f is the frequency (GHz). For most waveguides the value of a is normally about 0.8λ.

The narrow b dimension governs primarily the power-handling capacity of the guide. If the value of b is too low, the voltage established across the narrow dimension may exceed the breakdown potential of the air dielectric so that arcing will occur. We shall also see (Section 3-4) that the value of b, typically 0.4λ (or $a/2$), determines the upper frequency at which the waveguide can operate.

Example 3-1   A rectangular waveguide has internal dimensions of 16.5 and 8.25 cm. (a) Calculate the value of its cutoff frequency. (b) Suggest the designated band for which this particular waveguide would be suitable.

Solution   (a) The cutoff wavelength

$$\lambda_c = 2a = 2 \times 16.5 = 33 \text{ cm}$$

and the cutoff frequency

$$f_c = \frac{30}{33} \text{ GHz} = \textbf{0.909 GHz} \tag{3-1}$$

(b) If we assume that $a = 0.8\lambda$ for the center frequency f of the designated band,

$$f = \frac{30}{16.5/0.8} = 1.45 \text{ GHz}$$

This frequency is approximately at the center of the **L** band (1.12 to 1.7 GHz).

Example 3-2   A rectangular waveguide is required to operate over a band with a center frequency of 22.25 GHz. Suggest the values required for the guide's internal dimensions.

Solution   The wavelength corresponding to the frequency of 22.25 GHz is

$$\lambda = \frac{30}{22.25} = 1.35 \text{ cm}$$

Suggested internal dimensions are $a = 0.8\lambda = \textbf{1.08 cm}$ and $b = a/2 = \textbf{0.54 cm.}$

3-3 DEVELOPMENT OF THE FIELD PATTERNS IN A RECTANGULAR GUIDE

The E *Field*

Energy is transferred down a rectangular waveguide as an electromagnetic wave which consists of alternating electric (E) and magnetic (H) fields. Associated with the electric field is a voltage distribution across the guide's narrow dimension while "wall" currents flow along the inner surfaces to create the magnetic field. To further our understanding of the propagation process we must first derive the patterns of the E and H flux lines inside the waveguide.

Let us start by considering a lossless two-wire line which is one wavelength long and is terminated by a resistive load R_L equal in value to the line's surge impedance Z_0. As we learned in Section 2–3, this line is matched so that the voltage and the current waves move down the line in phase and their energy is totally absorbed by the load. Figure 3–3 shows such a line to which a number of $\lambda/4$ support stubs have been added. It must be emphasized that the voltage wave shown is an *instantaneous* distribution along the line. This wave would in fact travel down the line with its phase velocity, which is only slightly less than the speed of light.

At position A the voltage has reached its peak value, so that there is a maximum electric field between the wires; this is indicated by the high density of the flux lines. However, no voltage can ever exist at the shorted ends of the support stubs, so that as we move from the two-wire line to the ends of the support stubs at position A, the E field must decrease from its maximum value to zero.

Turning now to position B, the voltage is instantaneously zero, so that there

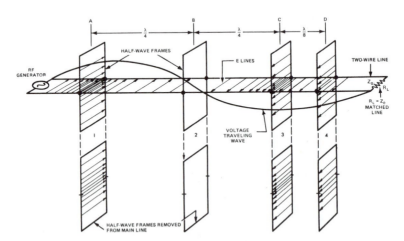

Figure 3–3 Instantaneous distribution of E lines on a matched line with support stubs.

are no E lines between the conductors and none exist on the stubs. Position C is separated in distance by one half-wavelength from, position A and by one quarter-wavelength from position B. At C the voltage is again at its instantaneous peak value but with opposite polarity compared with the voltage at position A. Consequently, the distributions of the E lines are the same at A and C but the directions of the two sets of lines are reversed. Position D is intermediate between position C and the load, so that the number of flux lines is correspondingly less.

We can now add further support stubs to create a hollow rectangular waveguide which is one wavelength long. The pattern of the instantaneous E lines inside the waveguide is illustrated by the various views of Figure 3–4. The symbol "×" is used to indicate a flux line whose direction is away from the viewer, while the direction of a line toward the viewer is shown by "·".

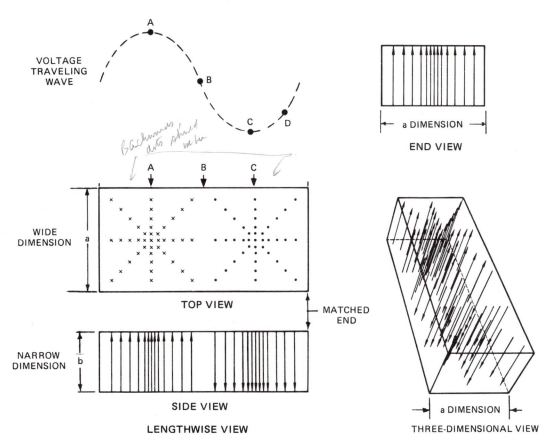

Figure 3–4   Instantaneous distribution of E lines in a one wavelength section of a rectangular waveguide.

The H *Field*

To derive the pattern of the *H* field inside a rectangular waveguide, we can start by considering a matched two-wire line (Figure 3–5) which is one wavelength long. The current wave is in phase with the voltage wave of Figure 3–3, and support stubs have been added between two positions, *A* and *E*, which are a half-wavelength apart. At these positions the current (electron flow) has instantaneously reached its peak value on the line. The directions of these currents and the currents on the support stubs are indicated by arrows. Each of these currents then contributes to the pattern of the *H* field inside the waveguide.

Around each conductor is a small individual loop whose arrow indicates the direction of the magnetic field surrounding that conductor (left-hand gripping rule). When two adjacent loops have arrows in opposite directions, their magnetic fields will tend to cancel, but when the arrows are in the same direction, the magnetic fields combine. Inside the waveguide the resultant magnetic field is in the form of an *H* loop which extends over a distance of half a wavelength; the various views of this *H* loop are illustrated in Figure 3–6. Outside the waveguide the individual loops cannot join to form a continuous flux path, so that there is no external magnetic field.

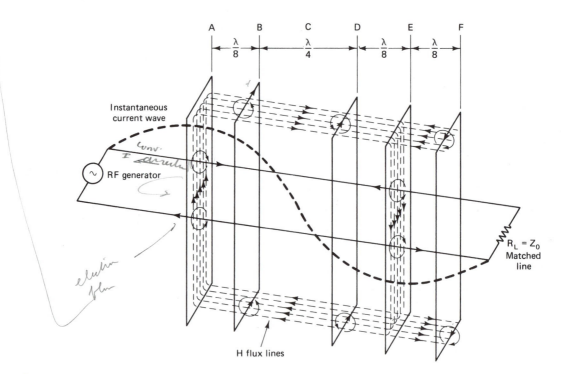

Figure 3–5   Instantaneous distribution of *H* lines on a matched line with support stubs.

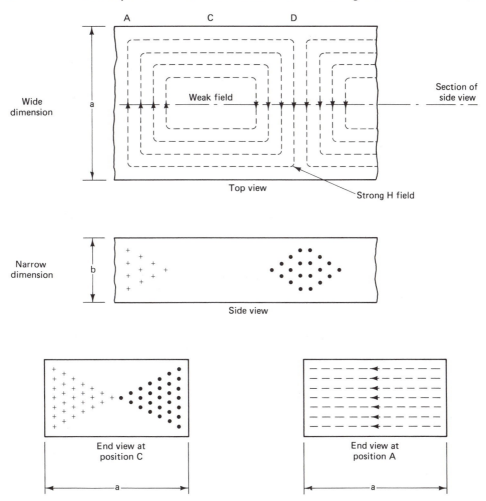

Figure 3–6   Instantaneous views of the magnetic flux lines in a rectangular waveguide.

When we combine the E and H fields over a distance of a wavelength, we obtain the patterns of Figure 3–7. The E field is entirely transverse to the direction of propagation and its peak coincides in position with the peak of the transverse component of the H field. However, the H field also has a longitudinal component in the same direction as the energy is being propagated.

The field patterns that we have established obey two important boundary conditions which were originally stated by Clerk Maxwell:

1. In order to exist, all *electrical* flux lines must be *perpendicular* to the walls of the guide. This also means that no electrical flux line whose direction is parallel to a wall can be positioned at the surface of that wall. Were such a line to

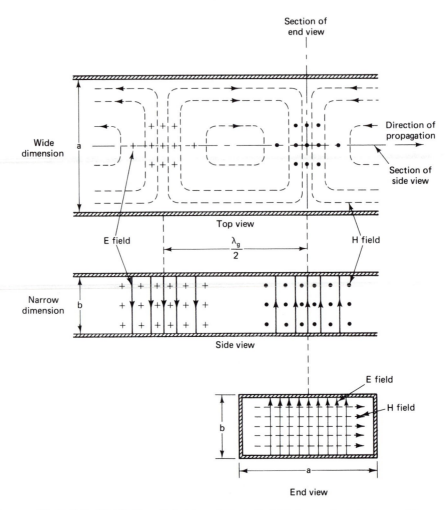

Figure 3–7 Distribution of electric and magnetic fields in a rectangular waveguide (TE$_{10}$ mode).

exist, the result would be a voltage stress at the wall; a surface current would then flow and the line would be eliminated.

2. In order to exist, all *magnetic* flux lines at a wall must run *parallel* to the surface. Consequently, no magnetic flux line can have a component that is perpendicular to the surface. Were such a component to exist, it would cut the wall, eddy currents would be induced, and the component would be removed.

We now know the field patterns that must exist inside the waveguide. The question is: How does the EM wave move down the waveguide in order to create the required field patterns and obey Maxwell's boundary conditions? The answer lies in the discussion in Section 3–4.

Example 3–3   A rectangular waveguide with internal dimensions of 4.8 cm and 2.4 cm is excited by a 2.9-GHz signal. Will the electric and magnetic fields be successfully propagated down the waveguide?

Solution   The free-space wavelength

$$\lambda = \frac{30}{2.9} = 10.3 \text{ cm}$$

and the cutoff wavelength

$$\lambda_c = 2 \times 4.8 = 9.6 \text{ cm}$$

Since $\lambda > \lambda_c$ (or $f < f_c$), the electric and magnetic fields will **not** be successfully propagated down the waveguide. The field configurations will, in fact, represent so-called evanescent modes, which will be rapidly attenuated.

3–4 PROPAGATION IN A RECTANGULAR WAVEGUIDE

In Section 1–2 we discussed the EM wave, which is radiated out into free space. Such a wave consists of alternating electric and magnetic fields that are in time phase although the two sets of flux lines are 90° apart in space. Both fields are entirely transverse in the sense that their directions are always at right angles to the direction of propagation. We then refer to a TEM (transverse electric, transverse magnetic) wave, which exists only on a transmission line or in free space.

Our next question: Does the energy move straight down the rectangular waveguide as a TEM wave? The answer is, no. If we enclosed a rectangular waveguide around a TEM wave, we could not obey Maxwell's boundary conditions. We also know that the H-field pattern contains a longitudinal component, and no such component exists in the TEM wave. To cut down on the suspense, the answer is: The electromagnetic energy progresses down the guide by a series of reflections off the internal surface of the narrow dimension (Figure 3–8). At each reflection the angles (θ) of incidence and reflection are equal.

Let us consider a TEM wave that approaches a plane conductor (flat metal surface) at an angle θ (Figure 3–9). The dark lines with their arrows represent the directions of the incident and reflected wavefronts, which are moving with the free-space velocity (virtually at the speed of light). The full lines are $\lambda/2$ apart in free

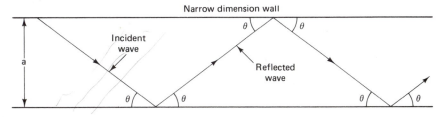

Figure 3–8   Path of an EM wave in a rectangular waveguide.

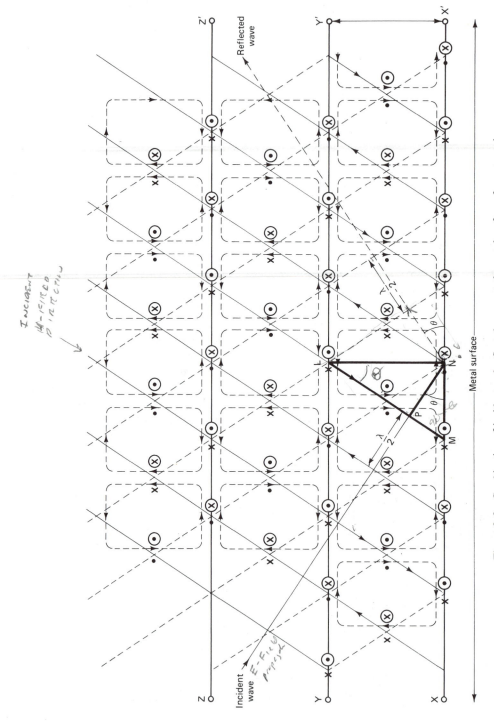

Figure 3-9 Combination of incident and reflected waves at the metal surface of a rectangular waveguide.

120

space and represent the incident H flux lines. The dashed lines are also $\lambda/2$ apart and are used to show the reflected H field. The symbols for the incident and reflected E fields are respectively "$\otimes$, $\odot$" and "$\times$, $\cdot$." These incident and reflected waves are of the same amplitude but have a phase reversal of 180°.

The fields of the incident and reflected waves are superimposed and must be combined to produce the resultant E- and H-field patterns. At the metal surface itself, the phase reversal causes cancellation between the incident and the reflected E fields, so that there are no resultant electric flux lines which are parallel to the surface. This obeys Maxwell's first boundary condition.

When the incident and reflected waves are combined, the resulting pattern of the E and H fields between the lines XX' and YY' is exactly the same as we derived previously from the support stub theory. We can therefore infer that the EM wave must progress down the guide by a series of reflections off the narrow dimension.

So far, so good. Now we must find the factors that determine the angle of incidence and the velocity with which the energy progresses down the guide. To do this we extract the triangle LMN from Figure 3-9 and display its magnified form in Figure 3-10a. As the wavefront moves from P to N (a distance of $\lambda/2$), the field pattern progresses a greater distance from M to N. The distance MN is a half wavelength of the field pattern as it exists inside the waveguide and is termed $\lambda_g/2$, where λ_g is the guide wavelength. For an analogy, think of sea waves approaching the shoreline at an angle (Figure 3-11). As the wavefront moves from one crest to the next through the distance CC' (one wavelength), the pattern of the crests at the shoreline covers the greater distance SS'.

In the right-angle triangle LPN,

$$\sin \theta = \frac{PN}{LN} = \frac{\lambda/2}{a} = \frac{\lambda}{2a} \tag{3-3}$$

For a given rectangular waveguide, the a and b dimensions are fixed. If the frequency is lowered, the wavelength is longer and the value of λ is increased (Figure 3-10b). In the limiting cutoff condition, $\lambda_c = 2a$ and $\theta = 90°$; the wave will then bounce back and forth between the narrow dimensions and will not progress down the guide. When the frequency is raised, the wavelength decreases and the value of θ is lowered (Figure 3-10c). This could be continued indefinitely; however, at the frequency of approximately $1.9f_c$, the narrow b dimension comes into play and limits the top frequency at which the waveguide can successfully operate.

In the right-angle triangle MPN,

$$\cos \theta = \frac{PN}{MN} = \frac{\lambda/2}{\lambda_g/2} = \frac{\lambda}{\lambda_g} \tag{3-4}$$

Therefore,

$$\tan \theta = \frac{\sin \theta}{\cos \theta} = \frac{\lambda/2a}{\lambda/\lambda_g} = \frac{\lambda_g}{2a} \tag{3-5}$$

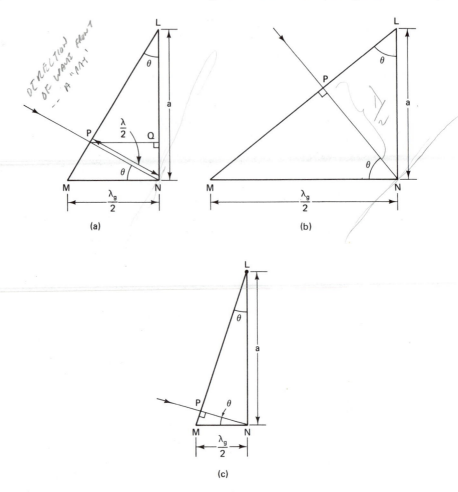

Figure 3–10   Effect of the frequency on the angle of incidence: (a) initial frequency; (b) lower frequency; (c) higher frequency.

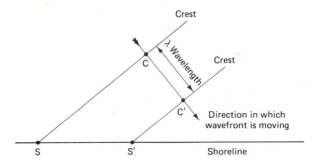

Figure 3–11   Sea-wave analogy.

Equation (3–5) shows that if we decrease the frequency, θ is raised and the guide wavelength λ_g is increased. In the cutoff condition, $\theta = 90°$ and the guide wavelength λ_g is infinitely long.

Using the trigonometrical relation $\sin^2 \theta + \cos^2 \theta = 1$, we can write

$$\left(\frac{\lambda}{2a}\right)^2 + \left(\frac{\lambda}{\lambda_g}\right)^2 = 1$$

This yields

$$\lambda = \frac{\lambda_g}{\sqrt{1 + (\lambda_g/2a)^2}} \tag{3–6}$$

and

$$\lambda_g = \frac{\lambda}{\sqrt{1 - (\lambda/2a)^2}} \tag{3–7}$$

$\lambda_g \geq \lambda$

Remembering that $\lambda = c/f$ and $\lambda_c = 2a = c/f_c$, we have

$$\lambda_g = \frac{\lambda}{\sqrt{1 - (f_c/f)^2}} \tag{3–8}$$

In the time that the wavefront moves from P to N at the speed of light c, the field pattern progresses from M to N (a distance of $\lambda_g/2$) with the *phase* velocity v_ϕ. With the same time interval the electromagnetic energy has moved down the guide a distance PQ; this physical movement of energy takes place at the *group* velocity v_g. Since $MN = PN \sec \theta$ and $PQ = PN \cos \theta$, the phase velocity

$$v_\phi = \frac{c}{\cos \theta} = c \sec \theta \tag{3–9}$$

and the group velocity

$$v_g = c \cos \theta \tag{3–10}$$

Therefore,

$$v_\phi \times v_g = \frac{c}{\cos \theta} \times c \cos \theta = c^2 \tag{3–11}$$

where v_ϕ, v_g, and c are all measured in meters per second.

At the cutoff condition the group velocity is zero and the phase velocity is infinite. This result does not contravene any physical laws since the phase velocity involves the *apparent* movement of a field pattern, not the movement of any physical quantity. As the frequency is raised, v_ϕ decreases and v_g increases as both these velocities approach the velocity of light, c. For an X-band waveguide the graphs of v_g and v_ϕ versus frequency are shown in Figure 3–12.

We are left to consider the waveguide's phase-shift constant, β. A phase shift

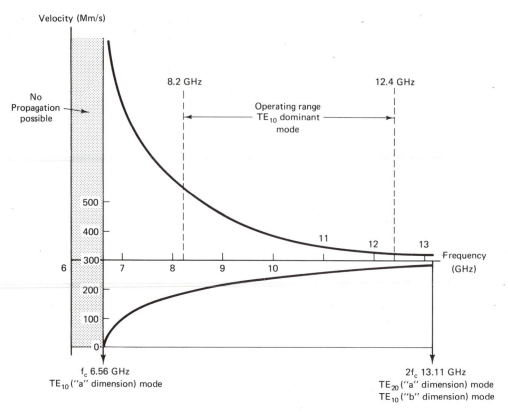

Figure 3–12   Variation of group velocity and phase velocity with frequency in an X-band rectangular waveguide.

of 2π radians occurs over a distance equal to the guide wavelength λ_g. Therefore, the phase-shift constant

$$\beta = \frac{2\pi}{\lambda_g} \qquad\qquad (3\text{–}12)$$

where

β = phase-shift constant (rad/m)
λ_g = guide wavelength (m)

Let us summarize this section by calculating the results for a practical WR 90 rectangular waveguide that is designed to operate over the X band, 8.2 to 12.4 GHz. The inner guide dimensions are 2.286 cm (a) and 1.143 cm (b). If the transmitted frequency is 10 GHz:

$$\text{Free-space wavelength } \lambda = \frac{c}{f}$$

$$= \frac{3 \times 10^8 \text{ m/s}}{10 \times 10^9 \text{ Hz}}$$
$$= 0.03 \text{ m}$$
$$= 3 \text{ cm}$$

$$\text{Cutoff wavelength } \lambda_c = 2a = 2 \times 2.286$$
$$= 4.572 \text{ cm}$$

$$\text{Cutoff frequency } f_c = \frac{c}{\lambda_c} = \frac{3 \times 10^8 \text{ m/s}}{4.572 \times 10^{-2} \text{ m}}$$
$$= 6.562 \text{ GHz}$$

$$\text{Angle of incidence } \theta = \sin^{-1} \frac{\lambda}{2a}$$
$$= \sin^{-1} \frac{3}{4.572}$$
$$= 41.01°$$

$$\text{Guide wavelength } \lambda_g = \frac{\lambda}{\cos \theta}$$
$$= \frac{3}{\cos 41.01°}$$
$$= 3.98 \text{ cm}$$

$$\text{Phase velocity, } v_\phi = \frac{c}{\cos \theta}$$
$$= \frac{300}{\cos 41.01°}$$
$$= 397.6 \text{ Mm/s}$$

$$\text{Group velocity } v_g = c \cos \theta$$
$$= 300 \cos 41.01°$$
$$= 226.4 \text{ Mm/s}$$

$$\text{Phase-shift constant } \beta = \frac{2\pi}{\lambda_g}$$
$$= \frac{2\pi}{3.98 \times 10^{-2} \text{ m}}$$
$$= 159 \text{ rad/m}$$

Dominant and Higher-Order Modes

The field patterns shown in Figure 3–13 represent only one of the infinite number of possible ways in which E and H fields can exist inside a rectangular waveguide. Each such field configuration is known as a mode of operation. For example, in

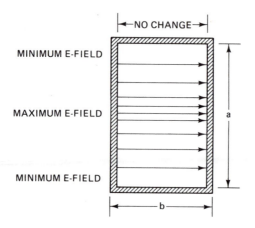

Figure 3–13 Dominant TE_{10} mode.

Figure 3–9 we could have inserted the upper metal surface at the line ZZ' rather than the line YY'. If this had been done, there would have been two H loops instead of one across the wide dimension. However, to accommodate this new mode, we would need to double the length of the wide dimension, a.

The field configurations described in Section 3–4 represent the dominant mode. For this mode the inner dimensions of the waveguide are the least possible so that all other modes require a larger waveguide—an obvious disadvantage. Furthermore, the dominant mode is the easiest to excite in the waveguide, is most efficient, and has the lowest cutoff frequency. Waveguides are normally designed so that only the dominant mode is propagated. For example, the WR 90 X-band waveguide with internal dimensions of 2.286 cm (a) and 1.143 cm (b) has a cutoff frequency of $f_c = 6.56$ GHz at the dominant mode. If the frequency is raised above $2f_c = 2 \times 6.56 = 13.1$ GHz, a higher-order mode will be excited in the a dimension and the dominant mode will at the same time appear in the b dimension (because in our example, $b = a/2$). Consequently, the energy introduced into the waveguide will be split between three possible modes; this is highly undesirable because (1) the methods of joining waveguide sections together may respond solely to the dominant mode, and (2) the means of removing the energy may be effective only at the dominant mode.

To classify the various waveguide modes, we use a numbering system which is similar to that for the cavity resonator (Section 1–3). We must first decide whether the mode is traverse electric (TE) or transverse magnetic (TM). In the TE mode all electric flux lines are at right angles to the direction of propagation, so that there are no longitudinal lines which are parallel to the direction of travel. This fits the description of the E lines in the rectangular waveguide which is operating at the dominant mode. For a TM mode the H lines are entirely transverse. Such is not the case with the dominant mode, where the H lines have a longitudinal component.

In addition to the designation TE or TM, we use subscript numbers to complete the description of the field patterns. In a rectangular guide, the first subscript is the *integral* number of half-wave patterns in the wide a dimension, while the second subscript is the number of half-wave patterns in the narrow b dimension. Figure 3-13 shows the cross-sectional distribution of the transverse E lines for the dominant mode. Clearly, there is one half-wave pattern along the a dimension but no change along the b dimension; the complete designation for the dominant mode is therefore TE_{10}. Two higher-order modes are shown in Figure 3-14 and you should verify their designations. Such modes might rarely be used under very special circumstances, but you should concentrate on the dominant mode.

(handwritten margin note: if wave velocity in guide is less than free space)
(handwritten margin note: $\lambda = \frac{v}{f}$ (down))

Dielectric Filled Waveguides

In the mode of propagation so far discussed the guide's wide dimension a had to exceed half of the free-space wavelength. However, if the waveguide is filled with a dielectric of high permittivity, the value of c falls, so that for a particular frequency, the corresponding wavelength λ is reduced. Consequently, the required value for the wide dimension a is decreased and a smaller waveguide is the result. However, this advantage of the dielectric filled waveguide is at the expense of the increased attenuation due to the dielectric hysteresis loss; as a result, this form of waveguide is normally used only over short distances.

Methods of Coupling

Fundamentally, there are three methods of coupling RF energy into or out of a waveguide: probe, loop, and aperture or slot; these methods were used with cavity resonators (Section 1-3). The E probe represents capacitive coupling and is illustrated in Figure 3-15a. Its action is similar to that of a $\lambda/4$ antenna, so that when the probe is excited by the RF signal, alternating E and H fields are established. The probe is normally positioned in the center of the a dimension and is a quarter wavelength, $\lambda_g/4$ (or an odd multiple of a quarter wavelength), from the sealed end, which may be in the form of an adjustable piston. At this position there is a maximum density of the E lines, so that there is maximum coupling between the probe and the guide. Usually, the probe is fed by a short length of a coaxial cable with its outer conductor connected to the wall of the waveguide. The inner conductor is joined to the probe, which extends into the guide but is insulated from the walls. The degree of coupling can be altered either by varying the depth of the probe's insertion or by shifting it from the position of the maximum E-line density or by partial shielding.

In a pulse-modulated radar system the bandwidth is large and is of the order of megahertz. So that the probe shall not discriminate appreciably against any of the numerous sidebands, we can use a variety of wideband probes (Figure 3-15a).

Figure 3-15b illustrates loop or inductive coupling. The loop is placed at a point of the maximum H field for which there are a number of possible locations; it is normally part of a coaxial cable whose outer conductor is connected to the waveguide. The inner conductor then forms the loop inside the guide and the end of the loop is connected to the internal wall. The degree of coupling can be varied by altering the position of the loop's plane relative to the direction of the H lines.

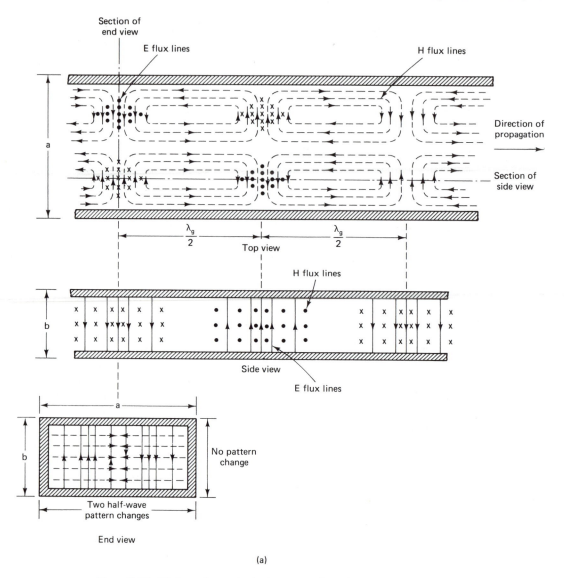

Figure 3–14 Higher order modes in the rectangular guide: (a) TE$_{20}$ mode; (b) TM$_{11}$ mode.

Aperture or slot coupling is shown in Figure 3–15c. Slot X is in an area where the *E* field is a maximum (electric field coupling), while slot Y is in a region of maximum *H*-field density (magnetic field coupling). The position of slot Z coincides with maximum *E* and *H* fields, so that the coupling is electromagnetic. X, Y, and Z are all radiating slots which are normally cut at right angles to the direction of the wall currents flowing along the inner surfaces of the guide; this produces maximum distortion of the wall currents so that radiation occurs (Figure 3–15d).

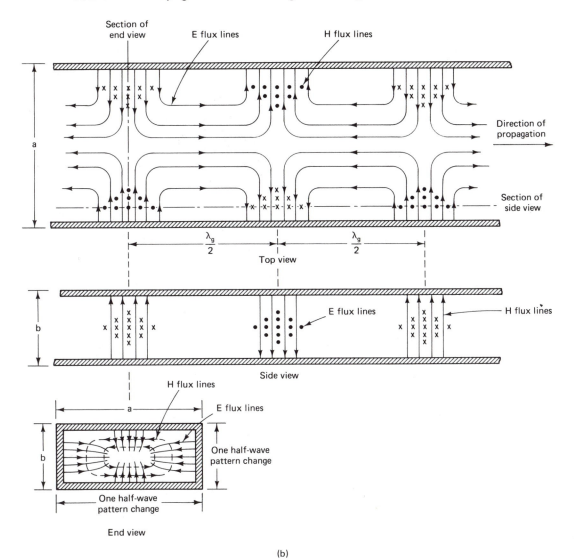

Fig. 3–14 (*cont.*)

The degree of coupling increases with the size of the slot, especially its width, and also depends on the angle of the slot in relation to the directions of the wall currents. Based on these principles a whole variety of radiating slots is illustrated in Figure 3–15e. For loose coupling the slots may be replaced by small circular holes. It should be emphasized that these methods of introducing energy into a waveguide are equally effective in removing it. By contrast, nonradiating slots produce minimum disturbance of the wall currents and are used for monitoring purposes, such as VSWR measurements; examples of such slots are shown in Figure 3–15f.

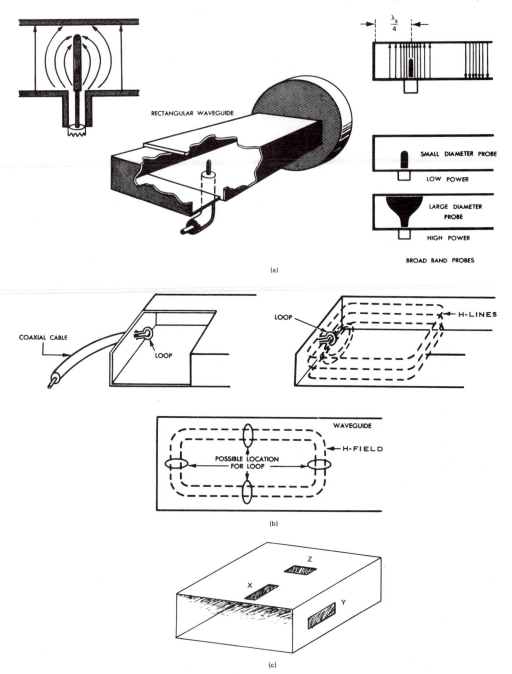

Figure 3–15 Methods of coupling to and from a rectangular waveguide.

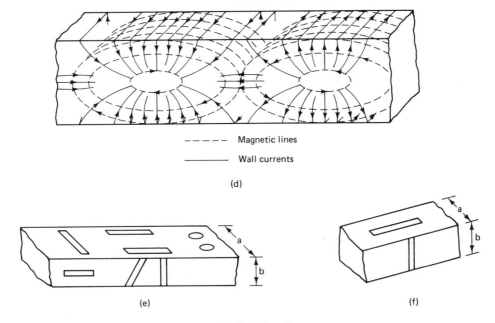

————— Magnetic lines

————— Wall currents

(d)

(e)   (f)

Fig. 3–15 (*cont.*)

Example 3–4   A WR 90 rectangular waveguide is designed to operate over the X band, 8.2 to 12.4 GHz. The inner guide dimensions are $a = 2.286$ cm and $b = 1.143$ cm. If the transmitted frequency is 8.5 GHz, calculate the values of the (a) free-space wavelength, (b) cutoff wavelength, (c) cutoff frequency, (d) angle of incidence, (e) guide wavelength, (f) phase velocity, (g) group velocity, and (h) phase-shift constant.

Solution   (a) The free-space wavelength

$$\lambda = \frac{30}{8.5} = \textbf{3.53 cm}\tag{1-3}$$

(b) The cutoff wavelength

$$\lambda_c = 2 \times 2.286 = \textbf{4.57 cm}$$

(c) The cutoff frequency

$$f_c = \frac{30}{4.57} = \textbf{6.56 GHz}$$

(d) The angle of incidence

$$\theta = \sin^{-1}\frac{3.53}{4.57} = \textbf{50.6°}\tag{3-3}$$

(e) The guide wavelength

$$\lambda_g = \frac{3.53}{\cos 50.6°} = \textbf{5.56 cm}\tag{3-4}$$

(f) The phase velocity

$$v_\phi = \frac{300}{\cos 50.6°} = 473 \text{ Mm/s} \qquad (3\text{--}9)$$

(g) The group velocity

$$v_g = 300 \cos 50.6° = 190 \text{ Mm/s} \qquad (3\text{--}10)$$

(h) The phase-shift constant

$$\beta = \frac{2\pi}{5.56 \times 10^{-2}} = 113 \text{ rad/m} \qquad (3\text{--}12)$$

This example has been used to show that all the results change when the frequency is altered in the problem illustrated in Section 3–4.

3–5 CIRCULAR WAVEGUIDES

Although rectangular guides are used almost exclusively in radar systems, there are specific cases in which circular waveguides find their application. A good example occurs in a radar antenna system which is required to revolve relative to the stationary transmitter and receiver. When one waveguide section is rotated relative to another, it is impossible to maintain continuity with rectangular guides, so that the only solution is to use a circular guide.

The dominant mode of a circular waveguide is illustrated in Figure 3–16a. For this mode the cutoff wavelength is 1.71 times the inner diameter d of the waveguide (quoted result). Remember that the cutoff wavelength of a rectangular guide is twice the wide dimension when operating in the TE_{10} dominant mode. For the same cutoff frequency, $1.71d = 2a$ or $d = 2a/1.71 = 1.17a$, so that the circular guide is larger than the corresponding rectangular waveguide—an obvious disadvantage. However, the circular waveguide suffers from a far more serious problem. In a rectangular guide the directions of the E and H lines can be referred to the directions of the narrow and wide dimensions. With a circular guide no such references exist and consequently the wave's plane of polarization tends to rotate as the wave moves down the guide. It can then be difficult to remove the energy from the guide since, for example, an E probe requires that the direction of the probe is parallel to the direction of the E lines.

To classify the various modes that can exist in a circular guide, we again divide the modes into TE and TM. On the numbering system the first subscript indicates the number of full-wave E patterns that we encounter as we move around the circumference, and the second subscript refers to the number of half-wave E patterns across the diameter. For the dominant mode the electric flux lines are entirely transverse. There is one full-wave pattern around the circumference and one half-wave pattern across the diameter; the full designation of the dominant mode is therefore TE_{11} (Figure 3–16a).

When the center conductor is removed from a coaxial cable to create a circular

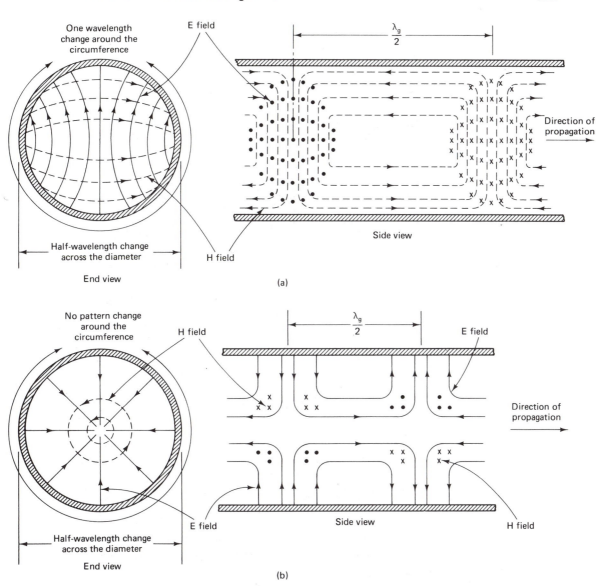

One wavelength change around the circumference

E field

$\frac{\lambda_g}{2}$

Direction of propagation

Half-wavelength change across the diameter

H field

End view

Side view

(a)

No pattern change around the circumference

H field

$\frac{\lambda_g}{2}$

E field

Direction of propagation

Half-wavelength change across the diameter

E field

Side view

H field

End view

(b)

Figure 3–16   Modes in the circular waveguide: (a) TE_{11} dominant mode; (b) TM_{01} mode.

waveguide, the mode of operation is designated as TM_{01} (Figure 3–16b), whose cutoff wavelength is only 1.31 times the guide's diameter. This mode is of particular interest because its E and H patterns are compatible with the fields that exist with the TE_{10} dominant mode of the rectangular guide. This allows us to convert from a rectangular guide to a circular guide, carry out a rotation, and then convert back to a rectangular guide (Figure 3–17).

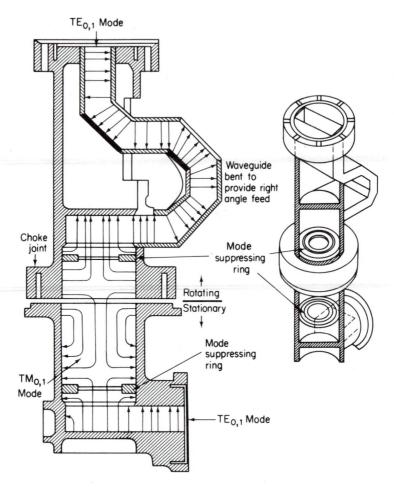

Figure 3-17 Conversion between a rectangular waveguide and a circular waveguide.

Summarizing, the use of a circular rather than a rectangular guide is limited by its increased size and the problem associating with the twisting of the wave's polarization plane.

Example 3–5 A circular waveguide has an inner diameter of 4.5 cm. Calculate its cutoff frequency when it is operated in (a) the TE_{11} mode and (b) the TM_{01} mode.

Solution (a) for the TE_{11} mode, the cutoff wavelength

$$\lambda_c = 1.71 \times 4.5 = 7.695 \text{ cm}$$

The cutoff frequency

$$f_c = \frac{30}{7.695} = \textbf{3.9 GHz}$$

(b) For the TM_{01} mode, the cutoff wavelength

$$\lambda_c = 1.31 \times 4.5 = 5.895 \text{ cm}$$

The cutoff frequency

$$f_c = \frac{30}{5.895} = \textbf{5.1 GHz}$$

3–6 MICROSTRIP AND STRIPLINE CONNECTIONS

At the present time we are in the process of miniaturizing microwave units; this involves the use of small printed boards, thin-film technology, and microwave integrated circuits (MICs). The problem is then to provide some form of connection between the miniature circuits. Clearly, a coaxial cable or a waveguide is not a practical solution; instead, we use short sections (less than 1 cm in length) of either stripline or microstrip (Figure 3–18a and b). Both these solutions may be regarded as modified forms of transmission lines, but they are included in this chapter because they represent a microwave alternative to the waveguide.

The simplest type of stripline (Figure 3–18a) consists of two grounded planes that behave as a single conductor. Inserted between these planes is a smaller conducting strip (the second conductor) which is separated from the grounded planes by a constant thickness (about 0.125 mm) of dielectric material. For microwave frequencies alumina is a suitable dielectric, with its low hysteresis loss and a relative permittivity of 9.6.

We can regard stripline as a form of coaxial cable in which the outer conductor is a rectangle with the narrow sides removed. Propagation over a section of stripline

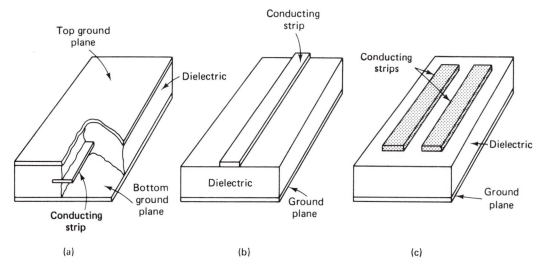

Figure 3–18   (a) Stripline; (b) single-conductor microstrip; (c) twin-conductor microstrip.

is carried out by means of a TEM wave. However, the radiation loss of stripline will exceed that of either the waveguide or the coaxial cable.

Microstrip is more commonly used and has a simpler construction than stripline since there is only a single grounded plane (Figure 3–18b). Consequently, microstrip can be compared with a two-wire transmission line and may easily be connected to microwave semiconductors and MICS. However, compared with stripline, there is more radiation loss from microstrip, which therefore has a lower Q value and power-handling capability. As a result, there is a greater mutual loading effect between two circuits that are connected by microstrip.

With both stripline and microstrip it is possible to mount one or more conducting strips on a single dielectric (Figure 3–18c). The configurations of these strips may be such that a section of each strip may represent a resistor (by using nichrome rather than copper), an inductor (of the order of nanohenries), or a capacitor (about 1 pF). Since ferrites may also be introduced, it is possible to manufacture complete circuits on either stripline or microstrip; examples of such circuits are filters (Section 4–12), isolators, and circulators (Section 4–10).

We already know about the approximate 3:2 ratio between the upper and lower frequency limits over which a waveguide can operate satisfactorily. No such limitation exists for stripline, where practical ratios of more than 3:1 are possible. However, above the X-band region the costs of microstrip and stripline components increase appreciably, so that their use is generally restricted to frequencies below 12.5 GHz. As a final point, stripline and microstrip components are manufactured with fixed values, whereas their waveguide equivalents are adjustable.

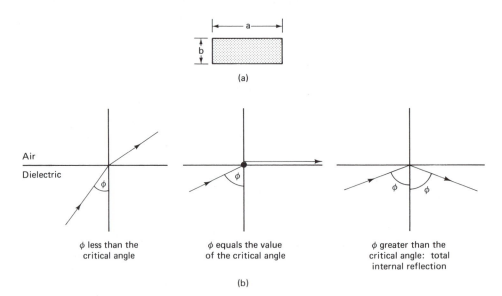

Figure 3–19 Principle of the dielectric waveguide: (a) dielectric slab; (b) behavior at various incident angles.

At frequencies above 25 GHz, microstrip losses become excessive and are of the order of 0.15 dB/cm. A practical alternative is to use a slab of dielectric material (alumina, silicon, gallium arsenide) whose value of relative permittivity is high compared with that of the surrounding free space (Figure 3–19a). The surface between the two media then acts as a discontinuity which allows EM waves to be propagated through the slab with an attenuation of only 0.05 dB/cm at a frequency of 50 GHz. This phenomenon is based on the optical principle of total internal reflection. Figure 3–19b shows a boundary existing between two dielectrics of different permittivities. When a wave strikes the boundary at an incident angle that exceeds a certain critical value, the wave is reflected and is then propagated along the slab. The critical angle corresponds to a certain critical frequency f_c, so that if the microwave frequency is less than this value, the slab will no longer behave as a form of waveguide.

In some applications a dielectric rod is used instead of a slab. For example, one type of end-fire antenna consists of a dielectric rod from which electromagnetic energy is allowed to leak out (Section 7–7).

3–7 OPTICAL FIBERS (FIBER OPTICS)

There are at the moment three transmission systems for carrying thousands of channels over long distances:

1. Microwave links (Section 7–8).
2. Coaxial cables, a number of which are placed in a single tube. To overcome the effects of attenuation over long distances, repeater units are inserted periodically to provide the necessary amplification.
3. Fiber optics, which is a relatively new technology with many advantages over the coaxial cable.

A single optical fiber is a thin, solid, circular cylinder manufactured from silica or glass with a high degree of purity. The diameter of this fiber is of the order of 0.1 mm, but it is surrounded by an equal thickness of ceramic cladding. The refractive index of the core is then made higher than that of the cladding. Based on the principle of total internal reflection (Section 3–6), an EM wave will be trapped inside the fiber and will be propagated in the dominant TE_{11} mode by multiple reflections off the side wall. The same result occurs if the cladding is discarded and the refractive index is gradually reduced from the fiber's center to its circumference.

A fiber-optics communications system involves a transmitting source such as a solid-state GaAs laser (Section 6–12). The information in the channels then modulates the infrared output of the laser. A typical frequency is 300 THz, which corresponds to a wavelength of 1 μm. This would allow a bandwidth of about 10 GHz, so that as many as 5000 channels may be carried by a single fiber which is less than 1 mm in diameter (compared with 1 cm for a coaxial cable). At the receiving end, demodulation is achieved by some form of photodiode which restores the original information.

In comparison with coaxial cables, optical fibers have additional advantages:

1. There is virtually no limit to the length of a fiber since sections can be joined with minimum reflection effects.
2. The fiber is immune to electromagnetic interference, which has no effect on energy in the infrared band.
3. The signal in one fiber does not interfere with the signal carried by a neighboring fiber. This compares with the considerable mutual interference between the signals carried by two adjacent copper wires.
4. Low attenuation: At the present time fibers can be manufactured with an attenuation of approximately 0.003 dB/m as opposed to 0.03 dB/m for a comparable coaxial cable; this reduces the number of repeater units required by increasing their allowable separation. Moreover, the cost of glass fibers is down to about $100.00 per kilometer and is clearly lower than that of a coaxial cable, with its relatively expensive copper conductors. The main constituent of glass is sand, of which there is a vast quantity available in the world; by contrast, the supply of copper is limited and is needed for the manufacture of low-resistance conductors. In addition, the weight of a coaxial cable is about 20 times greater than that of a comparable fiber optic tube.

New breakthroughs in fiber optics technology are in the research and development stage. These include:

1. Halide glass fibers, which are ultraclear filaments for the transmission of infrared waves. The attenuation for these fibers is extremely low, so that the black light may literally travel hundreds of miles without the need for optical repeater stations.
2. The transition from multimode to single-mode fibers. Multimode fibers have a minimum thickness of approximately 50 μm, can be easily spliced, and possess a wide bandwidth. However, single-mode fibers are only 10 μm thick (less than the thickness of a strand of human hair) and have a far greater bandwidth. The main disadvantages of single-mode fibers are the problems encounted in their manufacture and the difficulty experienced in joining such fibers together.

In the future a single fiber may carry enough information to control a number of robots working on an automated assembly line. A fiber-optic cable buried deep in the ground will be used to determine whether the cable is passing through valuable mineral deposits or worthless mud. Finally, a fiber-optic thermometer may be inserted into human tissue as an aid to hyperthermia treatment for the destruction of cancer cells.

To summarize, an optical-fiber communications system has the advantages of a large bandwidth, low cost, low attenuation, less weight, and minimum interference effects. It is not surprising that as existing coaxial-cable links become overloaded, they are being replaced by fiber-optic tubes.

PROBLEMS

BASIC PROBLEMS

3–1. A microwave signal has a frequency of 6.2 GHz. What is the wavelength of the signal in free space?

3–2. The internal dimensions of a rectangular waveguide are 4.755 cm and 2.215 cm. What are the values of the cutoff wavelength and the cutoff frequency for the TE_{10} dominant mode in the wide dimension? For which designated band would this waveguide be suitable?

3–3. In Problem 3–2 the frequency of the signal feeding the waveguide is 4.2 GHz. What is the angle of incidence inside the waveguide?

3–4. The internal dimensions of an S-band rectangular waveguide are 7.214 cm and 3.404 cm. The frequency of the signal propagating down the waveguide is 3.3 GHz. What is the guide wavelength for the dominant TE_{10} mode in the wide dimension?

3–5. In Problem 3–4, calculate the values of the group velocity, the phase velocity, and the phase-shift constant.

3–6. A rectangular K-band waveguide has internal dimensions of 1.067 cm and 0.432 cm. What is the cutoff frequency for the TE_{20} mode in the wide dimension?

3–7. A waveguide is sealed at one end and excited by an E probe. If the angle of incidence is 20° and the signal has a free-space wavelength of 12 cm, how far from the sealed end should the probe be positioned?

3–8. A circular waveguide has an internal diameter of 5.276 cm. What is the value of the cutoff frequency in the TE_{11} dominant mode?

3–9. A circular waveguide has an internal diameter of 3.815 cm. What is the value of the cutoff frequency in the TM_{10} mode?

3–10. The internal dimensions of a rectangular waveguide are 7.2 cm and 3.6 cm. Determine the frequency range over which only the dominant TE_{10} mode will be propagated down the waveguide.

ADVANCED PROBLEMS

3–11. The internal dimensions of an S-band rectangular guide are 7.214 cm and 3.404 cm. Calculate the cutoff frequencies for the TE_{10} and TE_{20} modes in the wide dimension and the TE_{10} mode in the narrow dimension.

3–12. Near the load the measured value of the VSWR on a rectangular guide is 1.8. Express this value in decibels. If the incident power being propagated down the line is 15 W, calculate the amount of power absorbed by the load.

3–13. The internal dimensions of a WR 90 rectangular X-band waveguide are 2.286 cm and 1.016 cm. If a 9.2-GHz signal is being propagated down the line in the dominant mode, what are the values of the angle of incidence, guide wavelength, group velocity, phase velocity, and phase-shift constant? What is the ratio of the signal frequency to the cutoff frequency?

3–14. A rectangular waveguide is excited by a 5.8 GHz signal. If the group velocity is to be 0.88 × c (the velocity of light), determine the length of the wide dimension.

3–15. Measurements show that 8% of a waveguide's incident power is reflected from its load. Neglecting any attenuation, calculate the value of the VSWR on the waveguide.

3–16. A rectangular waveguide is terminated by a brass plate that behaves as a short circuit. The internal dimensions of this C-band waveguide are 4.755 cm and 2.215 cm. If the signal frequency is 4.8 GHz, what is the separation in centimeters between two adjacent E-field nulls?

3–17. The angle of incidence is 25° for an EM wave propagating down a rectangular guide in the TE_{10} mode. If the wide dimension is 2.8 cm, calculate the frequency of the EM wave.

3–18. A circular waveguide and a rectangular waveguide have the same cutoff frequency and are both operated in their dominant modes. Assuming that the rectangular waveguide's wide dimension is twice the narrow dimension ($a = 2b$), what is the ratio of the circular waveguide's cross-sectional area to that of the rectangular waveguide?

3–19. A rectangular waveguide operating in the TE_{10} mode is terminated by a short circuit. Measurements show that the distance between two adjacent E-field nulls is 2.5 cm. If the frequency of the microwave signal is 9.8 GHz, what is the length of the guide's inner wide dimension?

3–20. A microwave source feeds a section of lossy waveguide that introduces a one-way attenuation of 2.2 dB. If the waveguide section at one end is terminated by a short circuit, calculate the value of the VSWR at the other end.

CHAPTER 4

WAVEGUIDE COMPONENTS AND ELEMENTS

4-1 INTRODUCTION

Since it is practically impossible to build a waveguide system in one piece, it is normally constructed in sections that must be connected by joints. Any irregularities in the joints cause reflection effects, create standing waves, and increase the attenuation. A proper permanent joint affords a good connection between the two sections of the waveguide and has very little effect on the *E*- and *H*-field patterns. During manufacture the waveguide sections are matched to within a few mils (0.001 in.) and then welded together. The result is a hermetically sealed and mirror-smooth joint. Moreover, in a circular waveguide system we sometimes require a rotating joint; one example is in a radar set, where the transmitter and the receiver are stationary but the antenna system is revolving.

It is normally difficult to run a waveguide in a straight line and therefore we are faced with a possibility of a number of bends or corners. The bends must be gradual with a limited radius of curvature, but by contrast, corners may be required to produce a 90° change in direction. Bends and corners as well as joints represent some degree of discontinuity, which will cause the production of standing waves and raise the value of the voltage standing-wave ratio. Devices such as irises (windows), posts, and screws are then used to reduce the SWR value to an acceptable level.

A twist in a waveguide is used to rotate a wave's plane of polarization through 90°. Such a twist must be gradual so that the discontinuity effect is kept to a minimum. In addition to twists, we sometimes employ a gradual taper as a transition between circular and rectangular waveguides.

In Chapter 2 we discussed the use of sliding or fixed stubs in matching the antenna load to the surge impedance of a line. A waveguide also possesses a certain impedance which may be matched to a terminating load by means of fixed stubs.

At a certain position in a waveguide system it may be necessary to split all or part of the microwave energy into particular directions. This is achieved by junctions that are combined to form coupler units, and these direct the energy as required. The same junction that is used to split the signal may alternatively be used to combine two or more signals.

A waveguide may, of course, feed its microwave energy into free space as in a radar system, but in other applications there are special matched terminations for which there is virtually total absorption of the energy, so that little reflection occurs.

The ancillary equipment used with waveguides is discussed in the following sections:

4–2. Joints

4–3. Bends and Corners

4–4. Twists and Tapers

4–5. Irises (Windows or Obstacles)

4–6. Posts and Screws

4–7. Wave Impedance

4–8. Matching Stubs

4–9. Waveguide Hybrid Couplers

4–10. Ferrite Characteristics

4–11. Attenuators

4–12. Microwave Filters

4–13. Fixed and Variable Phase Shifters

A variety of waveguide components is illustrated in Figure 4–1.

4–2 JOINTS

Where sections of a waveguide system must be taken apart for normal maintenance and repair, we obviously cannot use a permanent joint. To allow portions of the waveguide to be separated, the sections are connected by semipermanent butt joints of which the simplest is the "bolted flange" (Figure 4–2a). Here the two sections are merely bolted together, with a gasket to exclude moisture. However, care must be taken to provide perfect mechanical alignment to avoid the creation of a bend or step. Any such discontinuity would cause reflection effects and raise the value of the SWR. For the same reason it is important that the ends of the waveguides and the flanges themselves be provided with a smooth finish.

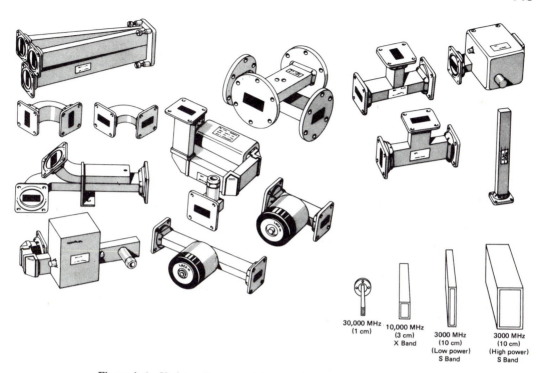

Figure 4–1   Variety of commercial waveguide components with size comparisons (figures in parentheses refer to the wavelengths). (Courtesy of MCL Inc.)

At the higher microwave frequencies the waveguide dimensions are reduced so that any given discontinuity becomes larger in relation to the guide dimensions and the wavelength involved. The discontinuity then presents a greater problem, so there must ultimately be a frequency limit for which the simple flange joint is no longer an adequate solution.

As an alternative to the mechanical connection, we can use the quarter-wavelength flange joint (Figure 4–2b), in which the open circuit at point X creates a short circuit at point Y; this is due to the standing-wave distribution between X and Y. As a result, there is no disturbance of the field patterns moving down the guide and consequently, no discontinuity exists. The disadvantage of such a joint is the leakage of energy through the open flange.

A superior solution to the flange or butt joint is the choke joint (Figure 4–2c and d), which is the one most commonly used. This consists of flanges that are connected to the waveguide at its center. In Figure 4–2c, the right-hand flange is flat and the one on the left is slotted one quarter-wave length, $\lambda_g/4$, deep from the inner surface of the waveguide. This slot is positioned at a distance of one quarter-wavelength, $\lambda_g/4$ from the point where the flanges are joined. Since the two quarter-

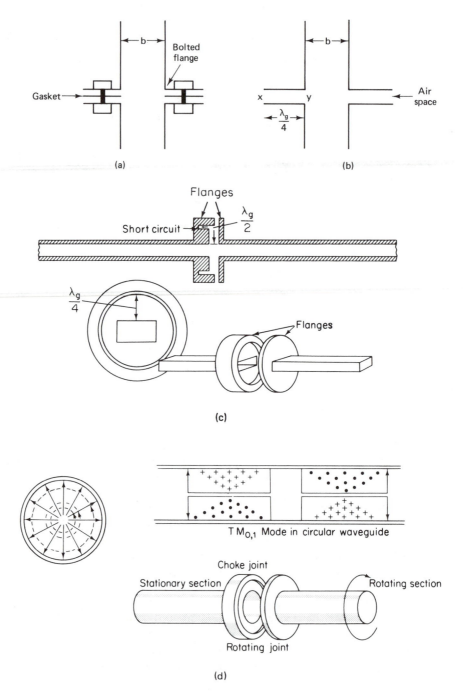

Figure 4–2 Waveguide joints: (a) bolted flange joint; (b) quarter-wavelength flange joint; (c) choke joint; (d) rotating choke joint and TM$_{01}$ mode.

wavelengths together form a half-wavelength section, this represents a short circuit at the place where the walls are joined together. The result is an *electrical* short circuit at the junction of the two waveguides. The two sections may actually be separated by as much as a tenth of a wavelength, which is 3 mm at 10 GHz in the X band. This separation allows us to seal the interior of the waveguide with a rubber gasket to exclude moisture. Any moisture introduces a discontinuity, creates standing waves, alters the guide's impedance, increases the attenuation, and may cause corrosion. For this reason long, level runs of a waveguide should be avoided; in addition, a small hole is sometimes drilled at the waveguide's lowest point so that any accumulation of moisture can be drained out. It follows that the waveguide run from a radar transmitter to its antenna should be kept to a minimum length.

The quarter-wavelength distance from the walls to the slot is modified slightly to compensate for the small reactance introduced by the short space existing between the slot and the periphery of the flange. The loss introduced by a well-designed choke joint is less than 0.03 dB, while a flange joint has a loss of 0.05 dB or more. However, since the principle of the choke joint depends on the half-wavelength distance, such a joint is frequency sensitive. Over a bandwidth that is about 10% of the center frequency, the SWR introduced by the choke joint should not be greater than 1.05.

We have already mentioned that rotating joints are usually required in a radar system where the transmitter and the receiver are stationary but the antenna system is revolving. A simple method for rotating one waveguide section relative to another is to use a mode of operation whose field distribution is symmetrical about the axis of rotation. This requirement is met by using a circular waveguide operating in the TM_{01} mode (Figure 3–17). The diameter of the circular waveguide ensures that higher-order modes cannot be propagated. At the same time, two circular irises (Section 4–5) attenuate the dominant TE_{11} mode by placing their metal diaphragms so that these surfaces are parallel to the electric field of the dominant mode. However, the same diaphragms will be perpendicular to the electric flux lines of the TM_{01} mode, whose propagation is therefore unaffected.

A choke joint separates the sections mechanically but joins them electrically. As one section rotates, the field distributions are fixed so that there are minimum reflections (Figure 4–2d). Since radar systems employ mainly rectangular waveguides, the circular rotating joint must be inserted between two rectangular sections (Figure 3–17). The joint then consists of two sections of circular guide; one section is rotating while the other is stationary. At the end of each of the sections there is a transition between the circular guide and the rectangular guide. At these junctions there may occur reflection effects which can be reduced by the insertion of irises.

In Figure 3–17 the rectangular sections are operating in the dominant TE_{10} mode. The *E* lines of the bottom rectangular section penetrate the circular section and excite the TM_{10} mode, which provides the required axial symmetry for rotating joints. At the top of the revolving circular guide the *E* lines couple the energy in the rectangular section that leads to the antenna; we are now operating once more in the dominant TE_{10} mode.

4–3 BENDS AND CORNERS

Any sudden change in the size, shape, or direction of the waveguide system will result in the introduction of reflection effects and an increase in the value of the SWR. However, if we look at the antenna system of Figure 7–19, we observe that such a system requires a bend, and this is in the direction of the wide dimension (Figure 4–3a), so that the H loops are primarily affected (H bend). The H-plane bend is therefore a piece of waveguide which is smoothly bent in a plane parallel to the magnetic field operating in the dominant mode. To reduce the amount of the reflection occurring in the bend, its mean length should be as long as possible and should preferably be a whole number of wavelengths.

As an alternative to the H bend in the wide dimension, the bend may be in the narrow dimension (Figure 4–2b), so that the E lines are mainly distorted (E bend). In order that either of these bends not produce an individual increase in the SWR of more than 1.05:1, the bend must have a mean radius of curvature of not less than $2\lambda_g$.

At the lower frequencies a bend would have to be excessively long, and a corner is normally preferred. It is clear that a sharp 90° bend (Figure 4–4a) is not permissible since total reflection would occur at such a discontinuity, and theoretically, the result would be an infinite value of SWR. In one solution (Figure 4–4b) the guide is bent through 45° twice with the two H bends one-quarter of the guide wavelength apart. The combination of direct reflection at one bend and inverted reflection from the

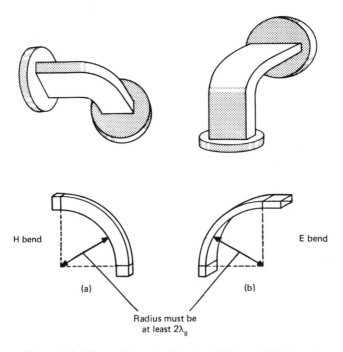

H bend

E bend

(a)

(b)

Radius must be
at least $2\lambda_g$

Figure 4–3   Waveguide gradual bends: (a) H bend; (b) E bend.

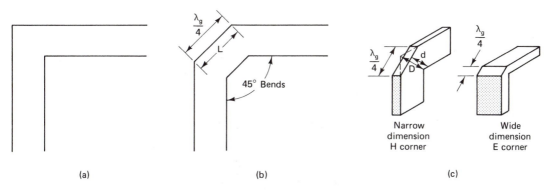

Figure 4–4   Waveguide bends: (a) 90° bend; (b) two 45° bends; (c) *H* corner, narrow dimension; (d) *E* corner, wide dimension.

other bend tends to cancel; it will then appear as if no reflection has occurred, although in practice the individual SWR introduced by the double bend is approximately 1.1. However, since this result depends on the distance L being equal to $\lambda_g/4$, the corner will behave in a frequency-sensitive manner. A superior solution to the two 45° bends is the mitered 90° bend of Figure 4–4c. Provided that $d = 0.65D$, the individual SWR introduced by the mitered *H* bend is only 1.05. These mitered bends do not change the mode of operation and are normally of the *H* variety. By contrast, the mitered *E* bend is less common because of the danger of arcing across the short distance, d.

Certain microwave systems may need special bends which are provided by flexible waveguides. Examples of such a guide are illustrated in Figure 4–5. The exterior is coated with rubber to avoid the dangers of oxidation and humidity. The waveguide itself is made of ribbon brass which is edge interlocked so that the internal dimensions are preserved in any angular position. Chromium plating is commonly used to reduce the skin effect on the interior surface.

4-4 TWISTS AND TAPERS

In Figure 3–17 the microwave energy which is coupled to the rectangular section leading to the antenna is vertically polarized. To convert to horizontal polarization, it is common practice to introduce a 90° twist, as illustrated in Figure 4–6a. Such

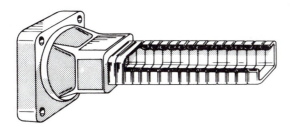

Figure 4–5   Section of a flexible waveguide. (Courtesy of Andrew Corporation.)

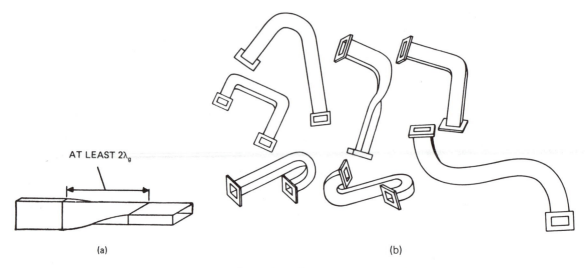

AT LEAST 2λ_g

(a)

(b)

Figure 4-6 Waveguide twists and bends.

a twist must occur over a distance of more than $2\lambda_g$ in order that the value of the SWR not exceed 1.1. However, shorter 90° twists are possible, provided that they are manufactured so that the rectangular dimensions do not vary too much. Twists may also be incorporated with bends as shown in Figure 4–6b.

On certain occasions we must join two waveguide sections that have different shapes for their cross-sectional areas. As an example, we might wish to connect a rectangular waveguide to a circular waveguide. This may be done with the aid of a gradual taper (Figure 4–7a) that extends over a distance of more than $2\lambda_g$. If the circular waveguide carries the TE_{11} mode, the rectangular waveguide will operate in the dominant TE_{10} mode, and vice versa. A taper may also be used to connect two rectangular waveguides with different dimensions (Figure 4–7b) provided that there is a match between their characteristic impedances (Section 4–7).

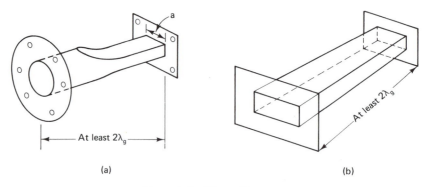

(a)

(b)

Figure 4-7 Waveguide tapers.

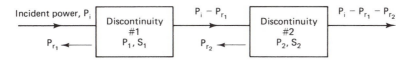

Figure 4-8 Combining of SWR values.

Combining SWR Values

So far we have discussed the value of the SWR introduced by an individual discontinuity. But how do we combine the various SWR values in determining the overall effect? In the example of Figure 4-8, there are two discontinuities, whose voltage reflection coefficient magnitudes are P_1 and P_2, respectively. The corresponding SWR values are

$$S_1 = \frac{1 + P_1}{1 - P_1} \quad \text{and} \quad S_2 = \frac{1 + P_2}{1 - P_2} \tag{2-38}$$

Then

$$P_{r_1} = P_1^2 \times P_i \tag{4-1}$$

and

$$P_{r_2} = P_2^2 (P_i - P_{r_1}) \tag{4-2}$$

Then the total reflected power

$$
\begin{aligned}
P_T^2 \times P_i &= P_{r_1} + P_{r_2} \\
&= P_1^2 \times P_i + P_2^2 (P_i - P_{r_1}) \\
&= P_i (P_1^2 + P_2^2 - P_1^2 \times P_2^2)
\end{aligned}
$$

where P_T is the magnitude of the overall reflection coefficient. Therefore,

$$P_T^2 = P_1^2 + P_2^2 - P_1^2 P_2^2$$

This yields

$$P_T = \sqrt{P_1^2 + P_2^2 - P_1^2 P_2^2} \tag{4-3}$$

where the overall value of the SWR,

$$S_T = \frac{1 + P_T}{1 - P_T}$$

For example, if $S_1 = 2.0$ and $S_2 = 3.0$,

$$P_1 = \frac{2-1}{2+1} = \frac{1}{3} \quad \text{and} \quad P_2 = \frac{3-1}{3+1} = \frac{1}{2}$$

Then

$$P_T = \sqrt{\left(\frac{1}{3}\right)^2 + \left(\frac{1}{2}\right)^2 - \left(\frac{1}{3}\right)^2 \times \left(\frac{1}{2}\right)^2}$$

$$= \sqrt{\frac{1}{9} + \frac{1}{4} - \frac{1}{36}} = \sqrt{\frac{1}{3}} = 0.577$$

and

$$S_T = \frac{1 + 0.577}{1 - 0.577} = \frac{1.577}{0.423} = \mathbf{3.7}$$

This method may be repeated if more than two discontinuities are involved.

4–5 IRISES (WINDOWS OR OBSTACLES)

No bend, joint, corner, or twist is ever perfect, so that in the presence of these discontinuities there is a certain amount of reflection, which causes a permanent susceptance to appear across the guide and presents some degree of mismatch. To overcome this mismatch and the production of standing waves, it is necessary to cancel out the susceptance by introducing another susceptance of the same magnitude but of opposite nature; the final purpose is to bring down the magnitude of the SWR until it approaches the ideal value of unity. This is the same principle as is used by the matching stub(s) described in Section 2–6; for example, a capacitive susceptance could be canceled by a parallel stub that offered an equal value of inductive susceptance.

Irises are sometimes referred to as windows, apertures, or obstacles. Each consists of one or more metal diaphragms which have a high conductivity and are small in thickness when compared to one wavelength. We must emphasize that the iris is a permanent obstacle, and no control over its effect is possible. The most common types of iris are illustrated in Figure 4–9.

The capacitive iris of Figure 4–9a is deliberately placed in a position where the electric field is strong. Assuming that the waveguide is being operated in the dominant mode, the voltage that previously existed between the top and bottom surfaces of the waveguide now appears between two metallic surfaces which are closer together. The capacitance, and therefore the capacitive susceptance, are increased.

The inductive iris of Figure 4–9b is placed where the magnetic field is strong and the electric field is relatively weak. The presence of the metal diaphragms allows currents to exist in surfaces where no current previously flowed. The effect is to intensify the magnetic field, increase the inductance, and reduce the inductive susceptance.

We can combine the capacitive and inductive irises to produce one which at a particular frequency is parallel resonant in the dominant mode; such an iris must be carefully shaped and positioned to achieve the desired effect. For the dominant mode the shunt impedance of this iris is extremely high and therefore its attenuation

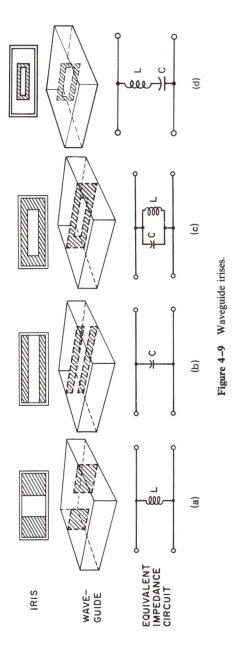

IRIS

WAVE-
GUIDE

EQUIVALENT
IMPEDANCE
CIRCUIT

(a) (b) (c) (d)

Figure 4-9 Waveguide irises.

effect can be neglected. However, at other frequencies or modes the impedance is comparatively low, and severe attenuation is the result. Consequently, the parallel resonant iris is used as a bandpass filter to suppress unwanted modes. The circular version of this iris (Figure 4–9c) was encountered in Figure 3–17, where it was used to suppress the unwanted TE_{11} mode.

The series resonant iris is shown in Figure 4–9d. It is supported by material that is nonmetallic and transparent to the flow of microwave energy.

4–6 POSTS AND SCREWS

The same results achieved by a capacitive or an inductive iris can be provided by a nonadjustable metallic cylindrical post which protrudes into the waveguide from a position in the wall of the guide's wide dimension (Figure 4–10). The effects that occur are due to the reradiation resulting from the voltage and current distribution associated with the post. If the post extends only a short distance (compared with the wavelength) into the guide (Figure 4–10a), the post behaves capacitively and presents capacitive susceptance which increases with the depth of penetration. When this depth is approximately equal to one quarter-wavelength (Figure 4–10b), the post is resonant in the series mode. Beyond one quarter-wavelength (Figure 4–10c) the post behaves inductively, with the susceptance decreasing as we move farther away from the center of the waveguide. When extended completely across the waveguide (Figure 4–10d), the post is inductive; however, the amount of susceptance

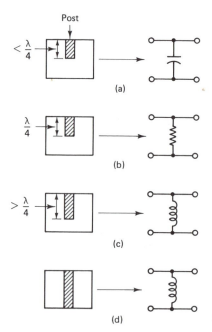

Figure 4–10 Use of posts.

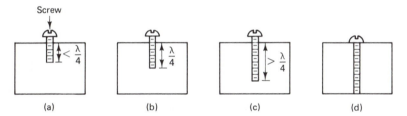

Figure 4-11 Adjustable screws.

decreases as the diameter of the post is reduced. Moreover, the thicker the post, the lower is its effective Q, so that the post, like the iris, may be used as a bandpass filter.

The screw or slug is an adjustable post and is illustrated in Figure 4-11. According to the amount of penetration, the screw may introduce either inductive or capacitive susceptance; this is similar to the action of a fixed-position stub. In fact, a combination of two screws $\lambda_g/4$ apart (Figure 4-12) may be used to match a waveguide to its load (compare the use of two fixed stubs in the Smith chart, Example 2-15). As we shall see in Section 4-8, the same result can be achieved with two H-plane stubs. A greater range of impedance match is achieved with three fixed screws with a separation of $3\lambda_g/8$ between adjacent screws.

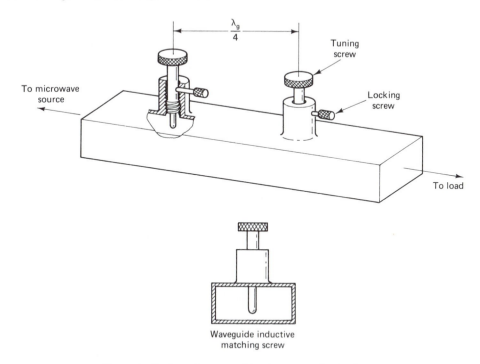

Figure 4-12 Impedance matching by means of twin screws.

When carrying out the matching procedure, all screws are initially fully retracted. One screw is then advanced, but if the SWR increases, that screw is returned to the retracted position and another is advanced. The screws are then subjected to a series of adjustments until no further reduction in the SWR is possible.

With a simple screw a large range of susceptance is possible by changing the depth of penetration and also varying its position on the line by sliding the screw along. This is similar in principle to the sliding stub of the Smith chart example (Example 2–13).

4–7 WAVE IMPEDANCE

Consider a transverse electric, transverse magnetic (TEM) electromagnetic wave, which is being propagated through a vacuum (free space). The electric field intensity $\mathscr{E}$ is measured in volts per meter (V/m), while the magnetic field intensity, H, is expressed in amperes per meter (A/m). The ratio of $\mathscr{E}$ to H must be measured in ohms and is called the intrinsic impedance, η_0, of free space. Therefore, the intrinsic impedance of free space

$$\eta_0 = \frac{\mathscr{E}}{H} \quad \Omega \tag{4-4}$$

The intrinsic impedance of free space is a constant and is given by

$$\eta_0 = \sqrt{\frac{\mu_0}{\epsilon_0}} \quad \Omega \tag{4-5}$$

where the permeability of free space, $\mu_0 = 4\pi \times 10^{-7}$ H/m, and the permittivity of free space, $\epsilon_0 = 8.85 \times 10^{-12}$ F/m.

The velocity with which all electromagnetic waves travel in free space is also a constant and is equal to the velocity of light, c (approximately 3×10^8 m/s). Clerk Maxwell showed that

$$c = \frac{1}{\sqrt{\mu_0 \epsilon_0}} \tag{4-6}$$

This yields

$$\frac{1}{\sqrt{\epsilon_0}} = \sqrt{\mu_0} \times c$$

and therefore the intrinsic impedance of free space is

$$\begin{aligned}\eta_0 &= \sqrt{\mu_0} \times \sqrt{\mu_0} \times c \\ &= \mu_0 \times c \\ &= 4\pi \times 10^{-7} \text{ H/m} \times 3 \times 10^8 \text{ m/s} \\ &= 120\pi \ \Omega \\ &= 377 \ \Omega\end{aligned} \tag{4-7}$$

For waveguides we refer to the wave impedance, which is comparable with the surge impedance of transmission lines. However, this wave impedance η depends on the particular mode of operation. For example:

All TE modes: The wave impedance

$$\eta = \eta_0 \times \frac{\lambda_g}{\lambda} \qquad \qquad (4\text{–}8)$$

All TM modes: The wave impedance

$$\eta = \eta_0 \times \frac{\lambda}{\lambda_g} \qquad \qquad (4\text{–}9)$$

Note that the value of the wave impedance depends on the values of η_0, the frequency, and the wide a dimension. This means that the concept of wave impedance could not be used to match two waveguide sections with different narrow b dimensions.

Waveguide Terminations

The primary use of wave impedance is in the correct design of waveguide terminations. Of course, on a waveguide there is no way of attaching a fixed resistive load as a termination. However, the end of the waveguide can be filled with graphited sand (Figure 4–13a), which will then dissipate the required energy. Virtually no energy is reflected back into the waveguide and the SWR is less than 1.01. Another method is to include a resistive rod (Figure 4–13b) which is positioned at the point where the density of the E lines is greatest. Yet a third method is to terminate the waveguide with a taper (Figure 4–13c) which is aligned with either the E or H lines. Such a taper is made from either powdered iron or carbon mixed with a binder which is deposited on a dielectric strip. When the flux lines cut the wedge, the induced currents create the required energy loss. For all these matched terminations there is virtually total absorption of the energy and little reflection occurs.

For test and monitoring purposes it is often desirable for nearly all of the energy to be reflected back from the end of the waveguide. This can be accomplished by permanently welding a metal plate (short circuit) at the end of the guide (Figure 4–13d). If it is necessary that the end of the plate be movable, the contact between the guide and the plate must be exceptionally good so that the H field will not be attenuated (Figure 4–13e).

If we wish to use a movable short, the required arrangement is similar to that of the choke joint discussed previously. Basically, it consists of an adjustable plunger that fits into the guide as shown in Figure 4–13f. The walls of the waveguide and the plunger form a half-wave channel. Since the half-wave channel is closed at one end, the other end also behaves as a short circuit. The result is a perfect connection between the wall and the plunger. The actual physical connection is made a quarter-wavelength from the short circuit, where the standing-wave current is at its minimum level. This makes it possible for the plunger to slide loosely in the guide at the point where the contact resistance is very low.

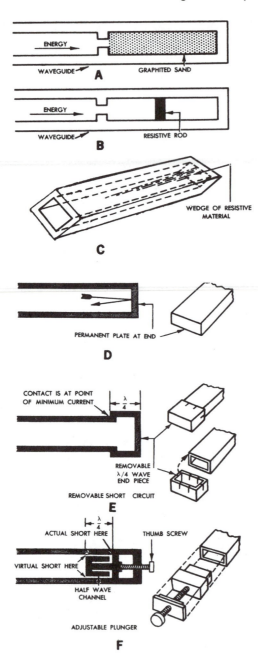

ENERGY

WAVEGUIDE **A** GRAPHITED SAND

ENERGY

WAVEGUIDE **B** RESISTIVE ROD

WEDGE OF RESISTIVE MATERIAL

C

PERMANENT PLATE AT END

D

CONTACT IS AT POINT OF MINIMUM CURRENT

$\frac{\lambda}{4}$

REMOVABLE $\lambda/4$ WAVE END PIECE

REMOVABLE SHORT CIRCUIT

E

$\frac{\lambda}{4}$

ACTUAL SHORT HERE

THUMB SCREW

VIRTUAL SHORT HERE

HALF WAVE CHANNEL

ADJUSTABLE PLUNGER

F

Figure 4–13 Waveguide terminations.

Other examples of terminating a waveguide are the crystal detector (Section 8–2) for signal demodulation and the thermistor or barretter mount for the measurement of microwave power (Section 8–3). Such terminations are designed so that they create an SWR of less than 1.1.

Characteristic Impedance

Consider two waveguide sections which have the same wide dimensions, frequency, and the same mode of operation. However, if the narrow dimensions are different, the two sections can be matched only if their characteristic impedances are made equal. The characteristic impedance of a waveguide is defined:

$$\text{characteristic impedance} = \eta_0 \times \frac{\pi b \lambda_g}{2a\lambda} \qquad (4\text{–}10)$$

The match can be achieved by using a taper (Section 4–4) to adjust the values of b and a until the characteristic impedances of the two sections are the same.

In another application it is possible to flare out the guide until the factor $\pi b \lambda_g / (2a\lambda)$ is equal to unity. The characteristic impedance is then matched to the intrinsic impedance of free space (Figure 4–14). The match may be further improved by including a dielectric baffle plate across the end of the face of the guide; such a plate will also assist in excluding moisture from the waveguide.

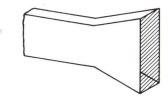

Figure 4–14   Waveguide flare and dielectric baffle.

4–8 MATCHING STUBS

In Section 2–6 we discussed the use of a single sliding stub or two fixed stubs in matching an antenna load to the surge impedance of the line. The stub itself was regarded as a limited section of line, about one quarter-wavelength long and terminated by a movable short. Such a stub is then placed across the main line and provides the required value of susceptance.

At microwave frequencies, stubs have pistons which use the principle of the choke joint's half-wave channel to achieve the necessary short circuit (Figure 4–15). When the two stubs are adjusted for their correct lengths, the combination of the stubs and the antenna load produces an overall impedance which is entirely resistive and is equal to the value of the waveguide's impedance.

Waveguide stubs may either be of the E or H variety. For an E stub the electric flux lines penetrate from the waveguide to set up standing waves on the

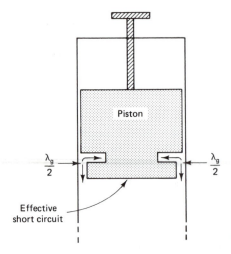

Figure 4–15 Cross section of waveguide matching stub.

stubs. Since the stub is across the wide dimension, its reactance is inserted in series with the equivalent line and is not connected in parallel (Figure 4–16a).

The H stubs are joined to the waveguide's narrow dimension so that the H loops can penetrate from the waveguide into the stub. By contrast with the E stub, the H stub is across the equivalent main line and represents a parallel susceptance (Figure 4–16b).

The stub positions which we have so far described are compatible with the main waveguide in the sense that their tee junctions do not create any field discontinuities.

4–9 WAVEGUIDE HYBRID COUPLERS

In Figure 4–17 we have compatible E and H junctions, but there is also an incompatible junction since any input signal at D will produce equal outputs at A and B but zero output from C. This is the principle behind some of the couplers used with waveguides.

Magic Tee Couplers

The magic tee coupler is a hybrid combination of E and H junctions which makes the unit a three-dimensional device, because we have positioned another half-wave stublike arm on the top side of the flat, H-plane junction. We can then show the wave and field relationships of the E and H planes by splitting the assembly and drawing a phantom view of them broken apart (Figure 4–18a).

As a further explanation of the waveguide junctions, we will recall that the stub equivalent of the input coupling to the H arm is a shunt-connected injection of the signal into the H field; the stub-equivalent coupling of the input to the E

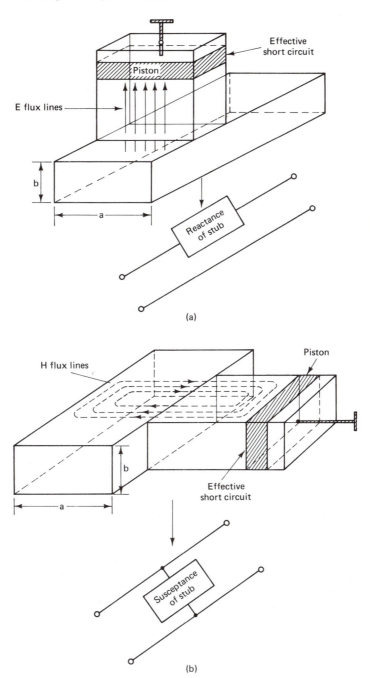

Figure 4-16   (a) *E*-plane and (b) *H*-plane stubs and equivalent circuits.

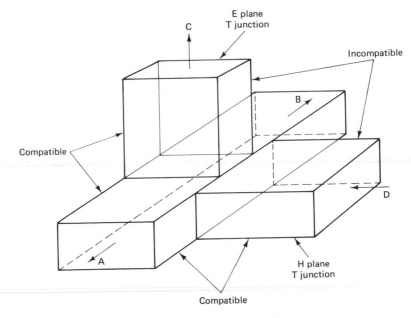

Figure 4–17   Magic tee directional coupler.

arm is a series-connected injection of the signal. Consequently, the microwave energy fed into the *H* arm (port *D*) divides equally, appearing at both of the collinear ports *A* and *B* while none appears at port *C* (Figure 4–18b). Similarly, the energy that is coupled into port *C* divides between the adjacent collinear ports *A* and *B* only (Figure 4–18c). Therefore, the energy entering port *D* cannot excite the dominant mode in port *C* because its *E* component lies in port *C*'s vertical plane, and no transfer can take place if, physically, the right-angle linkages are not made. The phase relationships for the inputs to ports *A* and *B* are precisely the same as when the inputs are fed to simple series or shunt tee junctions (Figure 4–18d).

Magic tee junctions are used as power dividers in balanced bridge circuits and for phase-related bridge measurements. In a balanced-mixer circuit, a matched detector is placed in each of the two collinear arms; microwave energy fed into the shunt arm then becomes completely decoupled from the power fed into the series arm. Figure 4–19 illustrates a commercial magic tee assembly.

The directional properties of the magic tee coupler may be summarized as follows:

Input signal	Output signals
D	*A* and *B* output signals are in phase; no output from *C*
C	*A* and *B* outputs are 180° out of phase; no output from *D*

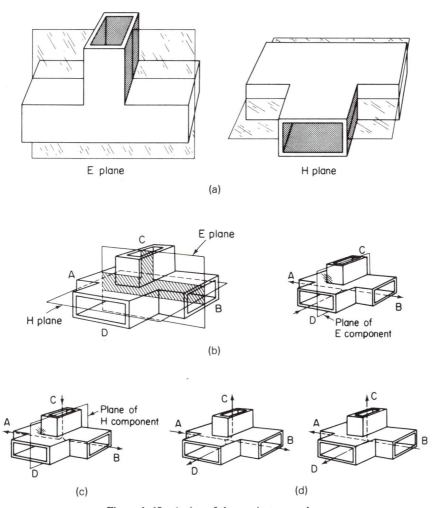

Figure 4–18   Action of the magic tee coupler.

Note that these directional properties are independent of the frequency.

Another purpose of the magic tee coupler is to match the A and B terminations to their guides. If the original signal is introduced at port D, there will be equal outputs from ports A and B but zero output from port C. If the distances from port C to the A and B terminations are equal, any reflected signals will emerge from port C. For a perfect match the terminations must be adjusted until the total output from port C is zero.

In yet another application the distances from port C to the two A, B terminations are made to differ by a quarter-wavelength. We can then use the magic tee to match these terminations to one another. If the match is correct, the two signals arriving back at C from the terminations will be equal in magnitude but 180° out of phase

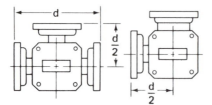

Figure 4–19   Commercial magic tee assembly.

(due to the half-wavelength difference between the two paths). The terminations can then be adjusted for zero output from C.

Rat-race Couplers

This device allows special coupling in waveguide systems and achieves its effect by folding a waveguide tee back upon itself. The result is a ringlike structure, so that mechanically we can place this structure flat using H-plane operation with shunt connections, or we can place it upright and use E-plane operation with series connections.

When conducting the folding operation and correctly locating the ports, the overall circular length must be compatible with the operating wavelength. We then see that the reentrant construction allows a fourth port to be added to the usual three-arm tee.

Figure 4–20 is a skeleton view of a typical rat-race. For proper operation it is necessary that the following two conditions are satisfied:

1. The mean circumference of the total rat-race must be 1.5 λ_g.
2. Each of the four ports must be separated from its neighbor by a distance of $\lambda_g/4$.

If a signal is fed into port A, the energy splits at the junction, so that half travels in the clockwise direction and the other half in the counterclockwise direction.

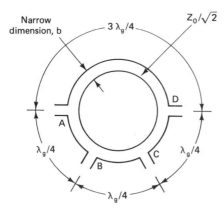

Figure 4–20   Principle of the rat-race junction, E-plane operation.

At ports B and D the outputs combine in phase but because of the $\lambda_g/2$ path difference, cancellation occurs at port C, whose output is zero. If an input signal is applied to port C, it is divided equally between ports B and D but the output at port A is zero.

If the rat-race uses E-plane operation with the impedance at each port equal to Z_0, the impedance of the complete ring is $Z_0/\sqrt{2}$ or $0.707Z_0$. With H-plane operation, the total ring impedance is $\sqrt{2}\,Z_0$ or $1.414Z_0$.

The rat-race can also be used either to combine two signals or to divide a single signal into two equal halves. If two unequal signals are applied at port A, an output proportional to their sum will emerge from ports B and D while a differential output will appear at port C.

A coaxial hybrid ring (Figure 4–21) is a simple coaxial equivalent of the rat-race circuit. In Figure 4–22 a phase inversion is introduced by inductive coupling which produces an effective path difference of $3\lambda/4$ between ports C and D. Consideration of the various path lengths shows the same directional properties as those for the magic tee.

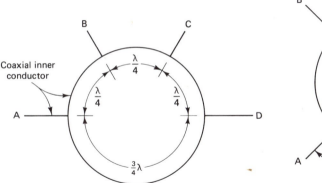

Figure 4–21 Coaxial hybrid ring.

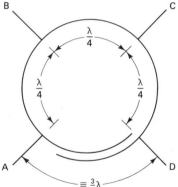

Figure 4–22 Coaxial hybrid ring with inductive coupling.

Directional Couplers

These are flanged, built-in waveguide assemblies which can sample a small amount of microwave power for measurement purposes or monitoring a system's performance. Such couplers may measure incident and/or reflected power, SWR values, provide a signal path to a receiver, or perform a variety of other operations.

Couplers may be unidirectional (Figure 4–23) and measure only the incident power, or bidirectional when measuring both incident and reflected power. The most common bidirectional coupler (Figure 4–24), consists of a length of main waveguide with two auxiliary sections mounted one on either side of its opposite walls. One section samples incident power while the other may be used to measure a small

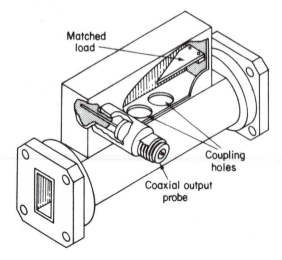

Figure 4–23   Unidirectional coupler.

portion of the reflected power. Figure 4–25 illustrates the placement of the auxiliary piece of waveguide in a simple directional assembly.

Three other features are involved with the construction of a coupler:

1. There must be coupling slots to allow the passage of the energy to the auxiliary section. These must pierce both the main and auxiliary walls so that the holes are common to both the main waveguide and the auxiliary section.

2. An internal probe must be used to detect the microwave power for indicating or measurement purposes; this usually extends through the side of the auxiliary waveguide and terminates by forming an output terminal to an externally located coaxial connection.

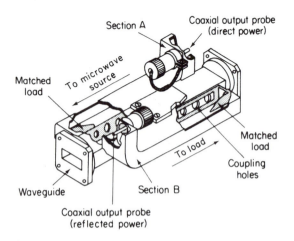

Figure 4–24   Bidirectional coupler.

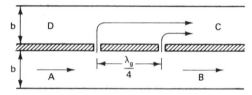

Figure 4–25   Basic directional coupler.

3. An internal load within the auxiliary section which absorbs the unwanted power from the coupling holes.

The various types of coupler are listed below, although the principle of operation is covered only for the two-hole or slot type.

1. Two-hole directional units with simple side-by-side, broad-wall H-field structures.
2. Bidirectional units with three holes which use double side-by-side wall structures with coupling holes or slots on each side (Figure 4–24).
3. Top-wall units containing narrow-wall, side-by-side, E-field couplers.
4. Two-hole crossed-guide units with common broad-wall sections (Figure 4–26a). The two amounts of energy passing through the small circular slots are 180° out of phase by virtue of the fact that the slots are set on either side of the waveguide's centerline. Some of the energy entering at port A is coupled to port C but not to port D.
5. Branching-guide couplers with a common wall instead of coupling holes (Figure 4–26b).

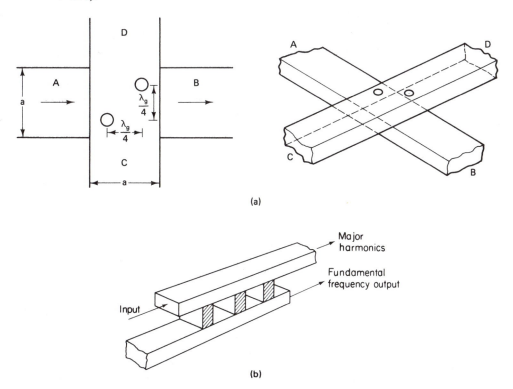

Figure 4-26 Two-hole crossed-guide and branching-guide couplers.

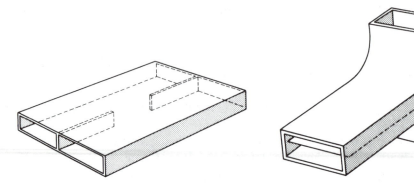

Figure 4–27 Short-slot coupler.   Figure 4–28 Bifurcated coupler.

6. Stripline couplers made from parallel, ground-plane, metallic strips running internally within the waveguide structure.

7. Short-slot couplers (Figure 4–27).

8. Bifurcated couplers (Figure 4–28).

Basic principle of operation.   Figure 4–29a illustrates the overall and internal structure of a bidirectional coupler with the auxiliary sections mounted on the opposite, flat, wide dimension walls of the main waveguide. The lower coupler diverts, by means of its coupling slots, a portion of the incident (forward) power; the upper section, by its slots, measures a sample of the reflected power. In the lower coupler, an internal probe is used to extract the sample of the incident power while a wedge acts as the load to absorb the reflected energy.

Figure 4–29b shows details of the slot locations which represent the H-field

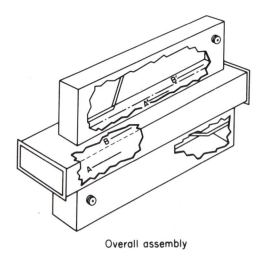

Overall assembly

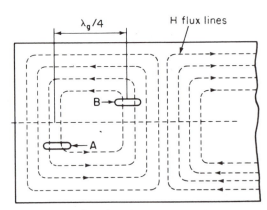

Coupling slot arrangement

(a)

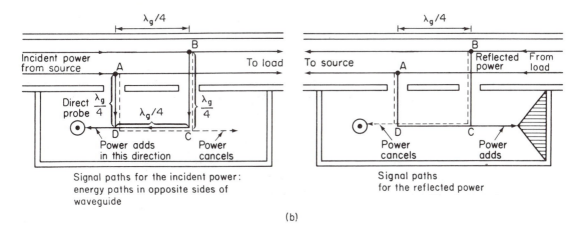

Signal paths for the incident power:
energy paths in opposite sides of
waveguide

(b)

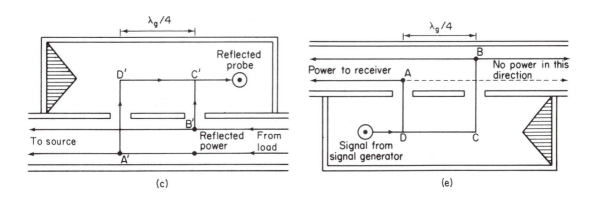

(c)   (e)

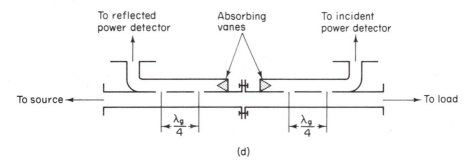

(d)

Figure 4-29 Bidirectional couplers: (a) basic structure; (b) coupler for sampling
the incident power; (c) coupler for sampling the reflected power; (d) arrangement
for measuring the incident and reflected powers; (e) basic principle.

coupling. Note that the coupling slots A and B are a quarter-wavelength ($\lambda_g/4$) apart and are on the opposite sides of the guide's centerline; this means that we are sampling two 180° out-of-phase components of the magnetic field.

When the incident microwave power, moving from left to right, is coupled through slot A, it travels through a distance of a quarter-wavelength to point D (this is represented by the solid-line power coupling). One quarter-wavelength farther on, an equal amount of energy is coupled through slot B to points C and D, thereby completing path BCD.

The difference in distance between paths AD and BCD is $3\lambda_g/4 - \lambda_g/4 = \lambda_g/2$, which is equivalent to a phase difference of 180°. However since A and B are on opposite sides of the guide, another 180° phase difference is introduced, so that there is a total 360° phase change between the two signals arriving at point D; these two signals therefore combine and a measure of the incident power is extracted from the direct probe.

The incident energy from slot A can also travel by way of the dashed-line path ADC (an equal split occurs at point D), while energy from slot B splits equally at point C so that half of the energy travels the path ABC. These paths are of equal length, but since the energy coupled from slot B is 180° out of phase with that from slot A, the fields cancel and there is no measure of the incident power at point C.

The sample of the reflected power reaching point C goes through a similar addition of combined signals and is absorbed by a dissipating load placed at C. Similarly, reflected power at the output probe is self-canceling. To respond to the reflected power, we must arrange to position such a coupler on the opposite wall of the main guide as shown in Figure 4–29c. A probe would then be installed to measure the reflected power.

We have seen that a coupler has directional properties and can be constructed so that it measures the amount of power transmitted in only one direction. However, two auxiliary waveguide sections, turned in opposite directions, can be used to measure separately the incident and the reflected power. For this purpose the physical arrangement is illustrated in Figure 4–29d, and is used in the reflectometer (Section 8–5).

Microwave power may also be fed into a waveguide by means of a coupler probe. Figure 4–29e illustrates this mechanism by using a signal-generator source which is applied to the lower waveguide; paths similar to those illustrated in Figure 4–29b deliver small amounts of power toward the receiver.

In designing a coupler, the main two factors are the degree of coupling and the amount of unwanted power which is absorbed by the dissipating load within the waveguide structure. The degree of coupling is determined by the size and location of the holes in the waveguide walls; it is normally the ratio (expressed in decibels) of the coupler's input power to its measured output power; for example, if this ratio is 100:1, we would have a 20-dB coupler.

The directivity of a coupler is the ratio in decibels of the coupler's measured output power to the amount of unwanted power present due to reflection effects. If a 20-dB coupler had a directivity of 20 dB, the unwanted power would be 1/10,000

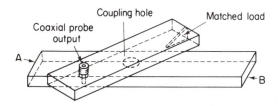

Figure 4–30 Bethe or single-hole crossed-guide coupler.

(−40 dB) of the input power. Although a high degree of directivity is relatively easy to achieve at a particular frequency, it is much more difficult to obtain over a relatively narrow band of frequencies. In this connection we must realize that the frequency determines the separation of the two slots as a fraction of the wavelength.

An improvement in the degree of directivity is possible with the Bethe or single-hole crossguide coupler pictured in Figure 4–30. Here the waves in the auxiliary guide are generated through a single hole which includes signals produced by both the electric and magnetic fields. Because of the phase relationships involved in the coupling process, the signals generated by the two types of coupling cancel in the forward direction but reinforce in the reverse direction. Consequently, the power entering at point A is coupled to the coaxial probe output, while the power entering at point B is absorbed by the matched load. If the two waveguides were paralleled, the magnetic component would be coupled to a greater degree than the electric component, so that the directivity would be poor. But by placing the auxiliary guide at the proper angle, the amplitude of the magnetically excited wave is made equal to that of the electrically excited wave; in this design the angle used depends on the frequency of operation. Basically, the directivity is improved because the Bethe coupler relies on a single hole rather than the separation between two holes.

Coaxial couplers are useful in high-power radar and communications equipment, where spurious transmitter harmonics are extremely difficult to control in rectangular guides because of their tendency to be sensitive to the excitation of several spurious modes. Slot coupling can result in the excitation of the higher TE_{11} mode in a rectangular guide when coupled from the TEM mode of a coaxial line. This slot relationship is shown in Figure 4–31a. The commercial design of simple directional and dual directional couplers is illustrated in Figure 4–31a. Stripline couplers with relatively simple configurations are commercially available in solid-state microwave switching and amplifier assembly (Figure 4–31b).

4–10 FERRITE CHARACTERISTICS

Ferrites have unusual magnetic properties which inspire their use in microwave equipment. The first difference from magnetic metal-like iron and nickel is that ferrites are oxide-based compounds containing iron, zinc, manganese, cobalt, aluminum, or nickel which are formed by firing (at 1100°C or more) powdered oxides of materials and pressing them into specific shapes. This processing gives the ferrite the added

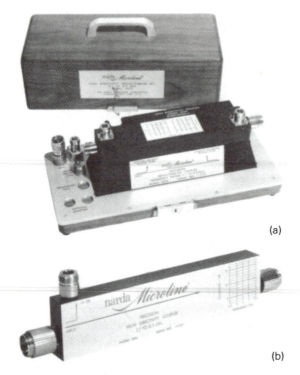

(a)

(b)

Figure 4–31 (a) Coaxial and (b) stripline couplers. (Courtesy of The Narda Microwave Corporation.)

characteristics of ceramic insulators so that they are usable at microwave frequencies. This is due to their high internal molecular resistance, which acts to reduce eddy-current losses.

Ferrites owe their characteristics to their atomic structure and in particular to the effects produced by the orbiting electrons as they spin on their own axes. When an electron spins, it creates a magnetic field along its own spin axis so that an associated current flows in a loop whose center is the electron axis. Consequently, those materials with atoms having a large number of spinning electrons have strong magnetic properties, and when placed in a static magnetic field, the electrons tend to be aligned and the material is magnetized. Any spinning body acts like a gyroscope and has a property called precession which is produced by applying another sideways magnetic force to the electron. This will cause the body to move in an off-center local orbital path at right angles to the sideways field so that it "wobbles" around its central axis. This wobble is another suborbital path, which usually has a resonant or natural precession frequency. The value of this frequency in the presence of a direct magnetic field lies between 3 and 9 GHz.

This result of combining a steady polarizing direct magnetic field with an RF field has directional properties on an electromagnetic wave which is applied to a ferrite positioned in a waveguide (Figure 4–32a). Ferrites can then be used in several devices to reduce reflected power, for modulation purposes, and in switching units.

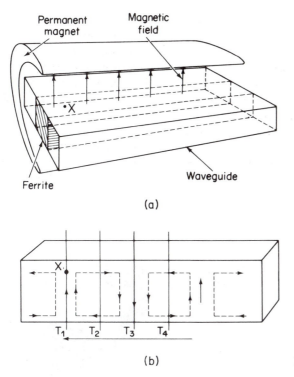

(a)

(b)   **Figure 4–32** Principle of the ferrite isolator.

Isolators

An isolator allows energy to flow in one direction but absorbs energy traveling in the opposite direction. This is accomplished using the rotational effects produced by combining a permanent direct magnetic field with an alternating electromagnetic field.

The principle is illustrated in Figure 4–32b with the apparent rotation of the magnetic wave component (without ferrite effects) as it travels from right to left. Let us consider the magnetic field at point X, which is off the centerline of the guide. At time T_1 this magnetic field component is pointed upward. When the condition at the time T_2 has reached point X, the magnetic field is pointed toward the right. For time T_3, the direction of the field is downward, while at time T_4 the direction is toward the left. Consequently, as the magnetic field passes point X in moving from right to left, its direction will appear to rotate in a clockwise manner; this is referred to as circular polarization. By contrast, if the wave is moving from left to right, the polarization will be counterclockwise.

Let us now place a section of ferrite at position X, where the RF magnetic field component is strongest and, at the same time, surround the waveguide with a permanent magnet. If the frequency of the wave is the same as the ferrite's gyro value and the field is passing from left to right, the precessing electrons will produce magnetic fields which rotate in the clockwise direction. This is the same direction as that of the circular polarization, so that the result is a resonant condition which produces severe attenuation. However, if the wave is moving from right to left, the direction of the circular polarization is opposite to the directions of the magnetic fields associated with the electrons and little attenuation occurs. In our example the typical attenuation for the right-to-left wave is usually less than 1 dB, while the attenuation in the opposite direction can extend from 10 to 30 dB.

Figure 4–33 illustrates various types of resonant isolator; these use rectangular, coaxial, or stripline construction with the permanent magnets mounted internally. Such units operate within the frequency range 1 to 10 GHz and with signal powers of less than 10 W. However, some units can be designed to operate with peak powers of 5 to 10 kW.

Faraday isolators.   This type of isolator uses the effect of a nonreciprocal phase shift created by a ferrite material. In 1848, Michael Faraday demonstrated that the

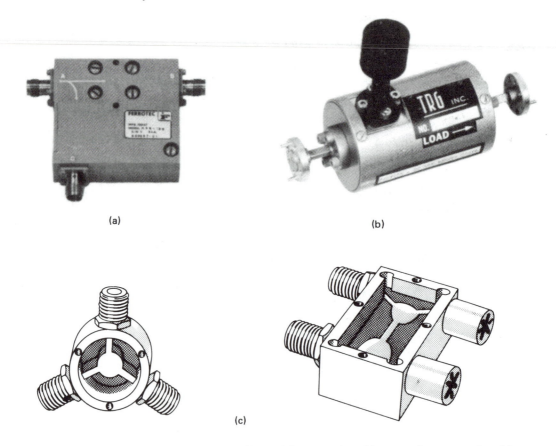

(a)

(b)

(c)

Figure 4–33 Isolator assemblies: (a) coaxial construction; (b) rectangular construction; (c) stripline construction.

plane of polarization of a linearly polarized light wave rotated when the light is passed through certain materials in a direction parallel to the flux lines of an external magnetic field. The same effect is produced in the microwave region when operating with ferrite materials. As stated previously, such materials are transparent to electromagnetic waves, have excellent magnetic properties, and have very high values of specific resistance.

The basic principle of the Faraday isolator is illustrated in Figure 4–34a. At the input end there is a rectangular waveguide which (relative to the plane of the paper) is fed by an EM wave with vertical electric flux lines. A transition is now made to a circular waveguide operating in the TE_{11} mode and a resistive attenuator vane is inserted parallel to the original rectangular guide's wide dimension. Since the E lines are perpendicular to the plane of the vane, no attenuation occurs.

The next step is to pass the wave through the ferrite specimen, which is situated in the presence of the external magnetic field. The result is to twist the plane of polarization through 45° in the clockwise direction. The emerging electric field is also at right angles to the second resistive vane so that the attenuation of the isolator in the *forward* direction has been kept to a minimum. Finally, there is another conversion from a circular to a rectangular waveguide whose orientation corresponds to the emerging wave's plane of polarization.

If we now attempt to send the same signal as a reverse wave back through the isolator, the plane of polarization will be twisted another 45° in the *same* direction as the shift of the forward wave (Figure 4–34b). Consequently, when the reverse

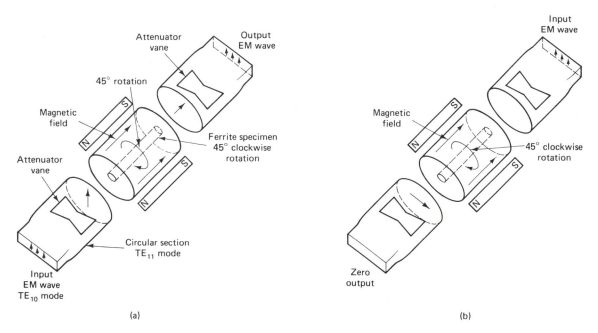

Figure 4–34 Rotation isolator in the (a) forward and (b) reverse directions.

wave emerges from the ferrite specimen, the E lines will be parallel to the first attenuator vane and a severe loss is the result. Typically, the low forward loss is about 1 dB or less while the high reverse loss is 20 dB or more. If the external magnetic field is supplied by an electromagnetic coil whose current can be varied, the characteristics of the isolator can be controlled.

Circulators

The properties of the ferrite specimen used in the isolator may be adapted to the circulator. There are many applications of circulators, but the principle is to establish various entry/exit points or "ports" where the RF power can either be fed or extracted. This is illustrated in Figure 4–35a where a circulator simultaneously connects a transmitter and a receiver to a single antenna. A variation of the same principle is shown in Figure 4–35b, where the transmitted energy circulates as indicated by the solid arrows. Transmitter 1 feeds the antenna while the received energy is directed to receiver 1. If the magnetic field is reversed, the energy will circulate as shown by the dashed arrows. In this instance the output of transmitter 2 is reflected from the filter of receiver 1 and fed to the antenna. At the same time the received signal is reflected from the filter of transmitter 1 and fed to receiver 2.

A circulator with one port terminated becomes an isolator with an exceptionally low loss. Two circulators, with one terminated by a matched load, can be used in tunnel-diode or parametric amplifiers (Figure 4–35c).

Stripline circulators can be constructed more simply; Figures 4–36a and b show how ferrite disks are used in coaxial line junctions, and Figure 4–36c illustrates typical field contours. Commercial circulators are illustrated in Figure 4–37.

Figure 4–38 shows a circulator which employs the principle of the 45° Faraday rotation. The results of this circulator are:

1. When a vertically polarized wave enters port 1, its plane of polarization is rotated through 45° by the ferrite specimen and the wave then leaves through port 2. Ports 3 and 4 represent incompatible junctions.

2. A 45° polarized wave entering port 2 is horizontally polarized after passing through the ferrite specimen and will therefore exit from port 3. Similarly, a wave entering port 3 has its plane of polarization rotated through 45° and only leaves through port 4.

3. A wave entering port 4 will be vertically polarized after leaving the ferrite specimen and will emerge from port 1.

These results are summarized in the circular network diagram of Figure 4–39. As a simple rule, a 90° rotation from the entrance port in the clockwise direction of the circular arrow will automatically locate the signal exit port. Such an arrangement could be used as the duplexer of a radar system which uses the same antenna for both the transmitter and the receiver; this is illustrated in Figure 4–40.

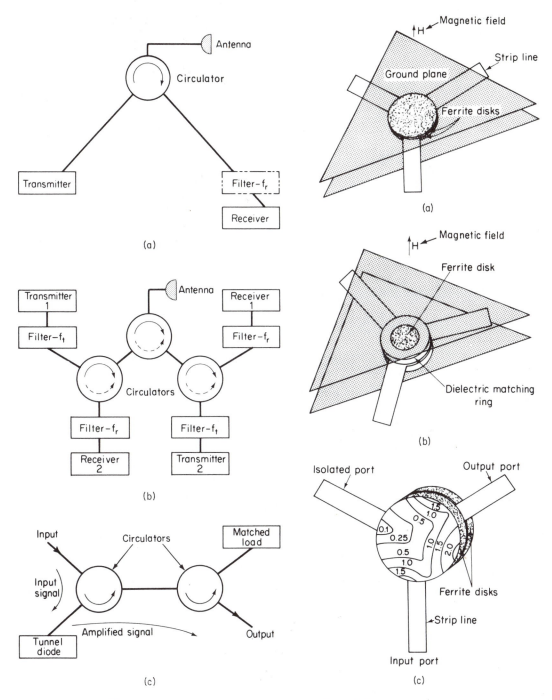

Figure 4-35 Circulator applications.

Figure 4-36 Commercial circulator designs.

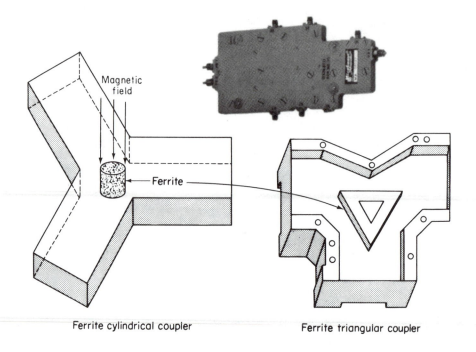

Magnetic
field

Ferrite

Ferrite cylindrical coupler

Ferrite triangular coupler

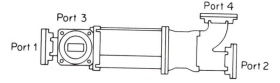

Port 3

Port 4

Port 1

Port 2

Figure 4-37 Commercial circulator assembly. (Courtesy of M/A-Com, Inc., and Alpha Industries, Inc.)

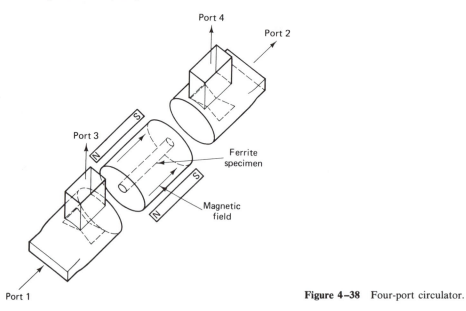

Port 4

Port 2

Port 3

Ferrite
specimen

Magnetic
field

Port 1

Figure 4-38 Four-port circulator.

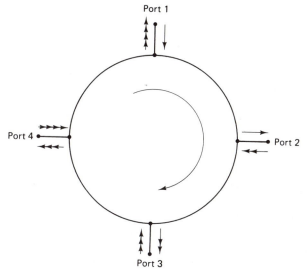

Figure 4-39   Principle of the circulator.

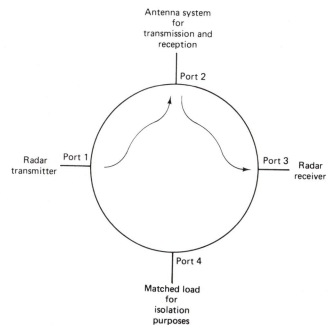

Figure 4-40   Circulator used as a duplexer for a radar antenna system.

4-11 ATTENUATORS

Attenuators in microwave circuits are commonly used to:

1. Measure power gain or loss in decibels
2. Provide signal generators with a means of calibrating their outputs accurately so that precise measurements may be made
3. Increase the power range of sensitive measuring instruments
4. Provide some degree of isolation between instruments so as to reduce their interaction
5. Reduce the power input to a particular stage to prevent overloading

The attenuators themselves fall into the three categories of fixed value, continuously variable, and step adjustable. In each category the various models provide wide ranges of accuracy and frequency coverage.

In selecting an attenuator to fulfill a particular purpose, the following specifications generally have to be taken into account:

1. Amount of attenuation in terms of the loss of decibels between the input and output power levels.
2. Degree of stability. This is measured by the change in the amount of attenuation due to the anticipated variations in temperature, humidity, frequency, and power level.
3. Value of the attenuator's characteristic impedance. This is important in order that the attenuator may be matched to the microwave units to which the attenuator is connected.
4. The accuracy of the attenuator in terms of the dial's maximum calibration error over its quoted frequency range.
5. The value of the SWR, which results directly from the insertion of the attenuator. This is quoted as the maximum value for the particular calibrated frequency or the highest value that occurs over the entire frequency range for which the attenuator is designed.
6. The degree of resolution, which measures the difference between the actual attenuation and the value indicated on the dial.
7. The amount of residual attenuation present when the dial reading is zero.
8. The variation in the attenuation value with the amount of the power level.

Coaxial Attenuators

Fixed-value attenuators for coaxial lines contain a glass tube center and a conductor on which is baked a thin metallic crust; this is similar to the construction of some precision resistors. A platinum/palladium or nichrome/chromium combination is used for this metallic film. The smooth glass base has the advantages of a high melting point and the ability not to twist or alter its shape.

The photograph of a fixed-value attenuator is shown in Figure 4–41a, and typical constructions are illustrated in Figures 4–41b and c. In Figure 4–41b the center conductor is the glass tube with a metallized circular disk in the middle, while on either side is a resistive film section. This attenuator is therefore equivalent to a T section pad and operates satisfactorily from dc up to 4 GHz. By contrast,

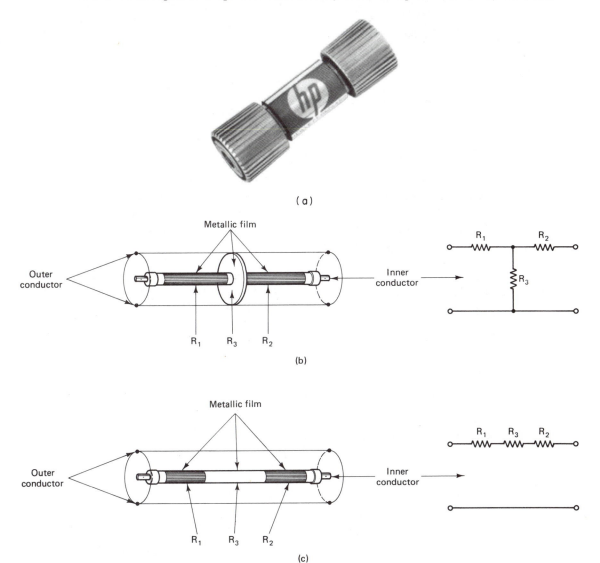

Figure 4–41   Coaxial attenuator pads. [(a) Courtesy of Hewlett Packard Company.]

Figure 4–42 Rotary step coaxial attenuators. (Courtesy of The Narda Microwave Corporation, and Hewlett Packard Company.)

the attenuator of Figure 4–41c may typically be used up to 20 GHz. Its main attenuation section in the center is highly resistive, while the lower-resistance sections on either side are used for matching purposes. The high resistance of the center section is achieved primarily by making the film's thickness less than its skin depth. Since this depth is dependent on the frequency, the degree of attenuation is also frequency sensitive. Moreover, the pad may contain lossy insulation material whose dielectric hysteresis effect increases with frequency. Consequently, if a precise value of attenuation is required, the pad must be calibrated for a particular frequency.

Step attenuators for coaxial lines may be controlled by a rotary switch which can select, in turn, particular attenuation values of a number of fixed pads (Figure 4–42). A variable attenuator may be formed by allowing the coaxial input to excite a waveguide which is operated at a frequency well below its cutoff value. Such a waveguide will automatically attenuate the signal and the loss can be continuously varied by moving the point of excitation (Figure 4–43).

Waveguide Attenuators

These are normally available in models whose attenuation can be continuously varied. Figure 4–44a shows a "flap" attenuator in which a resistive element is inserted into a longitudinal slot which is cut along the center of the wide dimension. The slot must normally be lined with an absorbing material to prevent excessive leakage.

Figure 4–43 Variable coaxial attenuator. (Courtesy of Hewlett Packard Company.)

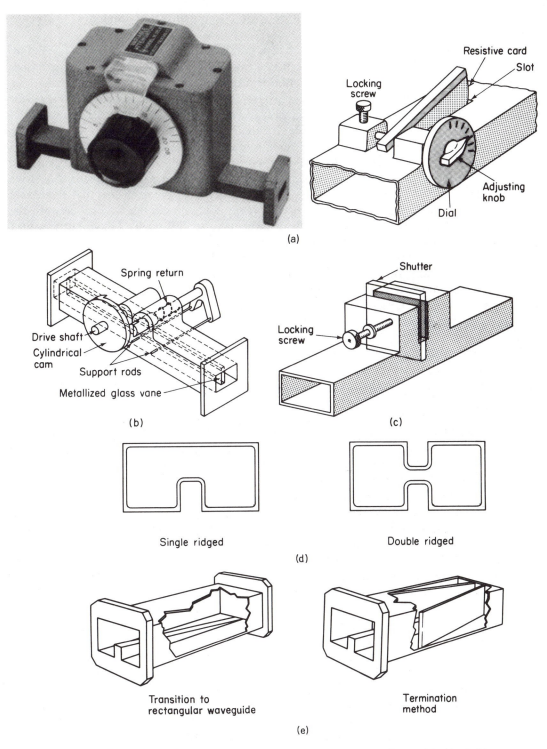

Figure 4-44 Waveguide attenuators: (a) flap; (b) vane; (c) shutter; (d) ridged; (e) ridged taper. [(a) Courtesy of Hewlett Packard Company.]

The resistive element is positioned where the electric field intensity is at its maximum, so that the degree of attenuation is determined by the flap's depth of penetration. However, there is no simple equation to determine the amount of the decibel loss, so that the dial of this type of attenuator has to be calibrated against a superior standard.

Another example of a variable attenuator is illustrated in Figure 4–44b. Basically it consists of a glass vane with a coating of either aquadag or carbon which acts as the "lossy" material. For maximum attenuation the vane is placed in the center of the guide's wide dimension, where the electric field intensity is greatest. A drive mechanism with a dial then shifts the vane away from the center so that the degree of attenuation can be varied. However, there is no simple relationship between the amounts of shift and the decibel loss, so that the dial must be calibrated against the reading of a precision attenuator. To match the attenuator to the waveguide, the vane can be tapered at each end; typically, a taper whose length is equal to $\lambda_g/2$ provides an adequate match. Although this type of attenuator has virtually no leakage, the amount of attenuation is frequency sensitive and the glass dielectric introduces an appreciable phase shift.

Attenuation may also be provided by inserting a resistive element through a shutter (Figure 4–44c). The plane of the element lies in the distribution of the electric field across the wide dimension of the waveguide and the result is a degree of attenuation which increases with the depth of the insertion. However, the element represents a discontinuity, so that a certain amount of the microwave energy is reflected.

With certain matching and coupling requirements it may be necessary to change the wave impedance and raise the attenuation before attaching the matching termination. This is commonly done by including a ridged waveguide section (Figure 4–44d and e) which is capable of varying the waveguide impedance by a factor of 20 or more and by multiplying the attenuation by a few hundred. These changes are brought about by gradually tapering the ridge.

Precision waveguide attenuators are illustrated in Figure 4–45. As we shall see, their loss in decibels can be expressed by a simple equation so that the dial reading has an accuracy of ±2% of the indicated attenuation over the operating frequency range. The maximum attenuation available is 50 dB, while the amount of phase shift introduced is less than 1°.

The principle of the precision attenuator depends on a resistive rotary vane which is placed in the center (movable) section of a circular waveguide arrangement which is tapered at either end toward rectangular sections (Figure 4–45b).

When all three vanes are aligned, their planes are at right angles to the direction of the electric field so that there is zero attenuation. The purpose of vane 1 is to eliminate any horizontal polarization so that the electric field emerging from this vane is entirely vertically polarized. When the center vane 2 is rotated through an angle of θ degrees, the $E\sin\theta$ component is totally attenuated while the $E\cos\theta$ component is unaffected. When the $E\cos\theta$ field approaches vane 3, its component, $E\cos\theta\sin\theta$, is absorbed so that the electric field output from the attenuator is $E\cos^2\theta$; this output has the same polarization as that of the input wave. The attenua-

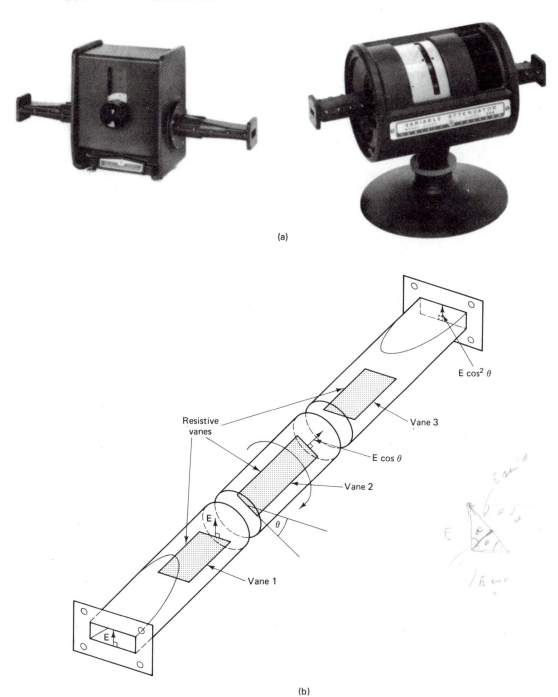

(a)

(b)

Figure 4–45 Precision waveguide attenuators.

tion is then $20\log\cos^2\theta = 40\log\cos\theta$ dB and is theoretically independent of the frequency.

4-12 MICROWAVE FILTERS

A filter is a frequency-responsive network of reactive components. There are four basic filter types, whose characteristics are illustrated in Figure 4–46.

1. *Low-pass filter* (Figure 4–46a). The attenuation is small at low frequencies which extend up to a certain cutoff frequency f_c; this range is referred to as the passband. Beyond the cutoff frequency the attenuation increases rapidly and then levels off; this region is called the stopband.

2. *High-pass filter* (Figure 4–46b). The action of this type is the reverse of the low-pass filter. The attenuation is high up to the cutoff frequency and afterward falls to a low level.

3. *Bandpass filter* (Figure 4–46c). This type has two cutoff frequencies, f_{c1} and f_{c2}, and provides low attenuation over the limited range of the frequencies between f_{c1} and f_{c2}. Frequencies below f_{c1} and higher than f_{c2} lie within the filter's two stopbands.

4. *Bandstop filter* (Figure 4–46d). The action of this type is the reverse of the bandpass filter. There is high attenuation between the cutoff frequencies, while passbands exist below f_{c1} and above f_{c2}.

The characteristics shown belong to microwave filters and contain many more irregularities than those of their low-frequency counterparts, which are constructed from lumped inductors and capacitors. By contrast, the basic element of the microwave filter is the cavity, which possesses a very high Q. In addition, the filter normally contains a number of such elements and it is difficult to arrange that all the elements function simultaneously. However the single resonant cavity is the major factor which determines the main properties of the filter. When such a cavity is inserted, the energy at its resonant frequency is passed through, while at other frequencies there is a mismatch, so that a reflection occurs. However, to obtain the desired response, it is necessary to couple together a number of cascaded cavities.

A typical multicavity filter with its characteristics is illustrated in Figure 4–47. Each section is coupled to its neighbor by an iris (Section 4–5) or a slot (Section

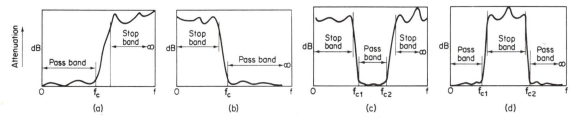

Figure 4–46 Attenuation characteristics of basic microwave filters: (a) low pass; (b) high pass; (c) bandpass; (d) bandstop.

3-4), while a capacitive screw is used for tuning each individual cavity. Since the screw effectively lowers the resonant frequency, each cavity is shorter than half of a wavelength; the screw can then be used to tune the cavity above or below the designed frequency. To avoid lowering the Q value of the cavity, the screw's depth of penetration is kept to a minimum and it is normally silver plated to reduce its insertion loss.

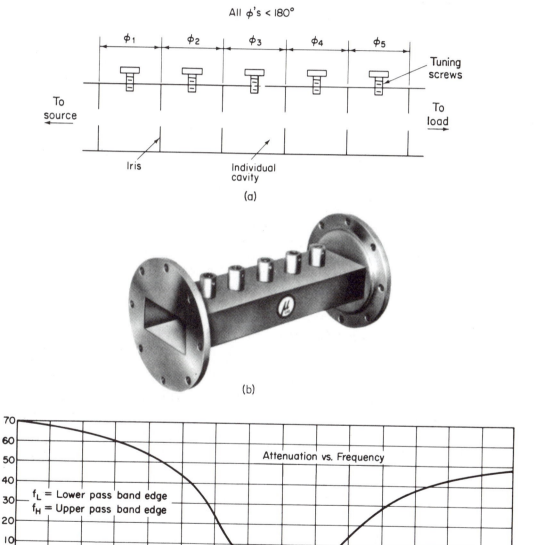

Figure 4-47 (a) Waveguide filter with its characteristics; (b) commercial microwave waveguide assembly; (c) characteristics. [(b) and (c) Courtesy of Microlab/FXR.]

In the construction of the filter we must bear in mind the following factors:

1. Resonant frequency of each cavity
2. Loaded Q of each cavity
3. Susceptance of each slot or iris
4. Degree of coupling between neighboring cavities
5. Filter's input and output impedances

Because of capacitive loading effects, the resonant frequencies of the end cavities differ slightly from the internal sections and we must compensate for this effect. The value of the loaded Q is difficult to control because it is determined by a variety of factors, such as the degree of coupling, as well as the cavity's skin depth, resonant frequency, internal dimensions, surface plating, and degree of smoothness.

The iris commonly used is of the inductive variety, so that Figure 4–48 represents the equivalent circuit of the cavities and the irises. Each iris is normally a round hole cut in a metal diaphragm (Figure 4–49a); this type allows a high Q to be maintained, but its design depends on so many factors that the performance of a fixed iris would be unpredictable. Consequently, it is common practice to use an adjustable iris to achieve optimum coupling (Figure 4–49b). The required frequency response for the filter is then obtained by varying the positions of the irises.

So far we have discussed microwave filters in relation to rectangular waveguides. However, similar filters can be formed from cylindrical waveguide sections which are separated by circular irises of the round-hole variety.

YIG Resonators and Filters

The letters YIG stand for yttrium-iron-garnet, which is a ferromagnetic oxide and has the properties of an insulator. It is, in fact, a ferrite material and behaves in a similar manner to the ferrites employed in isolators and circulators (Section 4–10). The YIG compound therefore possesses a gyromagnetic resonant frequency which is determined by the strength of the direct magnetic field in which the compound is placed.

A YIG [$Y_3 Fe_2 (FeO_4)_3$] resonator is a polished sphere which has a diameter of a few millimeters and rests in the center of two coupling loops with their axes at right angles. The complete assembly is then subjected to a direct magnetic field whose

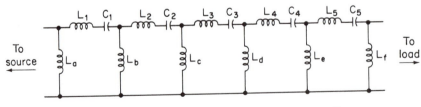

Figure 4–48 Lumped-circuit equivalent of a waveguide filter.

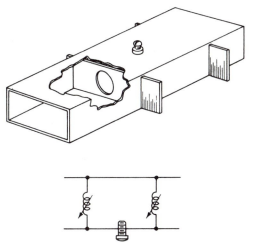

Figure 4–49 Waveguide filter section with adjustable iris.

flux lines lie in a plane at right angles to the coupling loops (Figure 4–50). If the direct field is produced by an electromagnet, the frequency may be changed by varying the current in the exciting coil. The complete mechanism has a high Q and may be used as the resonant circuit of an oscillator which will possess a high degree of frequency stability. Such an arrangement is shown in Figure 4–51, where the low output of a YIG oscillator can be varied between 1 and 4 GHz and also frequency modulated by an appropriate waveform. Figure 4–52 shows the frequency-conversion arrangement of a YIG-tuned microwave receiver.

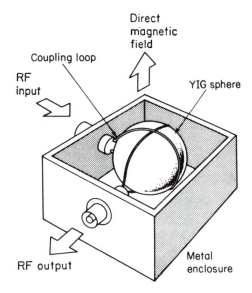

Figure 4–50 YIG resonator.

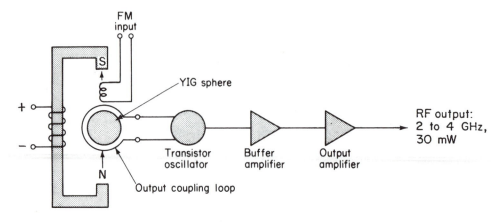

Figure 4-51   YIG oscillator.

As well as a YIG sphere behaving as an individual resonator, a number of such spheres may be constructed in tandem to form a bandpass filter. Each resonator has its own bandwidth, which is determined primarily by the ratio of the sphere's diameter to the diameter of the coupling loops. However, the degree of coupling between the resonators and their spacing is so arranged that there is a high degree of attenuation above and below the cutoff frequencies while a flat response exists over the passband. Figure 4-53a shows the tandem arrangement of four YIG units in a typical bandpass filter; each YIG resonator is capable of being individually tuned by the positioning of a dielectric rod. The entire unit is then inserted in the narrow dimension of a rectangular waveguide (Figure 4-53b) and is subjected to a direct magnetic field of constant intensity.

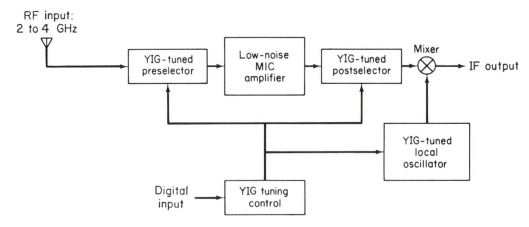

Figure 4-52   YIG-tuned microwave receiver.

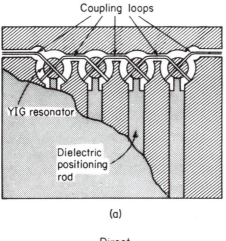

(a)

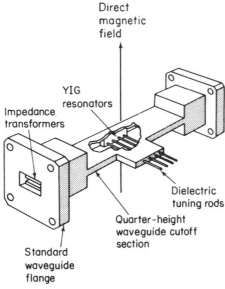

(b)

Figure 4-53 YIG bandpass filter.

Surface Acoustic Wave Devices

The conventional Miller, Pierce, and Butler oscillators employ quartz crystals which exhibit a marked piezoelectric effect. Because of the crystal's high Q (several thousand) and low-temperature coefficient, these oscillators possess a high degree of frequency stability. However, the fundamental frequency of a crystal wafer is inversely proportional to its thickness, and when the frequency is about 25 to 30 MHz, the required thickness is of the order of 1 mm or less. Such a crystal is liable to fracture easily, so that there is an upper limit to the fundamental frequency. However, the mode of a crystal's vibration is complex and contains overtone components which are virtually harmonics of the fundamental frequency. The output of an overtone oscillator can then be passed through a number of frequency multipliers, whose final output lies in the microwave region. Such an arrangement is complex, and the efficiencies of the multiplier stages are limited. It would be more effective if the microwave output could be generated at the oscillator stage.

The surface acoustic wave (SAW) device is capable of generating an output in the lower microwave region (S band). Its construction basically consists of two electrodes in the form of two interdigital metallic lines or fingers which are electrodeposited on a piezoelectric substrate (Figure 4–54); although only a few are shown, there are in practice hundreds of finger pairs.

When an RF voltage at the resonant frequency is applied, the piezoelectric effect will cause surface mechanical (acoustical) waves to travel in both directions and set up a standing-wave pattern whose acoustic half-wavelength is equal to the distance between two adjacent lines. This mechanical vibration is associated with a corresponding electrical oscillation.

The velocity of an acoustical wave depends on the medium through which it is traveling. In air this velocity is 340 m/s, but in quartz the wave travels nearly 10 times faster with a speed of approximately 3000 m/s. Using the normal equation $v = f\lambda$ [equation 1–1)], the wavelength corresponding to a frequency of 3 GHz (S band) is

$$\frac{3000 \text{ m/s}}{3 \times 10^9 \text{ Hz}} = 1 \ \mu\text{m}$$

The distance between two adjacent lines in the interdigital structure is then $\lambda/2$ or 0.5 μm and the width of each line is $\lambda/4$ or 0.25 μm. This establishes the upper-

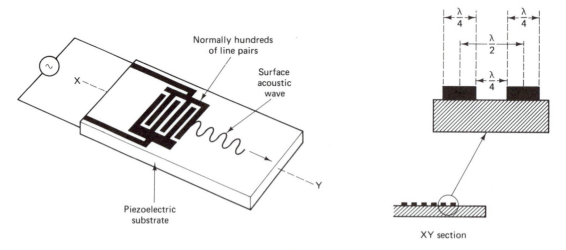

Figure 4–54   SAW device.

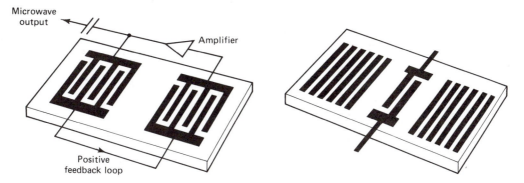

Figure 4–55 SAW oscillator.   **Figure 4–56** SAW resonator as a narrow-band filter.

frequency limit of the SAW resonator as about 5 GHz; the lower-frequency limit extends down to the region where it is possible to use conventional overtone oscillators.

To create an oscillator, two SAW resonators are connected to a microwave transistor amplifier (Figure 4–55) which is included in the positive feedback loop. The Q of the resonator is high but is less than the Q associated with the substrate. It is therefore possible to use the SAW resonator as a narrow-band filter (Figure 4–56). Here the surface acoustic wave is reflected by adjacent gratings to create a standing wave at the design frequency.

4-13 FIXED AND VARIABLE PHASE SHIFTERS

We observed in Section 3–4 that a waveguide's phase shift constant β was inversely proportional to the wavelength λ_g. Consequently, if we wish to introduce controlled amounts of phase shift between two given positions in a waveguide, it is necessary to change the magnitude of the guide wavelength, which in turn is determined by the value of λ. For a given frequency, λ can be lowered only by reducing the velocity of propagation, and this means that the medium must be changed.

Fixed amounts of phase shift may be created by:

1. Inserting dielectric rods across the diameters of circular waveguides. Such rods will be parallel to the directions of the electric flux lines, and this will introduce the required phase shift.

2. Restricting the cross section of a waveguide (Figure 4–57). By reducing the length a of a rectangular guide's wide dimension, the value of the guide wavelength will be lowered.

3. Inserting capacitive or inductive irises into the waveguide.

The most common methods of obtaining a variable phase shift are comparable with those which control the amount of attenuation. A dielectric slab or vane is specially

Figure 4–57 Restricting a waveguide's cross section to introduce a phase shift.

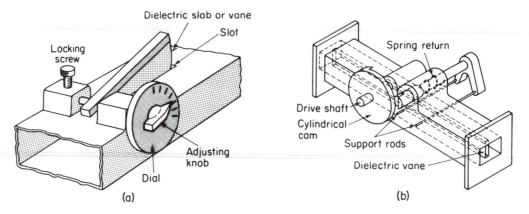

Figure 4–58 Dielectric vane phase shifters.

shaped to minimize reflection effects and is then inserted through a longitudinal slot which is cut in the waveguide's wide dimension (Figure 4–58a). The complete vane is therefore placed along these positions, where the electric field has its maximum intensity. If the vane is inserted deeper, there is more change in the medium and there is a greater phase shift. However, this is not a precision phase shifter since there is no simple mathematical relationship involved with its operation.

Another nonprecision variable phase shifter employs a dielectric vane which is placed inside the rectangular waveguide (Figure 4–58b), so that the vane's inside dimension is parallel to the direction of the electric flux lines. The vane is mounted on two support rods and is mechanically moved across the wide dimension. We can then assume that the amount of phase shift will decrease as the slab is moved from the position of maximum electric field intensity (at the center of the waveguide's wide dimension) to one or other of the side walls.

One type of precision waveguide phase shifter is illustrated in Figure 4–59a. It consists essentially of three circular waveguide sections, all of which contain one dielectric vane. Only the center section is capable of being rotated to provide the necessary phase shift. At each end there is a tapered transmission section which joins a rectangular waveguide operating in the dominant TE_{10} mode to the fixed circular waveguide which is excited in the TE_{11} mode.

In Figure 4–59b the direction of the linearly polarized E_1 field entering the fixed circular section is vertical and is at an angle of 45° to the planes of all three dielectric vanes. The E phasor may then be resolved into two in-phase components which are of equal magnitude and are mutually at right angles. One component

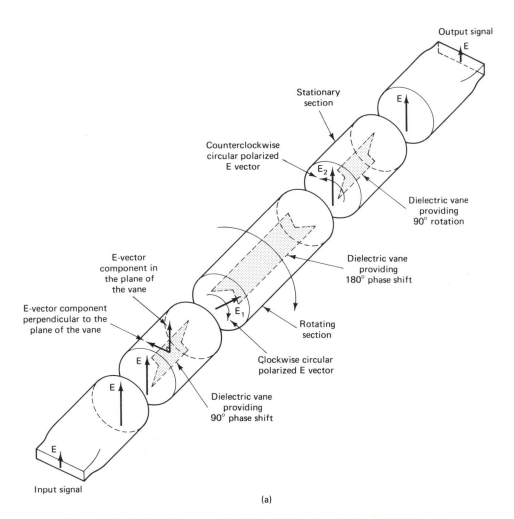

(a)

Figure 4–59 (a) Precision waveguide shifter; (b) linearly polarized E field of period T entering the first (fixed) circular section; (c) clockwise circularly polarized E field emerging from the first circular section and entering the second (variable) circular section; (d) counterclockwise circularly polarized E field emerging from the second central (variable) circular section and entering the third (fixed) circular section.

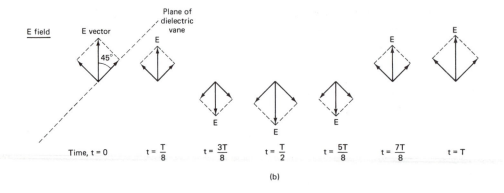

(b)

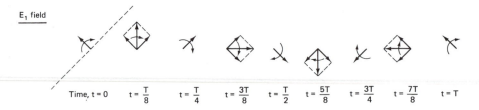

(c)

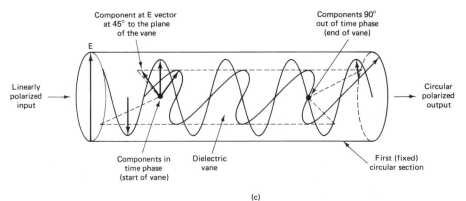

(d)

Fig. 4.59 (*cont.*)

will lie in the plane of the dielectric vane and will be subjected to a 90° phase shift delay (as determined by the length of the vane), while the other component will be perpendicular to the vane and will suffer no phase shift. Assuming that the two components are not unequally attenuated, the E_1 field emerging from the first section is circularly polarized in the *clockwise* direction (Figure 4–59c).

The circularly polarized E_1 field now enters the center variable section, whose vane is designed to provide a 180° phase delay for the component that lies in its plane. As a result, the E_2 field emerging from the center section will be circularly polarized in the *counterclockwise* (CCW) direction (Figure 4–59d).

Using the analysis we applied to the first 90° section, the third and final 90° section will convert the CCW polarized E_2 field to the linearly polarized output field E_3, which is in phase with the initial E field.

If the center circular section is now rotated through 180°, the arrangement of the vanes is essentially unchanged so that E and E_3 are still in phase. In other words, a 180° rotation of the center section has resulted in a 360° phase change, which is equivalent to a zero phase shift. Consequently, if the center section is rotated through an angle of θ degrees, the phase shift between E and E_3 will be 2θ degrees. Knowing this relationship, the precision phase shifter can be accurately calibrated and its dial can be used to provide a direct reading of the amount of the phase shift.

Ferrite Phase Shifters

The principle of the Faraday rotation (Section 4–10) is used in the following shifters:

The X-band phase shifter of Figure 4–60 is a miniature device used in large multielement phased antenna units (Figure 7–44). It contains a symmetrical circular waveguide which is filled with ferrite material so that an applied axial magnetic field creates the Faraday rotation effect and produces the necessary amount of phase shift.

In another application of the Faraday rotation, the gyrator of Figure 4–61 is a two-part device in which the phase shift in one direction differs by 180° from the phase shift in the reverse direction; this result is achieved by means of a 90° ferrite rotator section. In order that the input and output signals have the same plane of

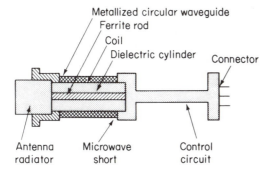

Figure 4–60 Miniaturized antenna phase shifter. (Courtesy of M/A-Com, Inc.)

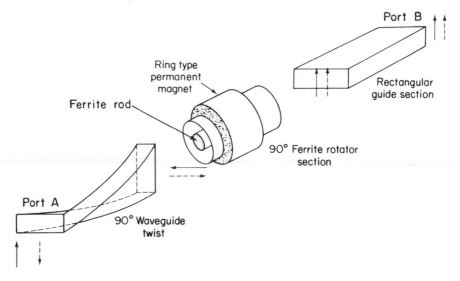

Figure 4–61 Gyrator phase shifter. Solid arrows indicate polarization of a wave traveling from left to right; dashed arrows indicate polarization of a wave traveling from right to left.

polarization, a 90° waveguide twist is normally used in connection with the ferrite section. Where an input signal (solid line) enters port *A*, its plane of polarization is subjected to a 90° *counterclockwise* rotation by the waveguide twist. However, when the ring magnet applies an axial field, the ferrite section rotates the wave through 90° in a *clockwise* direction so that the two effects cancel out and the output signal leaving port *B* is in phase with the input signal at port *A*. By comparison, when an input signal (dashed line) enters port *B*, its plane of polarization is subjected to *two* 90° *clockwise* rotations so that the emerging signal from port *A* is shifted through 180° with respect to the input signal at port *B*.

Two types of nonreciprocal ferrite phase shifters are shown in Figure 4–62. In each case the ferrite material is subjected to an axial magnetic field created by a biasing coil or wire. The level of the bias current then determines the amount of the phase shift introduced.

Duplexer Phase Shifters

A specialized phase shifter may be used as a duplexer for the common antenna system of a radar set. In the arrangement of Figure 4–63 the transmitted pulse must be transferred to the antenna, and virtually none of this pulse should be diverted to the receiver. By contrast, all the received echoes will pass to the receiver and will be unable to reach the transmitter.

In the transmit condition the initial rectangular guide carries the dominant

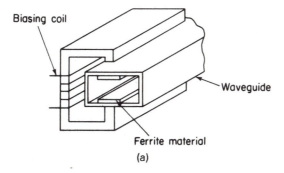

(a)

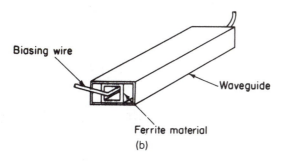

(b)

Figure 4–62 Ferrite phase shifters.

TE_{10} mode with the E lines in the horizontal direction. A transition section converts from the rectangular guide to a circular guide which is therefore excited in its dominant TE_{11} mode. Very little energy can then be coupled into the rectangular receiver arm because the two required directions for the E lines are at right angles. The transmitted signal now has its plane of polarization shifted through 90° in the counter-clockwise direction. This operation is achieved in a central circular phase shifter which is located between two transition sections. The phase shift is accomplished by a series of diametrically positioned gas-filled tubes which are manufactured from quartz. Each of the 16 tubes has its angular position successively advanced so that when they are ionized by the transmitted pulse, the guide's dielectric is changed and the E field is subjected to a total phase shift of 90°. When the signal emerges from the phase-shift section, its plane of polarization is correctly oriented for the final transition to the antenna.

When the received echoes arrive, they are far too weak to fire the tubes. Consequently, there is no phase shift in the circular section and the polarization in the TE_{11} mode is correct for the signal to pass through to the receiver arm. By contrast, the transmitter arm is transversely oriented so that the received echoes cannot reach the transmitter. This action is further reinforced by the introduction of shorting bars whose directions are parallel to the E lines of the received signal. These shorting bars act as the anti-TR (ATR) device.

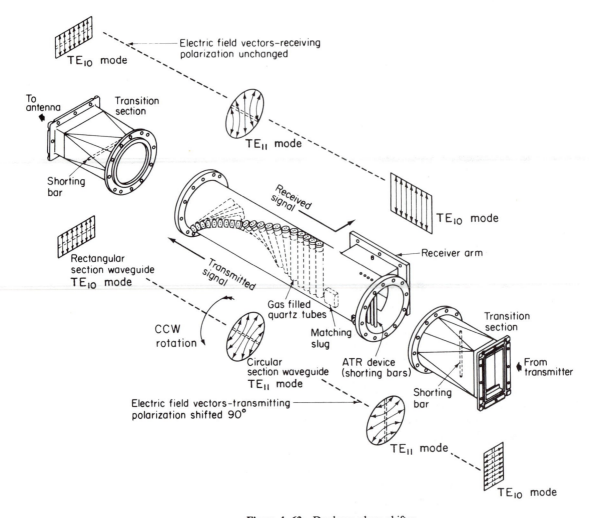

Figure 4–63   Duplexer phase shifter.

PROBLEMS

BASIC PROBLEMS

4–1. Discuss the advantages and disadvantages of the various types of waveguide joint.

4–2. Why are waveguide twists and bends constructed so that the direction of the propagated energy is gradually changed?

4–3. Describe the features of waveguide series and shunt tee sections. Explain the operation of the magic tee hybrid junction.

4–4. Explain the actions of the flap and the precision vane attenuators.

4–5. Describe in detail the operation of a directional coupler. Calculate the coupling value if the power in the main waveguide is 60 mW and the power delivered to the directional coupler is 0.4 mW.

4–6. Discuss the operation of posts and screws in obtaining an impedance match.

4–7. Explain the principles behind inductive and capacitive irises and discuss their uses.

4–8. What are the various methods of terminating a waveguide to (a) maximize and (b) minimize the reflections from the termination?

4–9. A precision rotary vane attenuator is set to 0 dB. The vane is now rotated through an angle of 20°. What is the value of the attenuation in decibels?

4–10. What is an isolator? Describe the action of the ferrite isolator.

ADVANCED PROBLEMS

4–11. A 10-GHz signal is being propagated in the dominant TE_{10} mode along a rectangular waveguide whose inner dimensions are 1.12 cm and 2.24 cm. Calculate the values of the wave impedance and the characteristic impedance.

4–12. In a rectangular waveguide a bend is followed by a 90° twist. If both the bend and the twist introduce an SWR of 1.1, what is their combined SWR value?

4–13. Two identical 20-dB directional couplers are attached to a waveguide in order to sample the incident and reflected powers.
(a) If the outputs of the incident and reflected couplers are 3 mW and 0.1 mW, respectively, what is the value of the SWR in the main waveguide?
(b) What are the values of the incident power, the reflected power, and the power dissipated in the terminating load?

4–14. A waveguide is terminated by a load with a reflection coefficient, P, of 0.3. The incident power is 100 mW and a directional coupler for the reflected power reads 1.0 mW. What is the value of the coupling factor?

4–15. An X-band waveguide has inner dimensions of 2.22 cm and 1.02 cm. What is its wave impedance at frequencies of 9, 10, and 11 GHz (assume that the dominant TE_{10} mode is used)?

4–16. What is the purpose of a circulator? Describe the action of a circulator that employs the Faraday rotation effect.

4–17. Describe the principles behind the operation of a microwave bandpass filter.

4–18. A rotary-vane precision attenuator is set to the 10-dB mark and the vane is then rotated through a *further* 25°. What is the *increase* in the amount of the attenuation?

4–19. A waveguide is joined to a fixed 4-dB attenuator which is terminated by a load. If the incident power in the waveguide is 200 mW and the reflected power in the waveguide is 15 mW, what is the value of the load's reflection coefficient, P?

4–20. Describe the action of the phase shifter which is used as a duplexer.

CHAPTER 5 _____

MICROWAVE TUBES

5–1 INTRODUCTION

We have already learned from the lighthouse triode (Section 1–5) that the ultimate limitation on conventional tubes is the transit time taken by an electron to cross from the cathode to the plate. It follows that an active microwave tube must actually make use of the transit time to achieve amplification and or oscillation. Examples of such devices follow.

Magnetron tube. This is normally the heart of a pulsed radar system. The magnetron is an oscillator which is capable of generating a short-duration RF pulse with a peak power output which is of the order of megawatts. However, there is a long time interval between pulses so that the duty cycle is low. For example, a magnetron generates 500 pulses per second but the time interval for each pulse is only 1 μs. The period between pulses is 1/500 s $=$ 2000 μs and therefore the duty cycle is only

$$
\begin{aligned}
\text{duty cycle} &= \frac{\text{active interval}}{\text{total period}} \\
&= \frac{\text{active interval}}{\text{sum of active and inactive intervals}} \\
&= \frac{1\mu\text{s}}{2000\ \mu\text{s}} = 0.0005
\end{aligned}
$$

200

If the peak power is 1 MW,

$$\text{average power} = \text{peak power} \times \text{duty cycle}$$
$$= 1,000,000 \times 0.0005 = 500 \text{ W}$$

The magnetron is also the active device in the microwave oven which operates at a frequency of 2.45 MHz. The space where the food is placed is continuously supplied with microwave energy.

The frequency range of magnetrons as a whole covers 0.6 to 30 GHz. For a particular magnetron the frequency is normally fixed, but there are methods of obtaining a limited tuning range as high as 10% of the center frequency.

Multicavity klystron tube.   This klystron is a stable microwave power amplifier which provides high gain (3 to 90 dB) at medium efficiency (30 to 50%). The frequency range of multicavity klystrons extends from 3 to 30 GHz.

Reflex klystron tube.   This tube is a microwave oscillator with a power output of only a few milliwatts and with a very low efficiency of 5% or less. Mechanically, its frequency can be tuned over a 30% range and electronically over a range of 2% or less. Its main use is in test equipment and as the local oscillator in a radar receiver.

Traveling-wave tube (TWT).   This is a high-gain, low-noise, wide-band microwave amplifier. Since no resonant cavities are employed, the upper-frequency limit of a TWT may be twice the lower limit. The primary use of the TWT is voltage amplification (although power TWTs with characteristics similar to those of a power klystron have been developed). The wide-bandwidth and low-noise characteristics have made the TWT ideal for use as RF amplifiers in microwave and electronic countermeasures equipment. For these purposes TWT have been designed for frequencies as low as 300 MHz but as high as 50 GHz.

Backward-wave oscillator (BWO).   There are two basic types of backward-wave oscillator: (1) The O type is frequently used as a low-power oscillator with an output of a few hundred milliwatts over a normal frequency range from 1 to 15 GHz. However, such oscillators have been made to operate as high as 200 GHz. (2) The M or crossed-field type is used primarily as a transmitting tube with a power output of a few hundred watts and an efficiency of about 30%. The output frequency is of the order of a few gigahertz. This type of BWO is also referred to as a carcinotron, which is commonly used as a noise generator.

Crossed-field amplifier (CFA).   These are primarily pulsed devices with peak powers up to 1 MW in the X-band region. At these frequencies the efficiency is about 50% and the duty cycle is of the order of 2%. The bandwidth available is normally about 20% of the center frequency. However, the principal disadvantage of the crossed-field amplifier is its relatively low gain of approximately 15 dB.

Gyrotron. There are various types of gyrotron tube that employ a helical beam in which the electrons revolve at a rate proportional to the intensity of a direct magnetic field. In the presence of an alternating electromagnetic wave there occurs a bunching process which is based on an electron's mass changing as a result of its acceleration to very high velocities. Although still in the developmental stage, some gyrotron tubes hold out the promise of large power capabilities (megawatts) at high microwave frequencies up to 100 GHz.

Our discussion of microwave tubes is covered by the following topics:

5–2 MAGNETRONS

Basically, the magnetron is a diode and has no grid. A magnetic field in the space between the plate (anode) and the cathode serves as the controlling mechanism. The plate of a magnetron does not have the same physical appearance as the plate of an ordinary electron tube. Since conventional *LC* networks become impractical at microwave frequencies, the plate is fabricated into a cylindrical copper block containing resonant cavities which serve as tuned circuits (Section 1–3). The magnetron's base differs greatly from the conventional base. It has short, large-diameter leads that are carefully sealed into the tube and shielded, as shown in Figure 5–1.

The cathode and the filament are at the center of the tube and are supported by the filament leads, which are large and rigid enough to keep the cathode and the filament fixed in position. The output lead is usually a probe or loop extending into one of the tuned cavities and coupled into the waveguide. The plate structure, shown in Figure 5–2, is a solid block of copper. The cylindrical holes around its circumference are the resonant cavities. A narrow slot runs from each cavity into the central portion of the tube and divides the inner structure into as many segments as there are cavities. Alternate segments are strapped together to put the cavities in parallel with regard to the output. These cavities control the generated frequency. The straps are circular metal bands which are placed across the top of the block at the entrance slots to the cavities. Since the cathode current is of the order of several amperes, the cathode must be large and must be able to withstand high operating

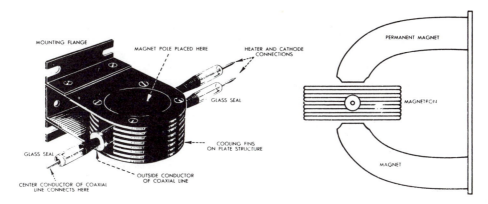

Figure 5–1   Construction of the magnetron.

temperatures. Such a cathode must also have good emission characteristics, particularly under back bombardment, because much of the output power is delivered from the large number of electrons emitted when high-velocity electrons return to strike the cathode. The cathode is indirectly heated and is constructed of a high emitting material. The open space between the anode and the cathode is called the interaction space because it is in this space that the electric and magnetic fields interact to exert a force on the electrons. The magnetic field is normally provided by a strong permanent magnet, mounted around the magnetron so that the direction of the magnetic field is parallel with the axis of the cathode. The cathode is mounted in the center of the interaction space. Since the anode is exposed while the cathode leads are carefully sealed, the anode is grounded and a large negative voltage of several kilovolts is applied to the cathode during the time of the magnetron's oscillation.

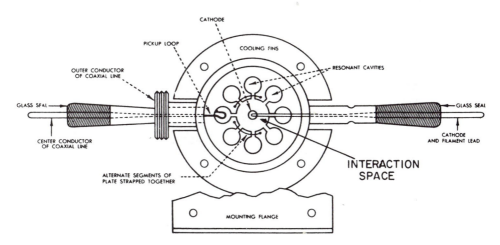

Figure 5–2   Cutaway view of the magnetron.

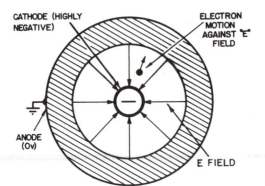

Figure 5-3 Motion of the electron in the magnetron's E field.

Basic Magnetron Principles

The theory of the operation of the magnetron is based on the motion of electrons under the influence of combined electric and magnetic fields. The following laws govern the motion. The direction of an electric field is from the positive electrode to the negative electrode. The law governing the motion of the electron due to an electric or E field states that the force exerted by the electric field on an electron is proportional to the electric field intensity (Section 5-1). Electrons tend to move from a point of negative potential toward a positive potential, as shown in Figure 5-3. In other words, the electrons tend to move against the direction of the E field. When an electron is being accelerated by an E field, as shown in Figure 5-3, energy is taken from the field by the electron.

The law of motion of an electron in a magnetic or H field states that the force exerted on the electron in a magnetic field is at right angles to both the directions of the field and the path of the electron. The direction of the force is such that the electron trajectories are clockwise when viewed in the direction of the magnetic field, as shown in Figure 5-4. This is the result of applying the right-hand motor rule. If the permanent magnetic field strength is increased, the electron path will take a

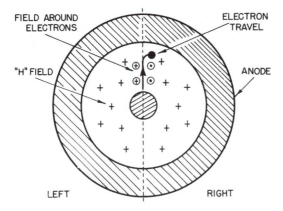

Figure 5-4 Motion of the electron in the magnetron's H field.

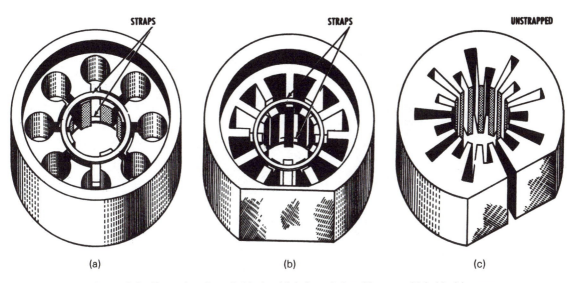

Figure 5–5 Examples of anode blocks: (a) hole and slot; (b) trapezoidal; (c) rising sun.

sharper bend. Similarly, if the velocity of the electron increases, the field around it increases and its path will again bend more sharply.

Electron Resonant Types of Magnetron

The first type shown in Figure 5–5a has cylindrical cavities and therefore possesses a hole-and-slot anode. The second type, called a vane anode, has trapezoidal cavities (Figure 5–5b). These first two anode blocks operate in such a way that alternate segments must be connected, or strapped, to ensure that each segment is opposite in polarity to its neighboring segment on either side (as shown in Figure 5–6). This also requires an even number of cavities. The third type, illustrated in Figure 5–5c, is called a "rising-sun block" because of its appearance. The alternative large and

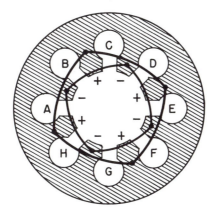

Figure 5–6 Strapping of the magnetron's alternate segments.

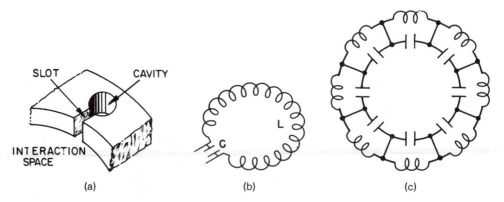

Figure 5–7 Equivalent electrical circuit of the hole and slot cavity.

small trapezoidal cavities in the block result in a stable frequency which lies between the resonant frequencies of the large and small cavities.

Figure 5–7a shows the physical appearance of the resonant cavities contained in the hole-and-slot anode which we will use when analyzing the operation of the electron resonant magnetron. Notice that the cavity consists of a cylindrical hole in the copper anode and a slot that connects the cavity to the interaction space.

The electrical equivalent circuit of the cavity and slot is shown in Figure 5–7b. The parallel sides of the slot form the plates of a capacitor, while the wall of the hole acts as an inductor. The hole and the slot then form a high-Q resonant LC circuit. As shown in Figure 5–5a, the anode of the magnetron contains a number of these cavities.

An analysis of the anode in Figure 5–5a shows that the LC tank circuits of the cavities are in series, as shown in Figure 5–7c. This is assuming that the straps have been removed. However, an analysis of the anode block after alternate segments have been strapped (Figure 5–6) will reveal that the cavities are now connected in parallel. The result of the strapping is shown in Figure 5–8.

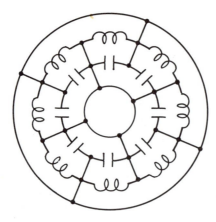

Figure 5–8 Cavities connected in parallel as the result of strapping.

Operation of the Magnetron

The resultant electric field in the electron resonant oscillator is a combination of the ac fields and a dc field. The dc field extends radially between the anode and the cathode, while the ac fields are due to the RF oscillations induced in the resonant cavities of the anode block.

Figure 5–9 shows the ac fields between adjacent segments at the instant of the peak value of one alternation in the RF oscillations associated with the cavities.

A strong dc field extends from the anode to the cathode and is due to a large negative dc voltage pulse applied to the cathode. This strong dc field causes electrons to accelerate toward the anode after they have been emitted from the cathode. These accelerated electrons take energy from the dc electric field. Oscillations are sustained in a magnetron because the electrons gain energy from the dc field and give up this energy to the ac fields as they come under their influence.

In Figure 5–9 consider an electron e_1 which is shown entering the field around the slot entrance to cavity X. The clockwise rotation of the electron path is due to the interaction of the magnetic field around the moving electron with the permanent magnetic field which is assumed to be entering the paper in Figure 5–9. Notice that electron e_1, which has entered the ac field around cavity X, is going against the direction of the ac field. This electron will therefore take energy from the ac field and will be accelerated, so that it turns more sharply as its velocity increases. Electron e_1 will therefore turn away from the anode and when it strikes the cathode, the energy it received from the ac field will be lost in the form of heat. This will force more electrons to leave the cathode and accelerate toward the anode. By contrast, electron e_2 is slowed down by the ac field and therefore gives up some of its energy to that field. Since electron e_2 loses some of its velocity, the deflection force exerted by the H field is reduced and the electron path deviates in the direction of the anode and not in the direction of the cathode as was the case with electron e_1.

The cathode-to-anode potential and the magnetic field strength (E-field to H-field relationship) determine the time taken by electron e_2 to travel from a position in front of cavity Y to a position in front of cavity Z. This time is equal to approximately one-half period of the RF oscillation of the cavities. When electron e_2 reaches a position in front of cavity Z, the ac field of that cavity will be reversed from that shown in Figure 5–9. As a result, electron e_2 will give up energy to the ac field of

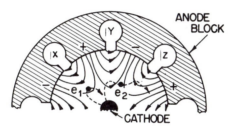

Figure 5–9 Motion of the electrons in the presence of the magnetron's E and H fields.

cavity Z and will slow down still further. Electron e_2 will actually give up energy
to each cavity as it passes and will eventually reach the anode, where its energy is
expended. Therefore, electron e_2 will have helped to sustain the oscillation because
it has taken energy from the dc field and given it to the ac field. Electron e_1, which
took energy from the ac field around cavity X, did little harm because it immediately
returned to the cathode. Electrons such as e_2 which give energy to the ac field as
they rotate clockwise from one ac field to the next, stay in the interaction space for
a considerable time before striking the anode.

The cumulative action of so many electrons, with some being returned to the
cathode while others are directed toward the anode, forms a pattern resembling the
spokes of a wheel, as indicated in Figure 5–10. This overall space-charge "wheel"
rotates about the cathode at an angular velocity of two anode segments for each
cycle of the ac field, and it also has a phase relationship which enables the electron
concentration to deliver energy continuously and therefore sustain the RF oscillation.
Electrons emitted from the area of the cathode between the spokes are, as previously
discussed, quickly returned to the cathode.

In Figure 5–10 it is assumed that alternative segments between cavities are at
the same potential at the same instant and that there is an ac field existing across
each individual cavity. This type of mode operation is called the pi (π) mode, since
adjacent segments of this mode have a phase difference of 180° or π radians. There
are, in fact, several other possible modes of oscillation, but the pi mode has the
greatest power output and is the one which is most commonly used.

To ensure that alternate segments have identical polarities, an even number of
cavities, usually six or eight, are used and alternate segments are strapped as pointed
out earlier. The frequency of the pi mode is separated from the frequencies of the
other modes by the strapping.

For the pi mode, all parts of each strapping ring are at the same potential,
but the two rings have alternately opposing potentials, as shown in Figure 5–11.

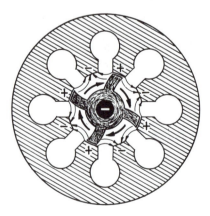

Figure 5–10 Concentration of the electrons
in the interaction space.

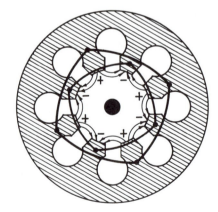

Figure 5–11 Separation of modes due to
strapping.

The stray capacitance between the rings then adds capacitive loading to the resonant mode; however, if there is a phase difference between the successive segments connected to a given strapping ring, current is caused to flow in that strap, which then has an inductance effect. As a result, an inductive shunt is placed in parallel with the cavity's equivalent circuit, thereby lowering the total inductance and increasing the frequency for all modes other than the π mode.

Coupling Methods

RF energy can be removed from a magnetron by means of a coupling loop. At frequencies lower than 10 GHz, the coupling loop is made by bending the inner conductor of a coaxial cable into a loop and soldering the end to the outer conductor, so that the loop projects into the cavity (Figure 5–12a). To obtain sufficient pickup at higher frequencies, the loop is located at the end of the cavity as shown in Figure 5–12b. Although it appears that the microwave energy is being taken from only a single cavity, all the cavities are in fact mutually coupled and therefore deliver their energy to the output coupling loop. The segment-fed loop is shown in Figure 5–12c. Here the loop intercepts the flux passing between the cavities. By contrast the strap-fed loop method (Figure 5–12d) intercepts the energy between the strap and the segment. On the output side, the coaxial cable feeds directly into a waveguide, with the vacuum seal at the inner conductor helping to support the line.

Aperture or slot coupling is illustrated in Figure 5–12e. This method allows the RF energy to be coupled directly to a waveguide with an iris feeding into the waveguide connector through the slot.

Tuning

A tunable magnetron permits the system to be operated at a precise frequency anywhere within a band of frequencies, as determined by the magnetron's characteristics. The resonant frequency of a magnetron may be varied by changing the inductance or capacitance of the resonant cavities. In Figure 5–13 an inductive tuning element is inserted into the hole portion of the hole-and-slot cavities. It changes the inductance of the resonant circuits by altering the surface-to-volume ratio in a region of high current. This type of tuner is illustrated in Figure 5–13 and is called a "sprocket" or "crown of thorns" tuner. All its tuning elements are attached to a frame which is positioned by means of a flexible bellows arrangement. The insertion of the tuning elements into each anode hole decreases the inductance of the cavity and therefore increases the resonant frequency. One of the limitations of inductive tuning is that it lowers the unloaded Q of the cavities and therefore reduces the efficiency of the tube.

The insertion of an element (ring) into the cavity slot as shown in Figure 5–14 increases the slot capacitance and decreases the resonant frequency. Because the gap is narrowed in width, the breakdown voltage will be lowered; the "capacity" magnetrons must therefore be operated with low voltages and hence low-power out-

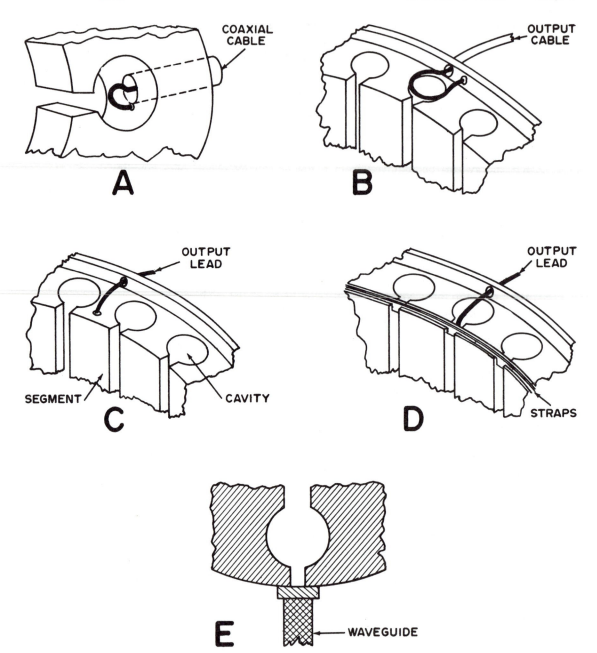

Figure 5–12   Methods of coupling from the hole and slot magnetron.

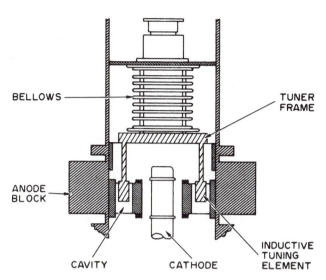

BELLOWS

TUNER FRAME

ANODE BLOCK

CAVITY   CATHODE

INDUCTIVE TUNING ELEMENT

Figure 5–13   Inductive magnetron tuning.

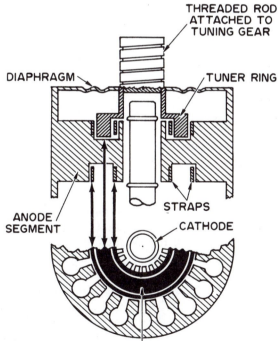

THREADED ROD ATTACHED TO TUNING GEAR

DIAPHRAGM

TUNER RING

ANODE SEGMENT

STRAPS

CATHODE

WIDE LINE REPRESENTS TUNER-RING POSITION BETWEEN THE MEGNETRON STRAPS

Figure 5–14   Capacitive magnetron tuning.

puts. The type of capacity tuner illustrated in Figure 5–14 is called a "cookie cutter" tuner. It consists of a metal ring inserted between two rings of a double-strapped magnetron, thereby increasing the strap's capacitance. Because of the mechanical and voltage breakdown problems associated with the cookie cutter tuner, it is more suited for use at the longer wavelengths. Both the capacitance and inductance tuners described above are symmetrical. Each cavity is affected in the same manner, and the angular symmetry of the pi mode is preserved.

A 10% frequency range may be obtained with either of the two tuning methods described. There is some indication that the cookie cutter tuner is more restricted than the crown of thorns tuner. The two tuning methods may be used in combination to cover a larger tuning range than is possible with either one alone.

Certain magnetrons may be tuned electronically by varying the anode voltage. These voltage-tunable magnetrons (VTMs) use a cold cathode with back-heating, low-Q cavities, and an additional injection electrode which aids the bunching process. Although such magnetrons are tunable over a wide range (maximum frequency: minimum frequency $\approx 2:1$), they have a low efficiency and are only suitable for CW operation as sweep oscillators.

Defects in the Magnetron

The main indication of a defective magnetron is a low reading of the magnetron current so that the undercurrent relay drops out. The CRT display will then show noise and the sweep, but the echoes will be unusually weak and fuzzy or will not be seen at all. In addition, the automatic frequency control (AFC) system will be ineffective.

If the magnetron itself is within its operating limits but the external magnet is too weak, the magnetron current will increase but the frequency of the oscillation will drift and the AFC system will not be effective. Under extreme conditions the oscillation may entirely cease. To avoid such weakening, the technician should not subject the magnet to extreme heat or any physical shocks. Furthermore, all metal tools should be kept well away from the magnet's presence.

In terms of general precautions, when servicing or maintaining a radar set, the technician should make certain that all power is shut off and that all capacitors are fully discharged. In particular, cathode ray tubes must be handled with great care. One further point: Arcing may occur in the modulator unit, the magnetron, the waveguide assembly, and other parts of the radar system. Such arcing may present a hazard if a ship is handling explosive or inflammable material.

Let us summarize the principles of the magnetron by considering a typical set of operating conditions. A magnetron is pulsed 500 times per second by an accompanying discharge line (Section 2–5) which consists of three LC sections each of inductance $L = 6.5$ μH and capacitance $C = 4000$ pF (Figure 5–15). The pulse length, $2N\sqrt{LC}$, is about 1.0 μs and the characteristic impedance $Z_0 = \sqrt{L/C} \approx 40$ Ω. An average magnetron requires an anode voltage of 26 kV and an anode current of 40 A, so that the magnetron impedance

$$Z_M = \frac{26{,}000 \text{ V}}{40 \text{ A}} = 640 \ \Omega$$

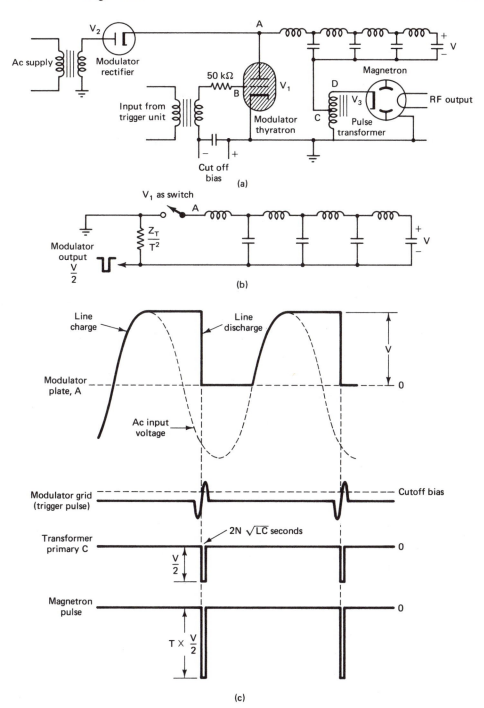

Figure 5–15 Discharge line modulator using thyratron tube: (a) modulator cutoff bias; (b) equivalent circuits; (c) waveforms of discharge line modulator circuit.

Hence the required turns ratio for the pulse transformer,

$$T = \sqrt{\frac{Z_M}{Z_0}} = 4:1$$

The primary pulse voltage = 26 kV/T = 6.5 kV. The primary pulse current = 40 A × T = 160 A. The required discharge line voltage = 2 × 6.5 kV = 13 kV and this is achieved by the ac supply and the diode D1. The peak of the pulse power input to the magnetron = 26,000 V × 40 A = 1040 kW. The average power input to the magnetron = 1040 kW × 1 μs/2000 μs ≈ 500 W. The typical RF peak power output is 400 kW, so that the magnetron efficiency is approximately 40%.

5-3 MULTICAVITY KLYSTRONS

The klystron tube is a stable microwave power amplifier which provides high gain at medium efficiency. Depending on the type of tube, klystron power outputs range from a few milliwatts to several megawatts peak power, and over 100 kW average power. The power gains vary from 3 to 90 dB. Klystron amplifiers are somewhat noisy and are therefore used mainly as power amplifiers. However, they have applications in many facets of microwave technology.

Operation of the Klystron Amplifier

The klystron tube makes a virtue of the very thing that defeats the triode—the transit time of the electrons as they cross from the cathode to the plate. By contrast, the klystron modulates the velocity of the electrons, so that as the electrons travel through the tube, electron bunches are formed (density modulation). These bunches deliver a positive feedback voltage to the output resonant circuit of the klystron. Figure 5-16 shows a cutaway representation of the basic klystron amplifier. The klystron amplifier consists of three separate sections: the electron gun, the RF section, and the collector.

Let us first consider the electron gun structure. It consists of a heater, cathode, grid, and anode. Electrons are emitted by the cathode and drawn toward the anode, which is operated at a positive potential with respect to the cathode. The electrons are formed into a narrow beam by either electrostatic or magnetic focusing techniques; this ensures that the electron beam does not spread out. The control grid is used to govern the number of electrons that reach the anode region. It may also be used to turn the tube completely on or off in certain pulsed amplifier applications.

The electron beam is well formed by the time it reaches the anode. The beam passes through a hole in the anode and on to the RF section of the tube, and eventually strikes the collector. The electrons are then returned to the cathode through an external power supply. It is evident that the collector of a klystron acts much like a plate of a triode insofar as the collection of electrons is concerned. However, there is one important difference. The plate of the triode is normally connected, in some fashion,

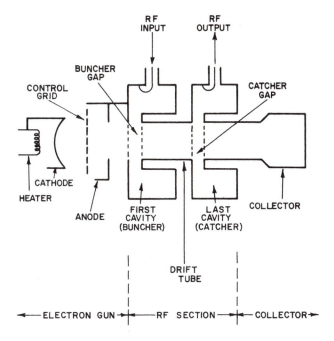

Figure 5-16 Basic klystron amplifier.

to the output RF circuit, while in a klystron amplifier, the collector has no connection to the RF circuitry at all.

Let us look at the RF section of the basic klystron amplifier. This part of the tube is quite different from a conventional triode circuit since the resonant circuits used in a klystron amplifier are reentrant cavities (Section 1-3).

Referring to Figure 5-16, the electrons pass through the cavity gaps in each of the resonators as well as the cylindrical metal tube between the gaps. This metal tube forms the so-called drift space. In a klystron amplifier the low-level RF input signal is coupled to the first resonator, which is called the buncher cavity. The signal may be coupled in through either a waveguide or a coaxial connection. If the cavity is tuned to the frequency of the RF input, it will be excited into oscillation. An electric field will exist across the buncher gap, alternating at the input frequency. For half a cycle, the electric field will be in a direction that will cause the field to increase the velocity of the electrons flowing through the gap. On the other half-cycle, the field will be in a direction that will cause the field to decrease the electron velocity. This effect is called velocity modulation and is illustrated in Figure 5-17a. Notice that when the voltage across the cavity gap is negative, the electrons will decelerate; when the voltage is zero, the electrons will be unaffected; and when the voltage is positive, the electrons will be accelerated.

After leaving the buncher gap (Figure 5-17b) the electrons proceed through the tube's drift region, then on to the collector. In the drift region, the electrons that have been speeded up by the electric field in the buncher gap will tend to overtake electrons that have been slowed down. Due to this action, bunches of electrons will

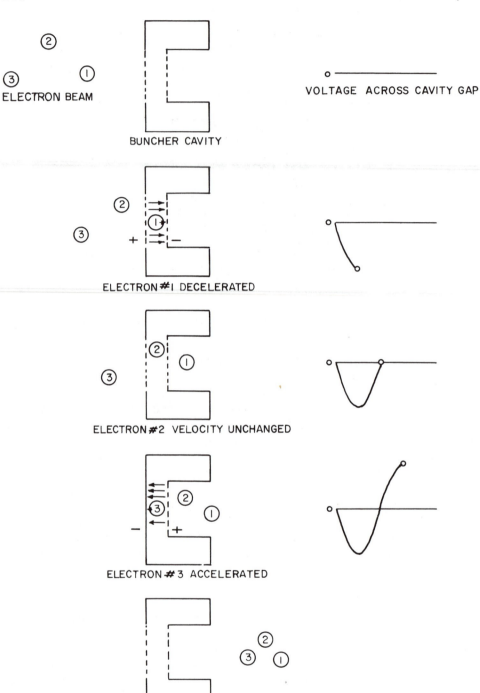

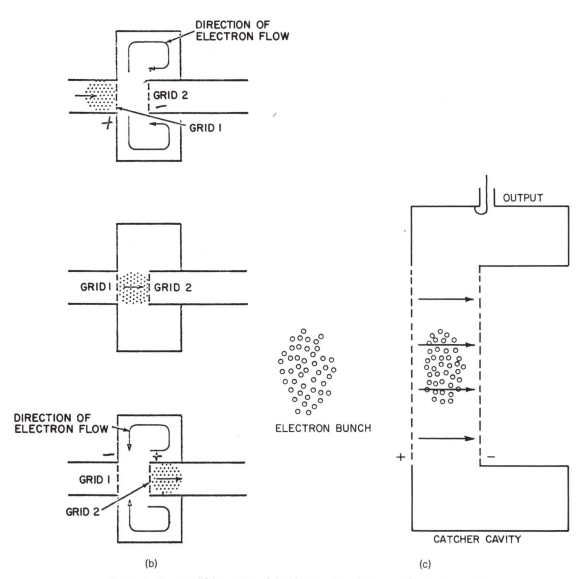

Figure 5–17 Amplifying action of the klystron: (a) velocity and (b) density modulation; (c) delivery of energy to the catcher cavity.

begin to form in the drift region and will be finally formed by the time they reach the gap of the second resonator, which is called the catcher cavity. Bunches of electrons periodically flow through the gap of this catcher cavity, and during the time between bunches, relatively few electrons flow through the gap. The time between the arrival of the electron bunches is equal to the period of the cycle of the RF input signal.

The initial bunch of electrons flowing through the catcher cavity will cause this cavity to oscillate at its resonant frequency. This sets up an alternating electric field across the catcher cavity gap, as illustrated in Figure 5–17c. With proper design and operating potentials, a bunch of electrons will arrive in the catcher cavity gap at the proper time to be retarded by the RF field, and therefore energy will be given up to the catcher cavity.

The RF power in the catcher cavity will be much greater than that applied in the buncher cavity. This is due to the ability of the concentrated bunches of electrons to deliver large amounts of energy to the catcher cavity. Since the electron bean delivers some of its energy to the output cavity, it arrives at the collector with less total energy than it had when it passed through the input cavity. This difference in beam energy is approximately equal to the energy delivered to the output cavity.

It is appropriate to mention here that velocity modulation does not form perfect bunches of electrons. There are some electrons which come through the input cavity with the wrong phase relationship and show up in the output cavity gap between the bunches. The electric field across the gap at the time these out-of-phase electrons come through is in a direction to accelerate them. This causes some energy to be taken from the cavity. However, much more energy will be contributed to the output cavity by the concentrated bunches of the electrons than will be withdrawn from it by the small number of out-of-phase electrons.

Multicavity Power Klystron Amplifiers

In the discussion above, only a basic two-cavity klystron has been considered. This simple type of klystron amplifier is not capable of high-gain, high-output power or suitable efficiency. With the addition of intermediate cavities and other physical modifications, the basic two-cavity klystron may be converted to a multicavity power klystron. This amplifier is capable of high-gain, high-power output and satisfactory efficiency. Figure 5–18 illustrates a typical multicavity klystron power amplifier.

In addition to the intermediate cavities, there are several physical differences between the basic and the multicavity klystron. The cathode of the multicavity power klystron must be larger, in order to be capable of emitting large numbers of electrons. The shape of the cathode is usually concave, to aid in focusing the electron beam. The collector must also be larger to allow for greater heat dissipation. In a high-power klystron, the electron beam may strike the collector with sufficient energy to cause the emission of x-rays from the collector. Many klystrons have a lead shield around the collector as protection against these x-rays. Most high-power klystrons are liquid cooled and must be constructed to facilitate the cooling system.

Klystron amplifiers have been built with as many as five intermediate cavities and therefore a total of seven cavities. The effect of the intermediate cavities is to improve the bunching process. This results in increased amplifier gain and to a lesser extent, increased efficiency. Adding more intermediate cavities is roughly analogous to adding more stages to an IF amplifier. The overall amplifier gain is increased and the overall bandwidth is reduced if all the stages are tuned to the same frequency. The same effect occurs with klystron amplifier tuning. A given klystron amplifier

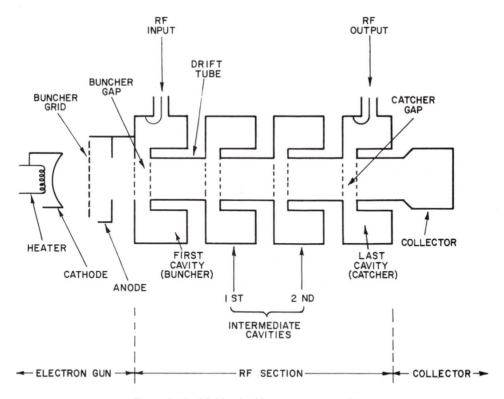

Figure 5–18   Multicavity klystron power amplifier.

tube will deliver high gain and narrow bandwidth if all the cavities are tuned to the same frequency. This is called synchronous tuning. If the cavities are tuned to slightly different frequencies, the gain of the klystron amplifier will be reduced and the bandwidth may be appreciably increased. This is called stagger tuning. Most klystron amplifiers that feature relatively wide bandwidths, are stagger tuned.

The klystron is not a perfect linear amplifier, so that the RF power output is not linearly related to the RF power input at all operating levels. Another way of stating this is that the klystron amplifier will saturate, just as a triode amplifier will have a limiting action if the input signal becomes too large. In fact, if the RF input is increased to levels above saturation, the RF power output will actually decrease. Figure 5–19 shows a plot of typical klystron amplifier performance for various tuning conditions. The RF output is plotted as a function of the RF input. Curve A of Figure 5–19 shows the typical performance for synchronous tuning. Under these conditions the tube has maximum gain. The power output is almost perfectly linear with respect to the power input, up to about 70% of saturation. However, as the RF input is increased beyond this point, the gain decreases and the tube saturates. As the RF input is increased beyond saturation, the RF output decreases.

To understand the reason for this decrease, you will recall that in the previous

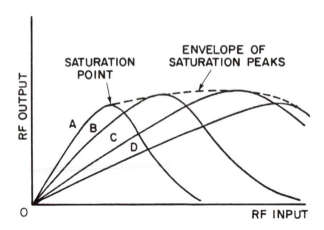

Figure 5–19 Comparison between synchronous and staggered tuning.

discussion, the electron bunches were formed by the action of the RF voltage across the buncher cavity gap. This RF voltage accelerated some electrons and slowed down other electrons, resulting in the formation of the bunches in the drift region. This speeding-up and slowing-down effect will be increased as the RF drive's power is increased. The saturation point shown in Figure 5–19 is reached when the bunches are most perfectly formed at the instant when they reach the output cavity gap. This results in the maximum power output condition. When the RF input is increased beyond this point, the bunches are perfectly formed before they reach the output gap. Consequently, they form too soon and by the time the bunches have reached the output gap, they are tending to debunch because of the mutual repulsion of the electrons and because the faster electrons have overtaken and passed the slower electrons. This is the reason for the decrease in output power.

If a multicavity klystron power amplifier is synchronously tuned, and the next-to-last cavity is then tuned to a higher frequency, the gain of the amplifier is reduced, but the saturation power output level may be increased. This effect is shown by curves B and C in Figure 5–19. Curve B represents a small amount of detuning of the next-to-last cavity, and curve C represents even more detuning. Note that the gain of the tube has been reduced but that the saturation output power is higher than that obtained with synchronous tuning. Many klystron amplifiers are stagger tuned because of the resulting higher output power capability with the same beam power input. This increases the efficiency, provided, of course, that enough RF drive power is available to operate under the stagger tuned condition. Also, as mentioned previously, stagger tuning results in a wider amplifier bandwidth. As might be expected, stagger tuning may be carried too far, at which point the saturation output power will drop. This is illustrated by curve D of Figure 5–19.

In the simple klystron amplifier the electrons were formed into a narrow beam by either electrostatic or magnetic focusing techniques. In a multicavity power klystron, it is even more important to focus the electron beam so that the spreading effect is kept to a minimum. This is normally achieved by an external permanent magnet or a number of electromagnets.

In a pulsed microwave system using a power klystron, one of three methods may be used to accomplish modulation. The first method is to switch the beam accelerating voltage on and off. The second is periodic interruption of the RF input signal. The last method is to turn the klystron beam current on and off.

When a klystron is pulsed by tuning the accelerating voltage on and off, the entire beam current must be pulsed as well. This action is similar to modulating a magnetron and requires a modulator capable of handling the full power of the beam.

If modulation is accomplished by switching of the RF input signal, the beam current must also be pulsed. If this is not done, some beam power will be dissipated (to no useful purpose) in the interval between the RF input pulses. This reduces the efficiency of the tube. Of the three methods, pulsing the modulating grid or anode is the most commonly used. For communications use, the klystron is usually modulated by applying the intelligence to the modulating grid.

The chief advantage of a klystron amplifier is that it is capable of high power output together with good stability, efficiency, and gain. Since a klystron is basically a power amplifier, it may be driven by a stable oscillator, operating at a low frequency, followed by a frequency multiplier chain. This arrangement results in a more stable operation than is possible with a self-excited power oscillator.

Another advantage of a klystron is that its dc and RF sections are separate. This allows the cathode and collector regions to be designed for optimum performance, without concern for their effect on the RF fields. As a result, the life of a klystron is increased over other types of microwave generator.

The chief limitations of klystron amplifiers are their large size, high operating voltages, and the complexity of their associated equipment, such as that required for cooling (Figure 5–20).

5–4 REFLEX KLYSTRONS

In a radar system, most receivers use 30 to 60 MHz as their intermediate frequencies. A highly important factor in receiver operation is the tracking stability of the local oscillator which generates the frequency that beats with the incoming signal to produce the IF. For example, if the local oscillator frequency is 3000 MHz, a frequency variation of as much as 0.1% would be equivalent to a 3-MHz frequency shift. This is equal to the bandwidth of most receivers and would cause a considerable loss in gain.

In receivers that use crystal mixers, the power required of the local oscillator is small, being only 20 to 50 mW in the 4-GHz region. Due to the very loose coupling, only about 1 mW reaches the crystal.

Another requirement of the local oscillator is that it must be tunable over a range of several megahertz. This is to compensate for changes in the transmitted frequency and in its own frequency. It is desirable that the local oscillator have the capability of being electronically tuned by varying the voltage applied to one of its electrodes.

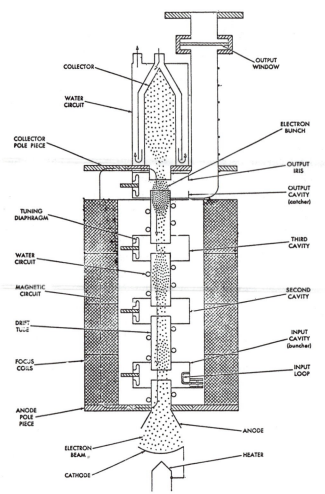

COLLECTOR

WATER
CIRCUIT

COLLECTOR
POLE PIECE

OUTPUT
WINDOW

ELECTRON
BUNCH

OUTPUT
IRIS

OUTPUT
CAVITY
(catcher)

TUNING
DIAPHRAGM

WATER
CIRCUIT

MAGNETIC
CIRCUIT

DRIFT
TUBE

FOCUS
COILS

ANODE
POLE
PIECE

ELECTRON
BEAM

CATHODE

THIRD
CAVITY

SECOND
CAVITY

INPUT
CAVITY
(buncher)

INPUT
LOOP

ANODE

HEATER

Figure 5–20 Construction of the multi-cavity power klystron.

Because the reflex klystron (Figure 5–21) meets all of these requirements, it is commonly used as a local oscillator in microwave receivers. Basically, this type of klystron consists of a source of electrons, a reentrant cavity, and a repeller or reflector plate.

The electrons that are accelerated by grid 1 will be velocity modulated as they pass through the cavity grids (grids 2 and 3). After moving through the cavity grids, the electrons will be traveling at different velocities. Since the reflector plate is made highly negative with respect to the cathode, these electrons will be repelled from the reflector and will reverse their direction. The high-velocity electrons will come physically closer to the reflector plate than either the medium- or low-velocity electrons. After repulsion, all electrons will be directed back toward the cavity grids. A bunching action then occurs on the return trip of the electrons. The distance that

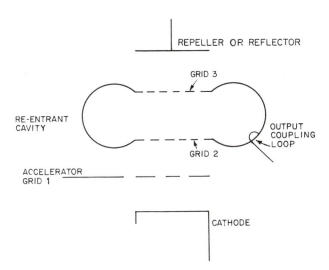

REPELLER OR REFLECTOR

GRID 3

RE-ENTRANT
CAVITY

OUTPUT
COUPLING
LOOP

GRID 2

ACCELERATOR
GRID 1

CATHODE

Figure 5–21   Principle of the reflex klystron.

the electrons move before they are repelled by the negative reflector is a function of the voltage on the accelerating grid, the negative dc voltage applied to the reflector, and the magnitude of the RF voltage existing on the cavity grids due to the oscillation in the cavity resonator. These applied voltages and the physical construction of the klystron should be such that the electrons will return to the cavity grids in bunches.

The bunching process itself is illustrated in Figure 5–22a, which is sometimes referred to as an Applegate diagram. Figure 5–22b shows a velocity–time diagram of the electrons during their transit. The electron at time 3 (center electron) passes the cavity as the RF field (bunching voltage) is zero, and its velocity is unaffected. Therefore, a bunch will form about this electron. The electrons at times 1 and 2 pass through the cavity with higher velocities, because they move through at a time when the RF voltage across the cavity grids is producing an accelerating field. There-

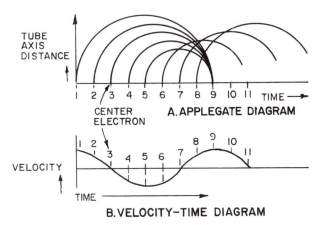

TUBE
AXIS
DISTANCE

1 2 3 4 5 6 7 8 9 10 11 TIME →

CENTER
ELECTRON

A. APPLEGATE DIAGRAM

VELOCITY

TIME →

B. VELOCITY-TIME DIAGRAM

Figure 5–22   Bunching process in the reflex klystron: (a) Applegate diagram; (b) velocity–time diagram.

fore, these electrons penetrate farther into the drift space and return to the cavity at essentially the same time as the center electron. Similarly, the electrons at times 4 and 5 leave with a lower velocity, penetrate a shorter distance, and return at the same instant as the previous three electrons.

When the bunch of electrons returns, the potential of the cavity grids is important. At the time of return the potential applied to the cavity grids must be such that some of the energy of the bunch is absorbed. The maximum absorption of energy will occur when the bunched electrons reach the midpoint between the cavity grids as the RF voltage between these grids reaches its peak value. As the electron bunch arrives at the midpoint, the grid nearest the repeller plate must be positive in relation to the other buncher grid. The electron bunch will then be decelerated in this field, so that some of its energy is expended in sustaining RF oscillations within the grids' cavity. For example, in Figure 5–22b, the center electron (time 3) remains in the drift space for three-quarters of a period and the bunch (electrons at times 1, 2, 3, 4, and 5) returns at time 9, when the cavity field has a maximum value in the direction that decelerates the returning bunch.

Under these conditions, electrons leaving the cathode will receive maximum acceleration from the cavity field, while the returning electron bunches will receive maximum deceleration. If the grids are separated by approximately one half-wavelength, the electron bunch will pass through the first grid (the one nearest the reflector) as its RF potential is zero and changing from negative to positive. The electron bunch would pass through the second grid when its potential is zero and is changing from negative to positive. After the returning electron bunches have given their energy up to the cavity, they are absorbed by the cavity grid nearest the cathode and are returned to the power supply.

The cavity grids perform a dual function—velocity modulation at the buncher grid followed by density modulation at the catcher grid. The output from the tube is extracted by means of the coupling loop shown in Figure 5–21.

By proper adjustment of the negative voltage applied to the reflector plate, the electrons that have passed through the bunching field may be made to pass through the resonator again at the proper time to deliver energy to this circuit. The result is the positive feedback needed to sustain oscillations in the cavity. Spent electrons are removed from the tube by the accelerating grid or by the grids of the resonator. The operating frequency of the tube can be varied over a small range by changing the voltage on the reflector plate. This potential determines the transit time of the electrons between their first and second passages through the resonator. However, the output power of the oscillator is affected considerably more than the frequency by changes in the magnitude of the reflector voltage. This is because the output power depends on the fact that the electrons are bunched exactly during the time of the decelerating half-cycle of the oscillating voltage. The volume of the resonant cavity is mechanically altered to change the oscillator frequency. The reflector voltage may be varied over a narrow range to provide minor frequency adjustments.

We mentioned that the electron bunches should arrive at the grids' midpoint when the RF swing is at its maximum positive value on the grid closest to the reflector plate. It is, however, not necessary for the electron bunches to return on the first positive half-cycle. They may be returned on the second, third, or fourth positive half-cycle. The positive half-cycle in which the electrons are returned and bunching occurs, determines the mode of operation.

The mode of operation is governed by the transit time of the electrons. This transit time means the time between which electrons leave the bunching grids and the time when the bunches deliver their energy to the cavity grids. Figure 5–23 shows the electrons being returned for the different operational modes. For the first mode, the bunching should occur $\frac{3}{4}$ of a period after the average velocity electrons leave the bunching grids, the second mode of operation occurs after $1\frac{3}{4}$ periods, the third after $2\frac{3}{4}$ periods, and the fourth after $3\frac{3}{4}$ periods. In practical operation, either the second, third, or fourth mode is used. Because of these time intervals we commonly refer to them as the $\frac{3}{4}$, $1\frac{3}{4}$, $2\frac{3}{4}$, $3\frac{3}{4}$ modes.

The mode of operation is determined by the electron transit time, which is the function of both the accelerating voltage and the reflector voltage. Consequently, the mode of operation is in fact controlled by the reflector voltage because the accelerating voltage is a fixed quantity.

The variations in the power output and frequency that occur as the reflector voltage is changed are particularly important characteristics of the klystron. Figure 5–24 illustrates curves that may be obtained if the power output and the frequency of the reflex klystron are measured as its reflector voltage is varied from a low to a high negative value while the accelerating voltage is kept constant. Oscillations occur only for certain values of the reflector voltage (corresponding to the various modes) for which the electron bunches return in the proper phase to deliver their energy to the cavity.

The center points of the modes, labeled A, B, and C in Figure 5–24, correspond

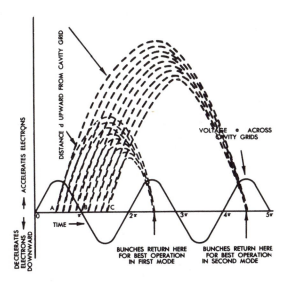

Figure 5–23 Formation of different modes in the reflex klystron. (Courtesy of Varian Associates, Inc.)

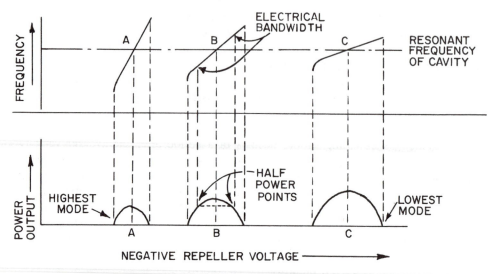

Figure 5–24 Frequency and power characteristics of the reflex klystron.

to those reflector voltages for which the time spent by the electrons in the drift space is a whole number of periods together with three-quarters of a period. At these points the oscillation frequency is the resonant frequency of the cavity, and the power output is the maximum power of the mode. Note that the power outputs for the various modes at the resonant frequency are not the same and the output is least in the highest mode. This can be explained by examining the factors which limit the amplitude of the oscillations and which, in turn, limit the output power.

The power limitation is due to overbunching as well as the usual losses in the oscillatory circuit. Overbunching occurs as oscillations build up and the bunching voltage becomes greater and increases the amounts of acceleration and deceleration. This causes bunching to occur in a shorter period of time (before the electrons reach the grid on the return trip), and this tends to reduce the magnitude of the oscillations. In the higher modes, where the bunches are formed more slowly, the electrons are more susceptible to overbunching.

As shown in Figure 5–24, the frequency of the oscillations is variable to a limited degree in any of the modes of operation by changing the reflector voltage. When the reflector voltage is altered, a bunch is caused to return either a little sooner or a little later than normal. Away from resonance the amplitude of the oscillation decreases by an amount depending on the Q of the cavity. The tuning range is small in comparison with the frequency of the oscillations and varies somewhat from one mode to another. It is greatest in the highest mode, because bunching and debunching take place at a slower rate and because a greater variation from the ideal time of return is possible without appreciable debunching, which would cause the amplitude of the oscillations to drop below the usable output level.

Another way to look at this is to consider that in the highest mode, the time taken by an electron in the drift space is greater, and the change in the period and its accompanying change in frequency occur in a relatively shorter interval of time. For example, in the third mode, the interval before the return must be about $2\frac{3}{4}$ periods (11 quarters of a period). A small change in the timing of the bunching voltage would therefore be only $\frac{3}{11}$ as great a portion of the interval as it would be if the operation were in the first mode, where the ideal time interval is only $\frac{3}{4}$ of a period.

The band of frequencies that can be obtained by varying the reflector voltage lies between the half-power points, as shown in Figure 5–24. This range of frequencies is known as the electronic tuning bandwidth (of the order of 1% of the cavity frequency).

The power output curve of the bandwidth is asymmetrical for the lower-order voltage modes. This results from the fact that as the negative reflector voltage is increased, not only does the bunching voltage decrease and cause the bunches to form at a later time, but the reflector voltage causes a quicker return. The effects of the two actions combine to cause poor bunching at the return of the electrons, resulting in a rapid drop in the output on the high side of the mode's peak. At lower voltages, however, even though the bunching voltage decreases and causes slower bunching, the decreased reflector voltage causes a later return to the cavity grids. In this way the two effects are counteracting and a greater change in the reflector voltage is possible before the output drops below the usable level. The asymmetry is not noticeable in the higher-order modes because the percentage change in the bunching that can occur in a higher-order mode is negligible.

As the local oscillator in a microwave receiver, a reflex klystron need not supply large amounts of power but should oscillate at a frequency that is relatively stable and easily controlled. The efficiency of a reflex klystron is normally on the order of 2% or less.

The need for a wide electronic tuning range suggests the use of a voltage mode of a high order. However, if a mode of excessively high order is selected, the power available is too small for local oscillator applications, and a compromise between the wide tuning range and the power output is necessary. The use of a very high order mode is also undesirable because the noise output of a reflex klystron is essentially the same for all voltage modes. Therefore, the closer coupling to the mixer required with high-order, low-power modes increases the receiver's noise figure. Usually the $1\frac{3}{4}$ or $2\frac{3}{4}$ voltage mode is found to be suitable. Since the modes are asymmetrical, the point of operation is usually a little below the resonant frequency of the cavity. Tuning above the operating frequency is then possible to a greater degree than if the precise resonant frequency were used.

In practice the reflex klystron is operated in conjunction with an automatic frequency control (AFC) circuit. Since the reflector voltage is effective in making small changes in frequency, the AFC circuit is used to control the reflector voltage to maintain the correct intermediate frequency. It should be noted that the coarse frequency of oscillation is determined by the dimensions of the cavity and there is, on most reflex klystrons, a mechanical adjustment which varies the cavity size. For

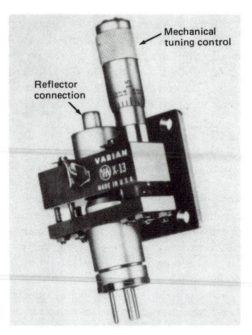

Mechanical
tuning control

Reflector
connection

Figure 5–25   X-band reflex klystron.
(Courtesy of Varian Associates)

example, a 10-GHz X-band klystron (Figure 5–25) may be mechanically tuned between
8.5 and 11.5 GHz.

5–5 TRAVELING-WAVE TUBES

The traveling-wave tube (TWT) is a high-gain, low-noise, wide-bandwidth microwave
amplifier. TWTs are capable of gains of 40 dB or more with bandwidths in which
the upper frequency is twice the lower frequency. TWTs have been designed for
frequencies as low as 300 MHz and as high as 50 GHz.

The primary use for the TWT is broadband voltage amplification (although
high-power TWTs, with characteristics similar to those of a power klystron, have
been developed). Their wide-bandwidth and low-noise characteristics make them ideal
for use as RF amplifiers in microwave equipment.

Figure 5–26 is a pictorial diagram of a traveling-wave tube. Notice that there
are no resonant cavities and this accounts for the wide bandwidth available. The
electron gun produces a stream of electrons which are focused into a narrow beam
by an axial magnetic field, much the same as in a klystron tube. This field is produced
by a permanent magnet or an electromagnet which surrounds the helix portion of
the tube. As it passes through the helix, the narrow beam is accelerated by a high
potential on the helix and the collector.

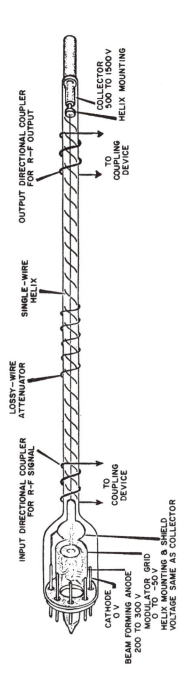

INPUT DIRECTIONAL COUPLER FOR R-F SIGNAL

TO COUPLING DEVICE

LOSSY-WIRE ATTENUATOR

SINGLE-WIRE HELIX

OUTPUT DIRECTIONAL COUPLER FOR R-F OUTPUT

TO COUPLING DEVICE

COLLECTOR 500 TO 1500V

HELIX MOUNTING

CATHODE 0 V

BEAM FORMING ANODE 200 TO 300 V

MODULATOR GRID 0 TO -50 V

HELIX MOUNTING & SHIELD VOLTAGE SAME AS COLLECTOR

Figure 5-26 Traveling-wave tube.

TWT Operation

While the electron beam in a klystron travels, for the most part, in regions free from RF electric fields, the beam in a TWT is continually interacting with an RF electric field which propagates along an external circuit surrounding the beam.

To achieve amplification, the TWT must propagate a wave whose phase velocity is nearly synchronized with the velocity of the electron beam. It is difficult to accelerate the beam to more than about one-fifth the velocity of light. Therefore, the forward velocity of the RF field propagating along the helix must be reduced to nearly that of the beam.

The phase velocity in a waveguide which is uniform in the direction of propagation is always greater than the velocity of light. However, this velocity may be reduced below the velocity of light by introducing a periodic variation of the circuit in the direction of propagation. The simplest form of variation is obtained by wrapping the circuit in the form of a helix which acts as a "slow wave" structure.

As explained previously, the electron beam is focused and constrained to flow along the axis of the helix. The longitudinal components of the input signal's RF electric field, along the axis of the helix or slow wave structure, continually interact with the electron beam to provide the amplification of the TWT. This interaction is pictured in Figure 5–27a, which illustrates the RF electric field of the input signal, as it propagates along the helix, and penetrates into the region occupied by the electron beam.

Let us first consider the case where the electron beam velocity is exactly synchronized with the circuit's phase velocity. The electrons then experience a steady dc electric force which tends to bunch them around position A and debunch them around position B. This action is due to the electric fields and is similar to velocity and density modulation which we previously discussed. In this case, as many electrons are accelerated as are decelerated; hence there is no net energy transfer between the beam and the RF electric field. To achieve amplification, the electron beam is adjusted to travel slightly faster than the RF electric field propagating along the helix. The bunching and debunching mechanisms just discussed are still at work, but the bunches now move slightly ahead of the fields on the helix. Under these conditions more electrons are in the decelerating field to the right of A than are in the accelerating field to the right of B. Since more electrons are decelerated than are accelerated, the energy balance is no longer maintained, energy is transferred from the beam to the RF field, and the signal is amplified.

The fields may propagate in either direction along the helix. This leads to the possibility of oscillation due to the reflections back along the helix. This tendency is minimized by placing some resistive material near the input end of the slow wave structure. This resistance may take the form of a lossy attenuator (Figure 5–26) or a graphite coating placed on insulators adjacent to the helix. Such lossy sections completely absorb any backward traveling wave. The forward wave is also absorbed to a great extent, but the signal is carried past the attenuator by the bunches of electrons. Since these bunches are not affected as they pass by the attenuator, they are capable of reinstituting the signal on the helix (Figure 5–27b).

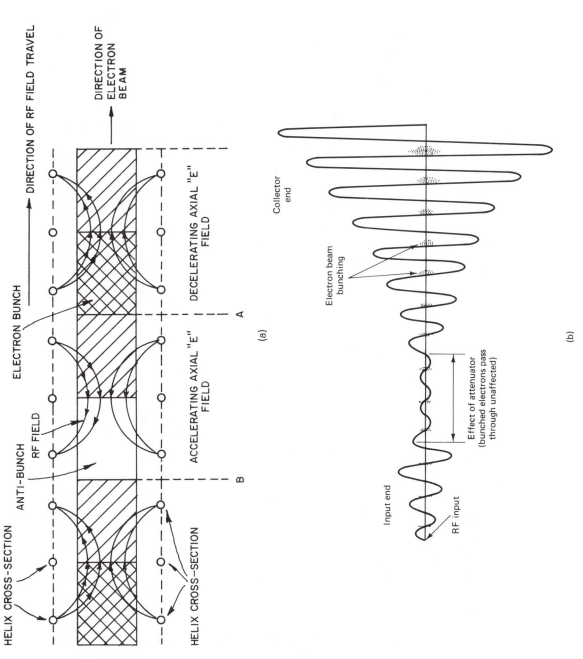

Figure 5–27 Interaction between the RF signal and the electron beam.

231

Methods of Coupling

Some means must be provided to apply the RF signal to one end of the helix and to remove it from the other end. Four methods of coupling are illustrated in Figure 5–28.

Figure 5–28a illustrates waveguide matching. The waveguide is terminated in a nonreflecting impedance, and the helix is inserted into the waveguide, as shown. The efficiency of the system is good, but the waveguide has a far higher Q than that of the traveling-wave tube. This means that the broadband characteristics of the TWT suffer in that the entire bandwidth is not available for amplification since the waveguide will not respond over such a wide spectrum.

The cavity match, illustrated in Figure 5–28b, is very similar to the waveguide match. Cavities may be made to resonate over wider ranges than waveguides, but they still have a high Q compared with the TWT. The helix is placed at the mouth of the cavity, thereby absorbing energy so that an E field is produced. The RF signal is then fed into the cavity by a coaxial cable.

Figure 5–28c illustrates a direct coaxial cable–helix match and is the simplest system of all. The center conductor of the input coaxial cable is connected directly to the helix. Although this method is used quite frequently, it has a disadvantage. A high VSWR is set up by this type of match, and this causes heating around the input connection. Since this connection passes through the glass envelope, the envelope is subject to heating and possible breakage at this point. However, this is a major problem only in the higher-power TWTs.

Figure 5–28d illustrates the coupled helix match. In this system, the coaxial center conductor is attached to a small helix. The major helix is inserted within this input helix, where it acts as the secondary of a transformer. This system has a good VSWR and is broader in bandwidth than cavities or waveguides, although it is unable to handle large amounts of power. We should note that any of the above may be used for output as well as input coupling.

The traveling-wave tube has also found application as a microwave mixer. By virtue of its wide bandwidth, the TWT can accommodate the frequencies generated by the heterodyning process (provided, of course, that the frequencies have been chosen to be within the range of the tube). The desired frequency is selected by the use of a filter on the output of the helix. Such a circuit has the added advantage of providing gain as well as simply acting as a mixer.

The TWT may be modulated by applying the modulating signal to a grid. This modulator grid may be used to turn the electron beam on and off, as in pulsed microwave applications, or to control the density of the beam and its ability to transfer energy to the traveling wave. The grid may be used to amplitude modulate the output signal.

A forward wave TWT may be constructed to serve as a microwave oscillator. Physically, TWT amplifiers and oscillators differ in two major ways. The helix of the oscillator is longer than that of the amplifier and there is no input connection

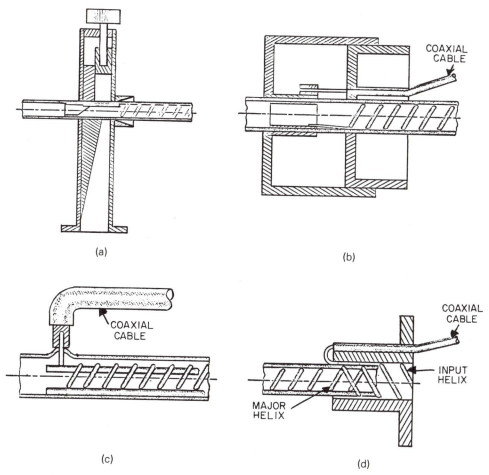

(a)   (b)

(c)   (d)

Figure 5–28   Methods of coupling the RF signal to and from the traveling-wave tube: (a) waveguide match; (b) cavity match; (c) direct coax-helix match; (d) coupled helix match.

to the oscillator. The operating frequency of a TWT oscillator is determined by the pitch of the tube's helix. The oscillator may be tuned, within limits, by adjusting the operating potentials of the tube.

The electron beam, passing through the helix, induces an electromagnetic field in the helix. Although initially weak, this field will, through the action previously described, cause bunching of succeeding portions of the electron beam. With the proper potentials applied, the bunches of electrons will reinforce the signal on the helix. This, in turn, increases the bunching of succeeding portions of the electron beam. The signal of the helix is sustained and amplified by this positive feedback resulting from the exchange of energy between the electron beam and the helix.

5–6 TWYSTRON AMPLIFIERS

The twystron tube is a hybrid combination of klystron and traveling-wave tube. It achieves a better overall performance than either tube can provide separately and consists basically of a multicavity klystron input section and a traveling-wave-tube output section. Figure 5–29 illustrates the cutaway view of a typical S-band version.

It is arranged that the four-cavity klystron driver section is stagger tuned and heavily loaded so that the gains at the ends of its band (2.7 to 3.3 GHz) are 5 to 10 dB higher than the gain at its midfrequency of 3 GHz. By contrast, the traveling-wave output section is designed primarily for the highest gain at the center frequency, while its gain at the ends of the band are 5 to 10 dB lower. It therefore follows that the twystron amplifier has a bandwidth of approximately 10% and a relatively flat gain response over the entire band. The amplifier's high efficiency is of the order of 30% and is the result of the more effective electron bunching in the klystron at either end of the frequency range, where the traveling-wave section gain is relatively low.

The normal operation of twystron amplifiers lies in the S and C bands. Peak power outputs of 1 to 10 MW are possible, with average powers ranging from 1 to 10 kW. The peak-power output response of a C-band twystron amplifier is illustrated in Figure 5–30.

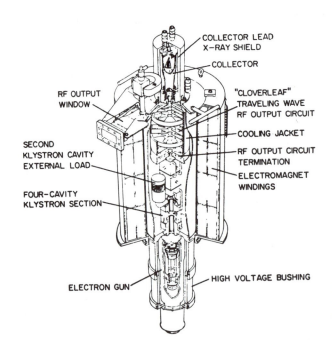

COLLECTOR LEAD
X–RAY SHIELD

COLLECTOR

RF OUTPUT
WINDOW

"CLOVERLEAF"
TRAVELING WAVE
RF OUTPUT CIRCUIT

COOLING JACKET

SECOND
KLYSTRON CAVITY
EXTERNAL LOAD

RF OUTPUT CIRCUIT
TERMINATION

ELECTROMAGNET
WINDINGS

FOUR-CAVITY
KLYSTRON SECTION

ELECTRON GUN

HIGH VOLTAGE BUSHING

Figure 5–29 S-band twystron amplifier. (Staprans, A., et al., "High-power linear-beam tubes." *Proc. IEEE,* 61, No. 3, 299–330, March 1973. © 1973 IEEE.)

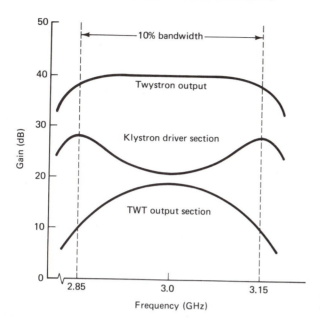

Figure 5-30 Gain response of S-band twystron amplifier.

5-7 BACKWARD-WAVE OSCILLATORS

O-Type Backward-Wave Oscillators

A typical O-type backward-wave oscillator is shown in Figure 5-31. The slow-wave circuit in the middle of the microwave region is a tape helix of such dimensions as to operate in the backward-wave fundamental mode or if the impedance is high enough to make its use practical, in a backward harmonic of the forward wave's fundamental component. The electron beam is shot into the helix, whose phase velocity is equal to the beam velocity.

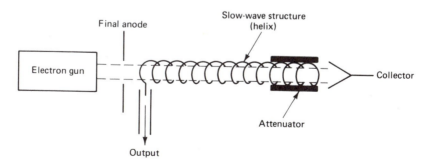

Figure 5-31   O-type backward-wave oscillator.

The oscillations begin in the backward-wave oscillator in much the same way as they begin in other oscillators. Noise components are established on the helix or other slow-wave structure, as a result of the shot noise coupled from the electron beam and from the thermal energy developed at the collector. The waves which travel backward on the tube, velocity modulate the beam and cause density modulation so that the electrons tend to bunch. These bunches of electrons reinforce the wave that exists on the line. In this way oscillations are built up at a single frequency determined solely by the electron-beam velocity, which in turn is a function of the accelerating voltage.

If a wave is propagated forward at the same speed as the electrons in the beam, an individual electron will experience a force of the same magnitude and direction at certain points on the periodic structure. Provided that the synchronism is maintained, some kinetic energy will be extracted from the beam and the wave will be amplified. This is the basic principle of the traveling wave tube. If, however, an electron travels in the opposite direction to the wave but this time with a reduced velocity, there is again the possibility of an individual electron experiencing a force of the same magnitude and direction at certain points along the periodic structure.

Assume that v (m/s) is the electron velocity required to synchronize with a forward wave whose phase change per segment is ϕ radians. Then to synchronize in the reverse direction, the electron velocity must be $v\phi/(2\pi - \phi)$ m/s.

If the phase difference per segment in Figure 5–32 is $\pi/2$, or four segments for one wavelength, then for an electron to synchronize with the wave at the gaps, it must travel forward one segment while the wave travels back three segments. It is then possible for the electron beam to be coupled to a component of the backward wave.

In producing oscillations, the electron beam gives up a proportion of its kinetic

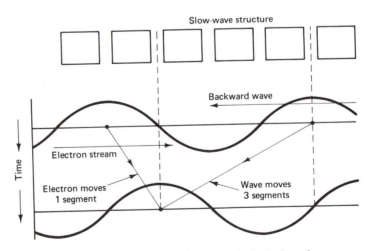

Figure 5–32 Electron-beam interaction with the backward wave.

energy of the RF field on the line, and this power may be taken from the line at the end nearest the beam's point of entry. At the collector end of the tube there is a matched load to absorb any reflected power.

As shown in equation (1–10), the electron velocity is proportional to the square root of the accelerating voltage. By contrast, the phase velocity of the wave on the periodic structure is a linear function of the wave's frequency. Therefore, the variation of the frequency with the voltage will tend to be rapid at the low voltages but will change more slowly at the higher voltages. Furthermore, unless the current is varied across the frequency band by the use of a grid, any shift in the frequency brought about by a change in the accelerating voltage will produce a considerable difference in the output power over the frequency band.

The O-type backward-wave oscillator is very useful as a low-power active device and is frequently used as a local oscillator or as a low-power modulator. The characteristics of this device are illustrated in Figure 5–33.

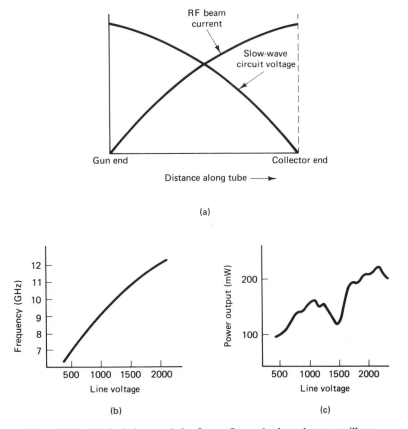

Figure 5–33 Typical characteristics for an O-type backward-wave oscillator.

M-Type Backward-Wave Oscillators

We have shown that not only does the traveling-wave tube possess a limited power output, but the actual output efficiency is low. This is because the electrons are continually slowed down by the wave as they give up kinetic energy and there comes a point where velocity synchronism between the beam and the wave no longer exists. Under these conditions there can be no further amplification, however long the tube, and therefore the tube saturates. Because the acceptable velocity range is so small (even when the tube is saturated), the emergent electron beam still possesses considerable kinetic energy, which in general is dissipated as heat at the collector. The tube is thus inefficient and the only way that the efficiency can be increased is (1) to enable the electrons to give up a higher fraction of their energy to the interaction process, or (2) to reduce the beam velocity after the interaction has taken place and before the electrons reach the collector. At the moment the best way of improving the efficiency is to modify the process of M-type interaction. This dates back to the early days of magnetrons, although its role in improving efficiency has been recognized only recently.

To make traveling-wave tubes and klystrons function, it is generally necessary to add some kind of magnetic focusing which operates in an axial direction; this limits the radial electron motion but does not interfere with the interaction mechanism. However, if a transverse rather than a longitudinal magnetic field is applied to the electron beam, the result is a new type of interaction.

Consider the linear crossed-field system shown in Figure 5–34. If the action of the RF field is ignored for the moment, an electron beam injected into the crossed-field region will drift to the right with a mean velocity, $v = \mathcal{E}/B$. The actual path described by the electrons depends on the injection velocity, but their mean drift velocity does not. Since the mean drift velocity is independent of the injection velocity, an electron beam traveling as shown will remain in synchronism with a wave propagating on the slow-wave structure over the whole of its length. It follows that no matter how much energy is extracted from the beam by the wave, the electrons will always maintain their drift velocity as a result of the transverse electric field.

Under dc conditions the magnetic field, provided that it is strong enough, prevents an electron from moving to the line of the slow-wave structure under the action of the transverse electric field. Therefore, an electron in the crossed-field region at

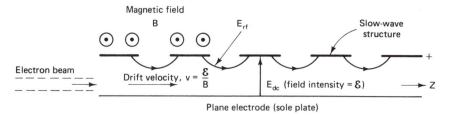

Figure 5–34 Linear crossed-field system.

any point between the electrodes possessses considerable potential energy. Without the magnetic field the electron would immediately move to the line of the slow-wave structure and arrive with a kinetic energy exactly equal to the potential energy it possessed when the magnetic field was present. Consequently, since the action of the RF wave is to draw energy from the electron beam (by endeavoring to slow it down) and since the drift velocity is compensated by the energy drawn from the transverse electric field, the electron trajectories must move out toward the line as the distance increases in the direction toward the collector. By moving toward the line, the electrons can give up their potential energy and make good the losses in the longitudinal drift which the slow wave is causing.

The basic device so far described would function as an M-type amplifier. However, such an amplifier is very noisy and this form of interaction finds its main use in the M-type backward-wave oscillator or carcinotron.

The slow-wave structure used is commonly an interdigital line (Figure 5–35a). Although the entire system is curved into an arc of a circle for convenience (Figure 5–35b), the electrical properties and operation are unchanged. The first bar on the line, from which the output is taken, is commonly called the anode, which serves as a grounded shield to the first fingers of the line and intercepts a large number of electrons which might otherwise melt these fingers.

The gun system used in the M-type device is completely different from that used in the O type. This is because of the effect of the magnetic field. The object is to inject electrons into the interaction space with their required drift velocity. Electrons are accelerated from the cathode by applying a voltage to the plate. The electrons start off in a direction which is normal to the cathode but are immediately bent into their cycloidal orbits by the magnetic field. At the top of their orbits the electrons pass into the region between the line and the sole plate and (in the absence of the RF field on the line) the electrons travel along the equipotential line with a mean velocity of $v = \mathscr{E}/B$. Even if the initial velocity is not quite correct, the electrons will still have the same average velocity, so that instead of following a linear path, the electrons will oscillate around the mean path. This allows a simple gun to be used and no compensation is required for the small variations in the line voltage.

When an RF field is excited on the interdigital line, it is so designed that the dominant mode is a backward wave. The tangential component of the RF field will alternatively force the electrons closer and farther from the line. When the electrons approach the line, they give up potential energy to the wave; but as they move away, energy will be extracted.

As the favorable electrons approach the line, they experience stronger fields and give up energy more quickly. By contrast, the unfavorable electrons, which extract energy, tend to move farther away from the line and therefore experience weaker fields.

As an approximation, no density modulation occurs in the M-type interaction; however, the beam is drawn up into the wave (Figure 5–36) while its space charge density and velocity remain the same. This is in direct contrast with the O type, where density modulation is essential.

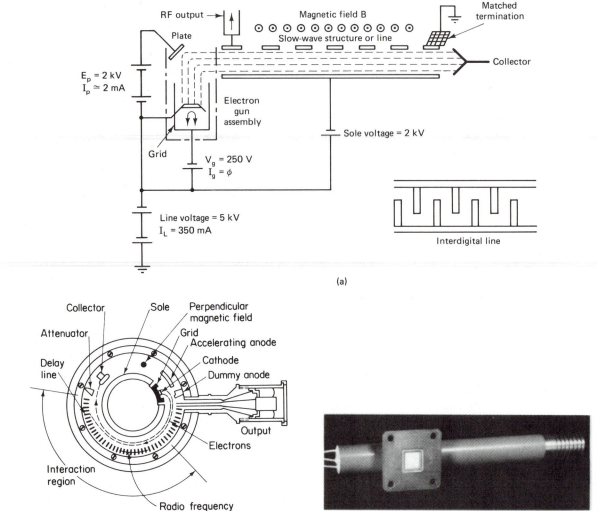

Figure 5–35 M-type backward-wave oscillator. [(b) Courtesy of Watkins-Johnson Company.]

The efficiency of the M-type backward oscillator is a function of $\mathscr{E}/B$ and can be increased by raising B or alternatively by reducing $\mathscr{E}$ for a given value of B. Theoretical efficiencies of about 80% should be possible, but in practice the efficiencies are of the order 20 to 30%. These characteristics are illustrated in Figure 5–37.

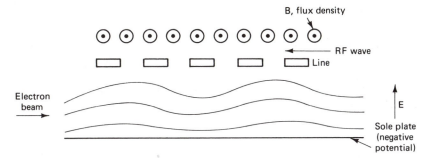

Figure 5-36 Electron beam formed into waves by the effect of the RF wave on the line.

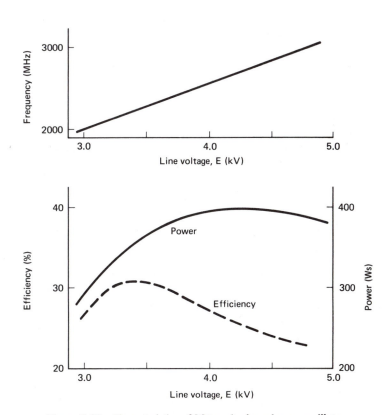

Figure 5-37 Characteristics of M-type backward-wave oscillator.

5–8 CROSSED-FIELD AMPLIFIERS

The crossed-field amplifier (CFA) was developed during the 1950s and its name implies the existence of direct electrical and magnetic fields whose directions are at right angles. Also present is an RF field, and the operation of the device depends on the interaction between the various fields in determining the motion of the electrons between the cathode and the anode. The same was also true of the magnetron described in Section 5–2, but the CFA and the magnetron, although similar in appearance, differ in a number of important ways:

1. The magnetron is an oscillator with a single RF output connection, while the CFA is a microwave power amplifier with both RF input and output connections.
2. In the CFA there is continuous interaction between the electron beam and a *moving* RF field, while in the magnetron the RF field is *stationary*.
3. Unlike the magnetron, the CFA contains a slow-wave structure similar to that of a traveling-wave tube.

Crossed-field amplifiers may be either forward wave or backward wave, as determined by the magnetic field in relation to the directions of the other fields. A forward wave CFA commonly uses the helix as the slow-wave structure while the bar line is preferred for the backward-wave type (Figures 5–38a and b).

The combined results of the various fields is to create a bunching effect which results in the formation of electron "spokes." In the forward wave CFA these "spokes" rotate in the same direction as the RF field moves. As in the TWT, this signal then receives energy from the bunched electrons, and therefore its strength increases as it progresses along the slow-wave structure. Consequently, the CFA provides a power gain between the output and input microwave signals; for CW operation this power ratio is of the order of 10. To prevent oscillation there is a drift space which has an isolating effect so that no feedback occurs.

Both the forward- and backward-wave CFAs of Figures 5–38a and b employ continuous cathode-emitting soles. Figure 5–38c shows an alternative arrangement (similar to that of the M-type BWO) in which a combination of cathode, grid and accelerating anode is used in an injected beam CFA. With this type the slow-wave structure may contain an attenuator section to prevent feedback and consequent oscillation.

The principal advantage of the CFA over other microwave amplifiers lies in its high efficiency, which permits a relatively small physical size. At frequencies around 2 GHz a pulsed CFA can produce a peak power of 4 MW with an efficiency of 60 to 70% and a duty factor on the order of 0.001. In the region of 10 GHz the efficiency falls to about 50% and the peak power available is approximately 1 MW. Bandwidths are normally 25% or less of the center frequency but can be as high as 50% for an injected beam CFA.

Compared with klystrons and TWTs, the major disadvantage of the CFA is its relatively low power gain (10 to 15 dB). Consequently, a CFA may require an

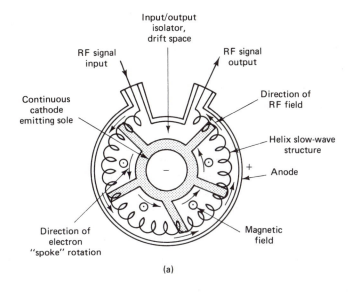

(a)

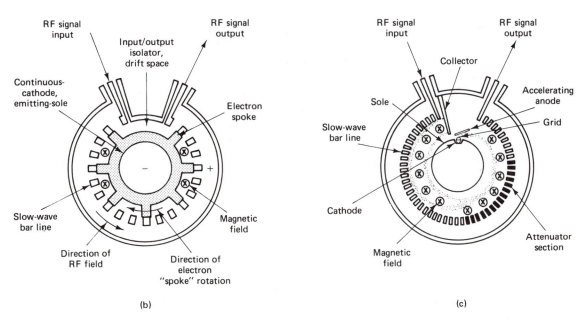

(b)   (c)

Figure 5–38   Crossed-field amplifiers: (a) forward wave; (b) backward wave; (c) injected beam backward wave.

additional driver stage. However, unlike the TWT, the strengths of both the electric and magnetic fields may be used to control the power output. The main application of crossed-field amplifiers is in radar equipment operating within the range 1 to 20 GHz.

5-9 GYROTRONS

The gyrotron is a tube which is still in the developmental stage but whose principle was independently discovered by American and Soviet engineers in 1958. You are no doubt familiar with the electromagnetic focusing techniques used in conjunction with a cathode ray tube. As the beam from the electron gun attempts to spread out due to the mutual repulsion of the electrons, the presence of a direct magnetic field causes the electrons to follow a helical path and spiral into a focus at the phosphor of the screen. This action is illustrated in Figure 5-39.

Electron e is being accelerated into the page by the direct electric field associated with the gun. However, if the electron has a transverse component of velocity v at position 1, the magnetic field will cause the electron to take a curved path and arrive at position 2; the direction of the path is found by the right-hand (motor) rule. Subsequent application of the same rule shows that the electron arrives at positions 3 and 4 and thus completes the full rotation. The number of times per second that the electron moves around the magnetic field is called the cyclotron frequency, whose value is proportional to the magnetic field intensity.

In addition to the direct electric and magnetic fields, the gyrotron action depends on the rotating electrons being placed in the presence of an alternating electromagnetic field whose frequency is also at the cyclotron rate. The idea is that the rotating electrons will deliver energy to the electromagnetic field either to sustain an oscillation

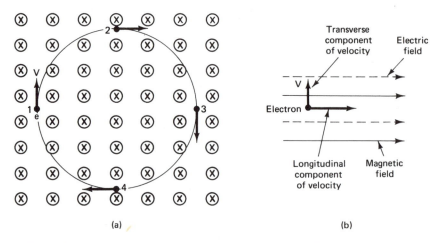

(a)   (b)

Figure 5-39 Spiral motion of an electron in the presence of direct electric and magnetic fields.

or to achieve amplification. However, this action depends on high electron velocities which result in an increase in an electrons' mass according to the principles of relativity. Such an increase in mass lowers the cyclotron rotation rate and results in energy being delivered to the EM field. The electron then loses mass and the phase of the cyclotron wave is shifted. The result is a form of electron bunching which can be compared to the action that occurs in a klystron or a traveling-wave tube. However, there is a major difference. The cyclotron action does not depend on a resonant cavity or a slow-wave structure such as a helix. Consequently, the dimensions of the interaction space are not limited and the power capability of a gyrotron in the millimeter band is much greater than those of the conventional tubes we have discussed previously.

Further, since the gyrotron action depends on the electron rotation, it is essential that most of the beam's energy is transverse to the direction in which the beam is traveling. This requires the development of a special type of electron gun which is capable of creating a hollow beam.

Bunching Mechanism

It must be emphasized that the gyrotron action produces angular velocity modulation rather than linear velocity modulation. Figure 5–40 shows an electron in the presence of an alternating magnetic field E and a direct magnetic field H. At position A the electron is accelerated from left to right by the instantaneous electric field. As a result of the magnetic field's presence, the motion of the electron follows a curved path and the electron arrives at position B. Provided that the timing is correct, the direction of the electric field will have reversed and the electron will again be accelerated. The electron therefore spins in a circle whose plane is at right angles to the direction of the magnetic field. However, those electrons which are accelerated by the electric field are less affected by the magnetic field and have a reduced angular

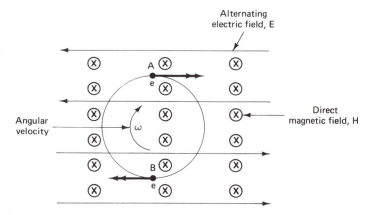

Figure 5–40 Spin motion of an electron in the presence of an alternating electric field and a direct magnetic field.

spin velocity. By contrast, other electrons which have been decelerated by the electric field will have a greater angular velocity. Consequently, a bunching process will take place and provided that the timing of the alternating electromagnetic field is correct, a bunch of electrons will deliver most of their transverse energy to the field so that amplification or oscillation is possible. In this process there is another important effect. When the electron is accelerated to a velocity which approaches that of light, the mass of the electron increases on the principles of relativity, and the angular velocity is reduced. However, when the electron bunches give up energy to the RF field, their velocity and mass are reduced. The result is an increase in the electron's velocity and a higher degree of bunching. Consequently, the relativistic effect is responsible for the final level of the gyrotron's RF power output.

The action so far described is illustrated in Figure 5–41. These represent idealized conditions which are separated by half of the period associated with the spin motion. In both situations the electron bunches are slowed down and energy is delivered to the alternating field. For this to occur, the periods of the alternating electric field and the spin motion must be equal.

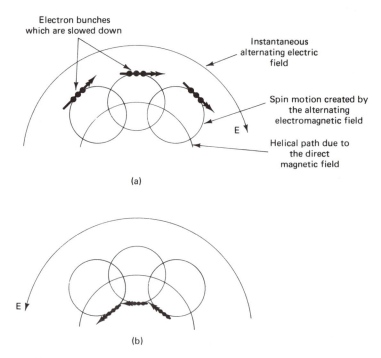

Figure 5–41 The bunching mechanism. The conditions in parts (a) and (b) are 180° out of phase (one-half period apart).

Types of Gyrotron Tube

A number of different types of gyrotron tube are being investigated. These are:

1. The gyro-monotron
2. The gyro-klystron
3. The gyro-traveling-wave tube
4. The gyro-backward-wave oscillator
5. The gyro-twystron (this is the gyrotron version of a tube which is a combination of a traveling-wave tube and a klystron)

The monotron is an oscillator which employs an electron gun, a single cylindrical cavity, and a collector. The electron beam is hollow and the electrons themselves travel in helical paths while interacting with the electromagnetic field, which exists close to the cavity wall. The energy of the resulting RF oscillations is then extracted through a waveguide arrangement. At the present time monotrons typically operate at 25 to 30 GHz and have achieved millisecond pulses with peak powers up to 1 MW. Research is being directed toward generating microwave frequencies of the order of 100 GHz; such monotrons may be used in nuclear fusion experiments.

The gyro-klystron uses the same gun and collector arrangements as the monotron (Figure 5–42). As compared with the conventional klystron, there are input and output cavities, but the main difficulty in the gyrotron version is to allow the RF energy to be propagated from the small output cavity while collecting the electron

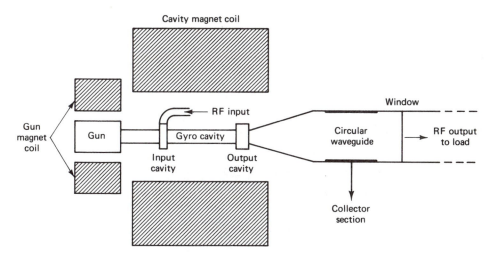

Figure 5–42   Principle of the Gyro-klystron.

beam and arranging for its power dissipation. The solution is to feed the microwave output to a relatively large diameter circular waveguide which is connected to the collector section. At a frequency of 28 GHz a typical power gain is of the order of 10,000 with a peak power output of 50 kW.

The Gyro-TWT, Gyro-BWO, and Gyro-twystron are still in the experimental stage and little is known about their capabilities at the present time.

5–10 MICROWAVE OVENS

This type of oven is a consumer application of one microwave tube, namely the magnetron, which commonly generates a frequency of 2.45 GHz (wavelength $\lambda = 30/2.45 = 12.24$ cm), although some ovens are operated at 915 MHz. As illustrated in Figure 5–43, a high-voltage transformer generates a secondary output of 2800 V rms, which is rectified by diode D_1. The diode current charges the high-voltage capacitor C_1, with the result that a maximum negative voltage of approximately 8 kV is applied to the cathode of the magnetron whose anode is grounded. A subsidiary winding provides an output of 3.2 V rms, 60 Hz for the cathode's heater circuit.

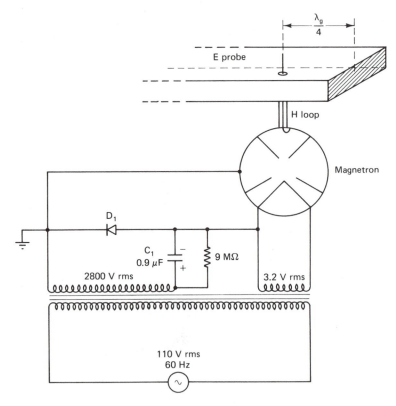

Figure 5–43   High-voltage power supply for a microwave oven.

When the magnetron oscillates, the energy is extracted from the resonant cavities by an H loop. At the other end of this loop is an E probe which protudes into a rectangular waveguide whose dimensions are $a = 6.83$ cm and $b = 3.81$ cm. The wide dimension therefore fulfills the requirement that a be greater than $\lambda/2 = 12.24/2 = 6.12$ cm. This end of the waveguide is closed, so that the probe is positioned at a distance of $\lambda_g/4$ from the sealed end (Figure 5-44). The other end of the waveguide is entirely open and the microwave energy is literally poured into a recess in which there is a metal stirring fan. A plastic ceiling shield then allows the microwave energy to pass through the main container in which the food is placed.

Microwave energy is reflected from metal surfaces so that the interior walls of the cooking oven are made from stainless steel or epoxy-coated metal. The reflections from the stationary walls and the moving stirring fan enable the microwave energy to be well distributed so that the food is evenly cooked. The typical input power to the oven is normally about 1.5 kW, which in turn produces a maximum microwave power of 700 W.

Some materials, such as glass, plastic, paper, and wood, are transparent to frequencies of the order of a few gigahertz and therefore do not absorb or reflect the microwave energy. Consequently, these substances are ideal for manufacturing the cooking containers. However, when the microwave energy comes into contact with the moisture contained in the food, the molecules associated with the moisture are set in vibration and the resulting friction produces the heat necessary for the cooking process. Although the waves can penetrate the food only to a depth of 2 or 3 cm, large items such as joints of meat are cooked by the conduction of heat to the center of the food.

With microwave the cooking process is fast and even, so that the food retains more of its natural flavor and nutritional value. However, the oven must be designed

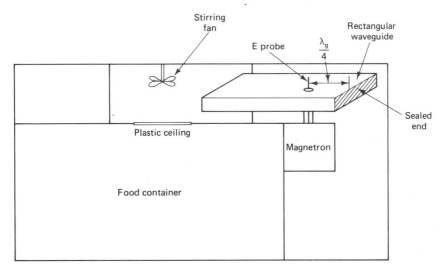

Figure 5-44   Principle of the microwave oven.

so that there is minimum leakage of microwave energy. For example, the oven door is surrounded by a special ferrite material which readily absorbs microwaves. In addition, the door cannot be opened without releasing a safety switch, which automatically shuts off all of the electromagnetic energy.

PROBLEMS

BASIC PROBLEMS

5–1. Explain how velocity modulation creates density modulation in a klystron amplifier. How does the reflex klystron differ from the amplifier klystron?

5–2. What are the advantages of the traveling-wave-tube amplifier over the klystron amplifier?

5–3. Describe the action of the slow-wave structure in a traveling-wave tube.

5–4. How are electron bunching and positive feedback achieved in the operation of a magnetron?

5–5. What is the purpose of the external magnetic field associated with the traveling-wave tube?

5–6. With the aid of a sketch, describe the operation of a crossed-field amplifier (CFA).

5–7. What are the advantages of a twystron amplifier compared with a TWT amplifier?

5–8. What is meant by π-mode operation in a magnetron? Describe how strapping separates the π mode from other possible modes.

5–9. Describe the basic principles behind the operation of a gyrotron tube.

5–10. With the aid of a sketch, explain the operation of a backward-wave oscillator (BWO).

ADVANCED PROBLEM

5–11. Discuss the following microwave tubes in terms of their frequency ranges, bandwidths, efficiencies, gains, power output capabilities, and applications.
(a) Two-cavity klystron
(b) Multicavity klystron
(c) Reflex klystron
(d) Magnetron
(e) Traveling-wave tube
(f) Twystron
(g) O-type backward-wave oscillator
(h) M-type backward-wave oscillator
(i) Crossed-field amplifier
(j) Gyrotron

CHAPTER 6

MICROWAVE SEMICONDUCTOR DEVICES

6-1 INTRODUCTION

Microwave solid-state technology has evolved enormously since the early 1960s. A large number of new semiconductor devices have been developed for the purposes of microwave oscillation, amplification, mixing/detection, frequency multiplication, and switching/limiting. Many of these devices have been greatly reduced in size and capable of being incorporated into microwave integrated circuits (MICs). Over the years improvements have resulted in higher degrees of performance, reliability, power capability, and flexibility. One of the major contributing factors has been the introduction of the semiconductor, gallium arsenide (GaAs), which has raised the operating frequency limit for many of the devices. Without such achievements the majority of today's microwave systems could not exist. The ultimate goal is to create a range of solid-state components whose use would extend up to the edge of the infrared region.

The study of modern microwave devices is an enormous field in itself; in fact, a number of books have been written on this subject alone. To keep this chapter within reasonable bounds, we will study the basic principles behind each device, then discuss its application, and finally indicate the possible research breakthroughs which are anticipated in the reasonably near future. In particular, we shall refer to such parameters as operating frequency range, bandwidth, gain, microwave power output capability, and noise figure (Section 8–8).

A "family tree" of microwave semiconductor devices is shown in Figure 6–1 and the topics for discussion may be conveniently grouped as follows:

6–2 MICROWAVE BIPOLAR TRANSISTORS

To review some basic principles, the action of bipolar transistors involves two sets of charge carriers; these are the positive charge carriers (holes) and the negative charge carriers (electrons). Large numbers of these carriers are created by doping pure germanium or silicon with certain impurities. However, in their intrinsic forms these semiconductors have only four valence electrons which are used in covalent bonding.

If the doping atoms are pentavalent, such as arsenic or phosphorus, four of the valence electrons are used in covalent bonding between neighboring atoms and the additional negative electron is the majority charge carrier, which is available for conduction purposes; such material, although electrically neutral, is referred to as *n* type. By contrast, with a trivalent impurity (indium, gallium), there is a gap or hole in the covalent bonding. Such holes attract electrons and therefore behave as positive majority charge carriers in the so-called *p*-type material.

A semiconductor can be doped to produce either *npn* or *pnp* crystals from which the bipolar transistors are manufactured. The three doped regions are referred to as the emitter, the base, and the collector. The emitter is heavily doped to produce a large number of majority carriers, the base is lightly doped and is very thin, while the collector is relatively lightly doped and is the largest of the three regions since it is required to dissipate the greatest amount of heat.

Figure 6–2a represents a *pnp* transistor which possesses an emitter–base junction and a collector–base junction. When a junction is formed, some electrons diffuse across from the *n* region to fill holes that exist in the *p* region. This has two effects:

1. In the vicinity of the junction there is a region that has been emptied of majority charge carriers. This is called the depletion layer, which contains negative ions on the *p* side and positive ions on the *n* side.

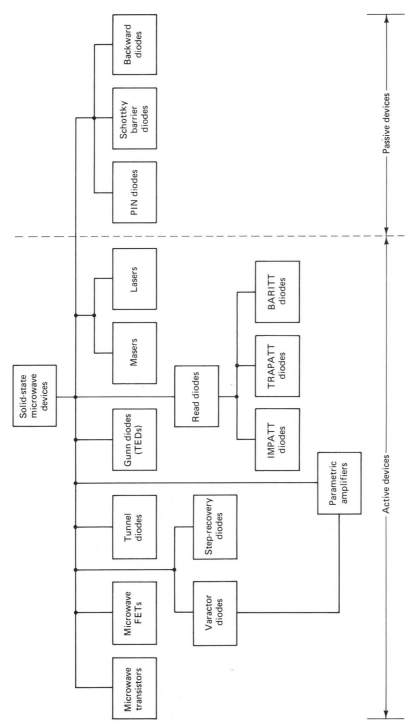

Figure 6-1 "Family tree" of microwave solid-state devices.

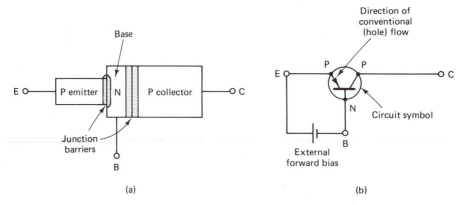

(a)   (b)

Figure 6–2 The *pnp* transistor.

2. Due to the presence of the charged ions, there is a potential barrier across the junction. At 25°C the potential barriers are 0.7 V and 0.3 V for silicon and germanium junctions, respectively.

Figure 6–2b shows an external voltage which is applied to the emitter–base *p-n* junction. The positive terminal is connected to the *p*-type emitter region while the negative terminal is joined to the *n*-type base region. Such an arrangement will oppose the potential barrier and is referred to as forward bias. Under these conditions a large emitter current will flow in the emitter–base circuit. By contrast, if we interchange the terminals of the external dc voltage to create a reverse bias, the potential barrier is reinforced, the depletion layer is increased, and the amount of the emitter current is extremely low. The fact that the current is not zero is due to the presence of minority charge carriers (electrons in the *p* region, holes in the *n* region), which are the result of thermal energy. It is clear that the *p-n* junction forms the basis for the semiconductor diode, since under forward-biased conditions its equivalent resistance is very low, but when reverse biased, the resistance is extremely high.

When the bipolar transistor is being operated under class A conditions, the emitter–base junction is forward biased while the collector–base junction is reverse biased. The transistor may therefore be regarded as equivalent to two diodes placed back to back. Referring to the *npn* transistor circuit of Figure 6–3, many of the

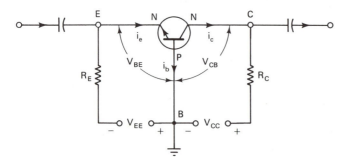

Figure 6–3 The *npn* transistor amplifier common-base arrangement.

conduction band electrons (majority charge carriers) in the emitter cross into the base under the influence of the forward bias; this action establishes the current i_e in the external emitter–base circuit. Since the base is so thin, a small number of the conduction electrons (less than 5%) will recombine with the base holes to form an external current i_b in the base lead. However, the majority of the electrons will drift across the base region into the depletion layer of the collector–base junction. The electrons will then be propelled into the collector region and will ultimately flow in the external circuit to form the collector current i_c. In this way the emitter current i_e directly controls the collector current i_c. Using Kirchhoff's current law gives

$$i_e = i_c + i_b \qquad (6\text{–}1)$$

In the common-base configuration the current gain h_{fb} (or α) is given by

$$h_{fb} = \frac{i_c}{i_e} \qquad (6\text{–}2)$$

and has a value which is slightly less than unity. In the common-emitter configuration, the current gain h_{fe} (or β) is

$$h_{fe} = \frac{i_c}{i_b} \qquad (6\text{–}3)$$

From equations (6–1), (6–2), and (6–3), it follows that

$$h_{fe} = \frac{h_{fb}}{1 - h_{fb}} \qquad (6\text{–}4)$$

So far we have discussed only the general operation of the bipolar transistor. But what particular difficulties do we encounter at microwave frequencies? Not surprisingly, the problems can be compared with those which occurred with the design of the lighthouse triode (Section 1–5):

1. The p and n regions act as conducting surfaces, while the depletion layer in the vicinity of the junction provides some degree of insulation. Consequently, there exist junction capacitances which limit the gain available at high frequencies. However, the values of these capacitances are not constant but depend on the depletion layer's width, which in turn is governed by the amount of bias applied to the junction. In addition, the junction capacitances provide feedback paths, but their effect is kept as low as possible by using the common-base arrangement, whose equivalent circuit at microwave frequencies is shown in Figure 6–4.

2. Lead inductances must be made as small as possible so that the signal loss across such inductances is very low. Since transistors are physically small, they can be packaged so that the lead inductance effect is reduced to a minimum.

3. The ultimate frequency limitation on a microwave bipolar transistor is the transit time taken for the majority charge carriers to cross from the emitter to the

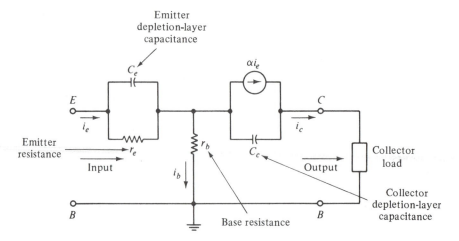

Figure 6-4 Equivalent circuit of microwave transistor. Common-base arrangement.

collector. Due to the limited mobility of the ions in the doped regions, the charge carriers have a lower velocity than the speed of the electrons moving through a tube. The total transit time is found by combining the following:

(a) The time T_1 for the majority carriers to move through the emitter region and cross the forward-biased emitter–base junction.

(b) The drift time T_2 through the base region. This is determined primarily by the degree of doping and the transistor's construction, which is aimed at making the base region as thin as possible.

(c) The time T_3 taken for the charges to move through the depletion layer of the reverse biased collector/base junction. This time is determined primarily by the width of the depletion layer and it can therefore be reduced by using a higher collector voltage. However, if too large a voltage is applied, the power dissipation would inevitably increase and damage might result. For this reason it is a common procedure to operate a microwave power transistor under class C conditions. Care must then be taken to avoid avalanche breakdown under the action of the reverse bias.

(d) The time T_4 for the majority carriers to cross the collector region.

The total transit time is the result of adding together T_1, T_2, T_3, and T_4; of these, T_2 and T_3 are the major contributors.

As the frequency is increased for an RF transistor, the current gains h_{fb} and h_{fe} become complex in nature and fall in value. There is ultimately a frequency f_T above which the current gain h_{fe} drops below unity. The value of f_T therefore reflects the high-frequency performance of the transistor and may be extended if the transit time is reduced by increasing the collector–emitter voltage V_{CE} to about 25 V, with a corresponding emitter current of 20 mA.

Summarizing, to optimize the high-frequency performance of a microwave tran-

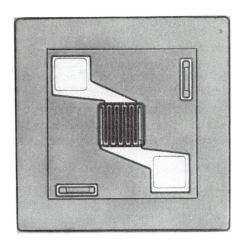

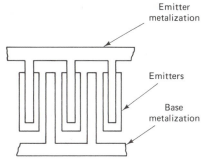

Figure 6–5 A 4-GHz silicon bipolar transistor. The five emitter stripes are each 0.5 μm wide and are separated by 5 μm. (Courtesy of Hewlett Packard Company.)

sistor, we need to cut transit time to a minimum by using narrow p and n regions and high values of V_{CE} and I_C. In addition, the values of the junction capacitances must be reduced as much as possible.

The practical solution is to use a very small emitter area and an extremely thin base; these construction features will both assist in reducing the transit time. In addition, the junction edges of both the emitter and the base must be large to provide a high current capability. The diffused epitaxial planar silicon transistor exhibits the best performance at microwave frequencies. The emitter and the base normally have interdigital structures (Figure 6–5) which are locked together. The total area of the chip so formed is about 5000 μm² and its typical gain–frequency response is illustrated in Figure 6–6.

Further development of the bipolar transistor for the higher microwave frequencies was preempted by the introduction of the gallium arsenide (GaAs) field-effect transistor, with its greater ion mobility. As a result, bipolar transistors are limited to about 5 GHz with a noise figure of about 3 dB and a power output of 1 W.

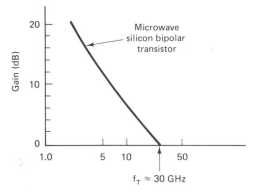

Figure 6–6 Maximum available gain from a microwave silicon bipolar transistor.

6–3 FIELD-EFFECT UNIPOLAR TRANSISTORS

Junction Field-Effect Transistors

The action of the *pnp* bipolar transistors involved two types of charge carriers: the electron and the hole. The FET is a unipolar transistor since its operation requires only one charge carrier, which may be either the electron or the hole. The two types of FET that we will discuss are the junction field-effect transistor (JFET) and the metal–semiconductor field-effect transistor (MESFET); the latter is sometimes referred to as the Schottky barrier FET.

The JFET is essentially a doped silicon bar which is regarded as a channel and behaves as a resistor. The doping may either be *p* type or *n* type, thereby creating either a *p*-channel or an *n*-channel JFET. At the ends of the channel are two terminals, referred to as the source and the drain. When a particular drain–source voltage (V_{DS}) is applied between the end terminals, the amount of current flow (I_D) between the source and the drain depends on the channel's resistance. The value of this resistance is controlled by a gate, which may consist of either two *n*-type regions diffused into a *p*-type channel or two *p*-type regions diffused into an *n*-type channel. In either case the two regions are commonly joined to provide a single gate. Cutaway view of both types with their schematic symbols are shown in Figure 6–7. In the

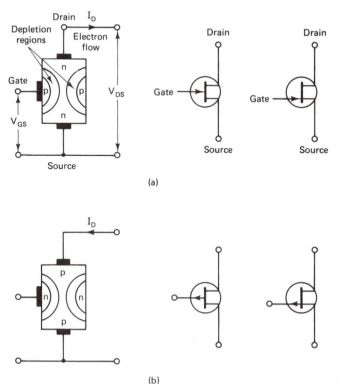

Figure 6–7 *n*-Channel and *p*-channel JFETs with their symbols.

schematic symbols the vertical line may be regarded as the channel; the arrow then points toward an n channel but away from a p channel. The gate line may either be symmetrically positioned with respect to the source and the drain, or drawn closer to the source.

If a reverse-biased voltage is applied between the gate and the source of an n-channel JFET, depletion layers will surround the two p regions that form the gate. If the reverse bias is increased, the depletion layers will spread more deeply into the channel until they almost touch. The channel's resistance will then be extremely high, so that the corresponding I_D is very small.

The reverse biasing of the gate–source junction may be compared with applying a negative voltage to a triode's grid relative to its cathode. Like the tube the FET is a voltage-controlled device in the sense that only the input voltage to the gate controls the output drain current. This is in contrast with the bipolar transistor, where the base–emitter junction is forward biased; the input voltage then controls the input current, which in turn determines the output current.

With reverse biasing of the gate–source junction very little gate current flows. This means that the input impedance to a JFET is of the order of several megohms; this is a definite advantage of the FET over the bipolar transistor, whose input impedance is relatively low. However, compared with the JFET, the output current of a bipolar transistor is much more sensitive to changes in the input voltage; the result is a lower voltage gain available from the JFET.

JFET drain and transconductance curves.   In Figure 6–8a the voltage (V_{DS}) between the source and the drain of the n-channel JFET is gradually increased from zero. At the same time the voltage between gate and source, $V_{GS} = 0$ V; this is referred to as the shorted gate condition. Initially, the available channel is broad, so that the drain current I_D is directly proportional to V_{DS} and rises rapidly as V_{DS} is increased. However, the drain voltage creates a reverse bias on the junction between the channel and the gate. The increase in V_{DS} causes the two depletion regions to widen until finally they almost come into contact. This occurs when V_{DS} equals a value called the pinch-off voltage V_p (Figure 6–8b); the available channel is then very narrow, so that the drain current is limited (or pinched off). Further raising of V_{DS} above the pinch-off point will produce only a small increase in the drain current. This relationship between I_D and V_{DS} is measured by the ac drain-to-source resistance

$$r_{ds} = \frac{\Delta V_{DS}}{\Delta I_D} \qquad (6\text{–}5)$$

No change in the situation occurs until the drain voltage equals $V_{DS\,(max)}$; at this point an avalanche effect takes place and the JFET breaks down. Over the operating range between V_p and $V_{DS\,(max)}$, the approximately constant value of the drain current with the shorted gate is referred to as I_{DSS} (drain-to-source current with shorted gate).

For each different negative value of V_{GS} a different drain current curve can

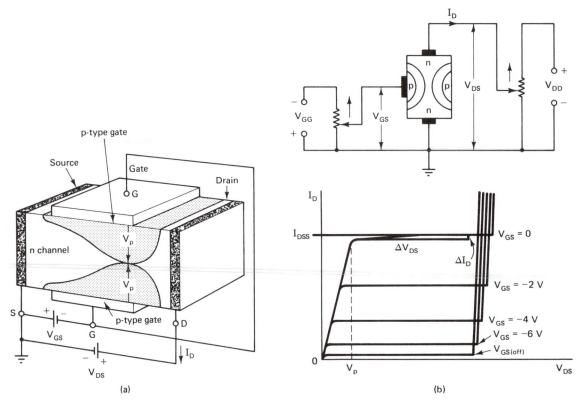

(a)   (b)

Figure 6–8   Drain characteristics of an n-channel JFET.

be obtained. This family of curves is illustrated in Figure 6–8b. Ultimately, V_{GS} can be sufficiently negative so that the drain current is virtually cut off and equal to zero; this value of the gate–source voltage is therefore referred to as $V_{GS\,(\text{off})}$.

At the cutoff condition the depletion layers nearly touch; this also occurred when V_{DS} was equal to V_p. Therefore, V_{GS} has the same value as V_p, although V_{GS} is a negative voltage while V_p is positive. The transconductance curve is the graph of drain current I_D plotted against the gate-to-source voltage V_{GS} while maintaining the drain-to-source voltage at a constant value. For example, in Figure 6–9a, the drain voltage is set to 12 V while the gate is initially shorted to the source so that $V_{GS} = 0$. The recorded drain current would then equal the value of I_{DSS}. If the reverse gate voltage is now increased from zero, the drain current will fall until ultimately cutoff is reached when $V_{GS} = V_{GS\,(\text{off})}$.

The shape of the transconductance curve is considered to be a parabola (Figure 6–9b), so that there is a mathematical relationship between I_D and V_{GS}:

$$I_D = I_{DSS}\left[1 - \frac{V_{GS}}{V_{GS\,(\text{off})}}\right]^2 \tag{6–6}$$

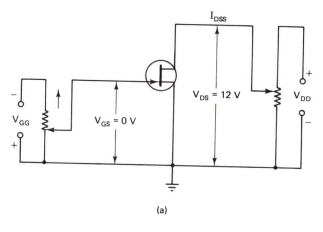

(a)

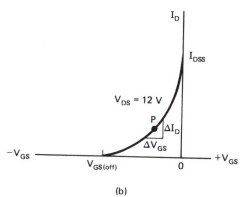

(b)

Figure 6–9   JFET transconductance curve.

In this equation both V_{GS} and $V_{GS\,(off)}$ are regarded as negative voltages.

In a FET amplifier the control which the gate voltage exercises over the drain current is measured by the transconductance g_m. At point P on the curve the transconductance is defined by

$$g_m = \frac{\Delta I_D}{\Delta V_{GS}}$$

(6–7)

and is normally measured in microsiemens (μS). By differentiating the expression for I_D with respect to V_{GS},

$$g_m = g_{mo}\left[1 - \frac{V_{GS}}{V_{GS\,(off)}}\right]$$

(6–8)

where g_{mo} is the transconductance for the shorted gate condition. The value of g_{mo} is $-2I_{DSS}/V_{GS\,(off)}$; since $V_{GS\,(off)}$ is a negative voltage, g_{mo} is a positive quantity.

Metal–Semiconductor Field-Effect Transistors

The most popular field-effect transistor for microwave frequencies has a Schottky barrier (metal-semiconductor) gate and is commonly referred to as a MESFET. With this arrangement the potential barrier is created at the metal–semiconductor interface by establishing stable charges in the doped material, thereby avoiding the need for a chemical junction. The resulting drain characteristics are illustrated in Figure 6–10.

In its construction the MESFET uses a GaAs mesa epitaxial film which is about 0.25 μm thick and is deposited on a GaAs semi-insulating substrate (Figure 6–11). The GaAs n channel is doped with either tin or sulfur and the Schottky barrier gate is evaporated aluminum. The source and drain contacts are alloys of Au/Te, Au/Ge, or Au/Te/Ge.

At microwave frequencies the n-type channel length is extremely short and the charge carriers must reach their ultimate velocity before arriving at the pinched region. The transistor's performance then depends on its parameters, which are displayed in the FET's cross section and its associated equivalent circuit (Figure 6–12).

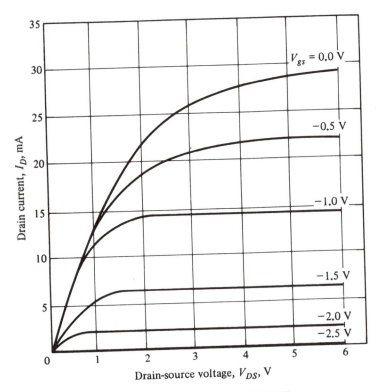

Figure 6–10   Drain curves of a MESFET.

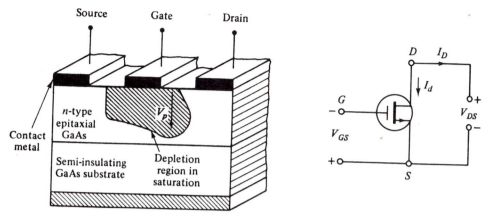

Figure 6–11 Construction and circuit symbol of a Schottky barrier **MESFET**.

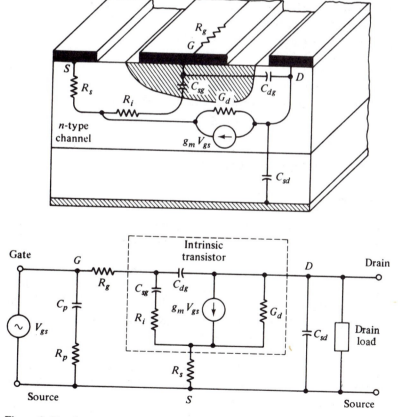

Figure 6–12 Cross section of a **MESFET** and its equivalent circuit. g_m, FET transconductance; G_d, drain conductance; R_i, input resistance; C_{sg}, source–gate capacitance; C_{dg}, drain–gate capacitance; R_g, gate metallization resistance; R_s, source–gate resistance; C_{sd}, source–drain capacitance; R_p, gate bonding-pad parasitic resistance; C_p, gate bonding-pad parasitic capacitance.

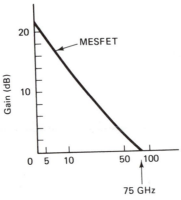

Figure 6–13   MESFET's gain–frequency response.

As a result of their short transit times, MESFETs can function at frequencies exceeding 30 GHz, but their normal operation is around 5 to 20 GHz. However, it is anticipated that their frequency range will ultimately be extended to 100 GHz (see the gain–frequency response of Figure 6–13).

MESFETs have an extremely high input resistance with a low noise figure, such as 1 dB at 5 GHz and 2 dB at 20 GHz. Their power capabilities range from several watts at 5 GHz to a few hundred milliwatts at 20 GHz. Finally, compared with bipolar power transistors, the individual GaAs FET can operate over much wider ranges of frequency and power.

6–4 MICROWAVE INTEGRATED CIRCUITS

In Section 3–7 we referred to the use of microstrip lines as the connectors for microwave integrated circuits (MICs). We are now going to discuss the actual integrated circuits themselves, whose commercial development has occurred since 1970. It should be immediately emphasized that MICs are quite different from conventional ICs. In particular, conventional ICs are manufactured with high packing densities which are not needed at microwave frequencies since these circuits do not contain arrays of identical devices. Consequently, the packing density for MICs is relatively low.

The first type of MIC was the hybrid variety, which was fabricated on a high-quality ceramic, glass, or ferrite substrate. The passive circuit elements were deposited on the substrate and then the active devices such as chips or transistors were mounted on the substrate and either soldered or bonded to the passive circuitry. An example of a passive element structure is illustrated in Figure 6–14a and its equivalent circuit is shown in Figure 6–14b. In general, the passive elements may be formed from either distributed microwave lines or lumped components (capacitors, inductors, resistors). In addition, we may use thick- or thin-film techniques. With thin films the conducting material for the resistor or the nonconducting dielectric for a capacitor is vacuum deposited on a substrate of ceramic or glass. By comparison, the pattern

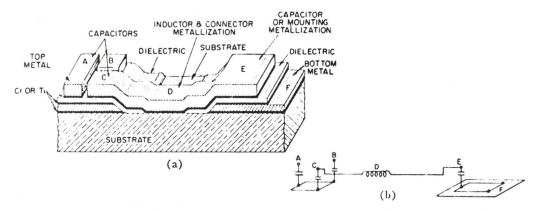

Figure 6–14 Passive element microwave integrated circuit. (Caulton, Martin, et al.:
"Status of lumped elements in microwave integrated circuits—present and future."
IEEE Trans. on Microwave Theory and Techniques, MTT-19, No. 7, 1971. © 1971
IEEE.)

of films whose thickness exceeds several thousand angstrom units is usually defined
by silk screening.

For the manufacture of an MIC we need four basic materials to act as the
resistive films, the dielectric films, the conducting surfaces, and the substrate. Examples
of these materials are:

1. *Resistive films*: chromium, nickel/chromium, tantalum, titanium
2. *Dielectric films*: silicon oxide, silicon dioxide
3. *Conductor materials*: silver, copper, gold, aluminum
4. *Substrate materials*: gallium arsenide, sapphire, alumina, beryllia

In the fabrication process the basic techniques used are those of epitaxy, diffusion,
deposition, and etching.

A typical hybrid MIC with an external volume of only 3 cm³ and hermetically
sealed to contain dry nitrogen might consist of a two-stage GaAs FET amplifier
which operates over a frequency range of 5 to 20 GHz. The input power would be
of the order of 1 mW with a gain of 10 dB and a noise figure of less than 8 dB.

Only the hybrid type was available until 1980, at which time the monolithic
MIC made its appearance. In this form of integrated circuit only a single wafer is
used and the active devices are grown in or on a GaAs semiconducting substrate.
This is achieved by injection or doping of the substrate to produce the required
unipolar transistors and diodes. The passive elements are either grown in the substrates
or deposited as films. Initially, although the monolithic MIC was physically much
smaller than its hybrid counterpart, its processing difficulties, low output power,
and poor overall performance severely limited its application. However, these difficulties have been largely overcome, and an existing monolithic GaAs FET power amplifier
with a volume of about 1 cm³ has a gain of 30 dB at 8 GHz with an output power

of 1200 mW over the frequency range 7.0 to 9.0 GHz; its efficiency is on the order of 25 to 30%.

Compared with discrete circuits, MICs provide the following advantages:

1. Small size and light weight
2. Low cost
3. High degree of reliability
4. Better overall performance

MICs have largely been successful because of the recent developments in microwave solid-state devices. As a result, MICs are used extensively in space and military equipment because they meet the specifications for severe vibrations, extreme shock, and wide temperature variations.

6–5 VARACTOR AND STEP-RECOVERY DIODES

Varactor Diodes

In Section 6–2 we discussed the capacitance that exists at a *p-n* junction due to the presence of the depletion layer. If a diode is reverse biased and the amount of the bias is increased, the depletion layer will widen and the junction capacitance will decrease in accordance with the relationship

$$C_T = \frac{\epsilon A}{W_d} \tag{6–9}$$

where

$$\epsilon = \text{permittivity of the depletion layer}$$
$$A = \text{effective area of the } pn \text{ junction (m}^2)$$
$$W_d = \text{width of the depletion layer (m)}$$
$$C_T = \text{junction, depletion region, or transition capacitance (F)}$$

The semiconductor devices which are used to behave as voltage-controlled variable capacitances are referred to as varactor or varicap diodes. The characteristic and constructions of typical varactor diodes with their accepted symbols are illustrated in Figure 6–15. The range of the reverse bias V_r is limited between zero and the point at which avalanche breakdown occurs. For a particular value of V_r, the expression for the capacitance C_T is

$$C_T = \frac{C_0}{(1 + V_r/V_0)^n} \tag{6–10}$$

where

$$C_0 = \text{value of capacitance corresponding to zero bias (pF)}$$
$$V_0 = \text{value of the knee voltage (0.7 V for silicon)}$$

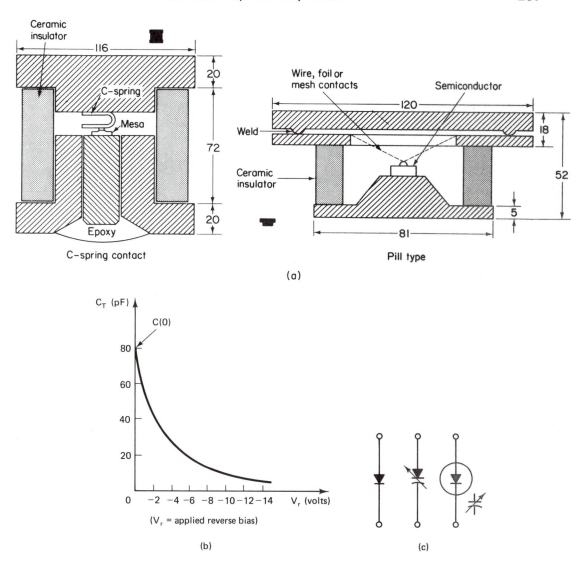

Figure 6-15 Construction, characteristic, and symbols of the varactor diode. [In part (a), dimensions are given in thousandths of an inch.]

$n =$ an exponent whose values are 0.50 and 0.33 for alloy and diffused p-n junctions, respectively

As a result of the exponent factor, there is a sharp initial drop in the value of C_T and the characteristic of C_T versus V_r has a marked curvature. It is this degree of curvature which allows the varactor diode to be used for frequency multiplication in the microwave region.

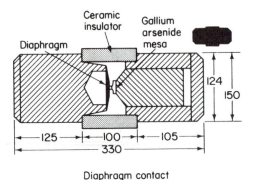

Diaphragm contact

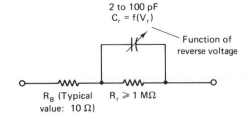

Figure 6–16 Construction of a GaAs varactor diode. (Dimensions are given in thousandths of an inch.)

Figure 6–17 Equivalent circuit of a varactor diode under reverse bias conditions.

The original microwave varactor diodes were of the silicon-diffused junction mesa type. However, silicon has largely been replaced by gallium arsenide with its higher mobility of charge carriers. As a result, GaAs diodes can operate at frequencies of the order of hundreds of gigahertz and at extremely low temperatures which are only a few degrees above absolute zero (such low temperatures can be required for parametric amplifiers, as described in Section 6–6). The construction of a typical GaAs varactor diode is illustrated in Figure 6–16.

In microwave bands we must consider the effects of lead inductance and stray capacitance between the anode and cathode sections. These factors are included in the varactor diode's equivalent circuit of Figure 6–17. Typical values are included for the equivalent reverse resistance R_r and the diode's bulk resistance R_B.

The diode must have a wide capacitance variation, so that the range of C_T $[C_{T\,(max)} : C_{T\,(min)}]$ is typically 5:1 or more. With regard to frequency, the cutoff value $f_c = \dfrac{1}{2\pi R_B C_{T\,(min)}}$, which for GaAs varactor diodes can exceed 500 GHz. However, as a result of skin effect, R_B increases as the frequency is raised, so that varactors are not normally operated at frequencies exceeding $0.2f_c$.

Varactor diodes represent a mature technology and were originally developed in the 1950s as simple voltage-variable capacitances (VVCs) and were subsequently used for the direct frequency modulation of oscillator stages and as the control devices for AFC circuits. However, for microwave purposes their main applications lies in frequency multiplier chains and parametric amplifiers.

Step-Recovery Diodes

The step-recovery diode is also known as a snap-off varactor and consists of a graded *p-n* junction of gallium arsenide or silicon. During a portion of the positive input signal cycle, the diode is forward biased and the result is a high current of short duration which stores charge in the epitaxial diffused junction. If a suitable forward bias is applied, the diode is designed to conduct heavily for a short period of time and the related charge is then stored in its depletion layer. Subsequently, when the bias is reversed, the diode rapidly discharges, to create a sharp output pulse whose duration is of the order of only a few hundred picoseconds or less. Such a pulse will naturally be extremely rich in harmonics, but to achieve efficient frequency multiplication, the duration or snap time of the pulse must be much shorter than the period of the output frequency. As an example, if the output frequency is 10 GHz, the snap time must be considerably less than $10^{12}/(10 \times 10^9)$ ps = 100 ps. This factor limits the output of step-recovery diodes to below 20 GHz.

Figure 6-18 shows a typical frequency multiplier circuit employing a step-recovery diode. During a part of the positive half-cycle the diode conducts and a reverse bias is created through the time-constant circuit which contains the resistor R_1; at the same time the junction's depletion layer is charged. Then as soon as the input voltage reaches the point where reverse bias is applied to the diode, the discharge takes place and the short-duration pulse is created. The output circuitry then contains

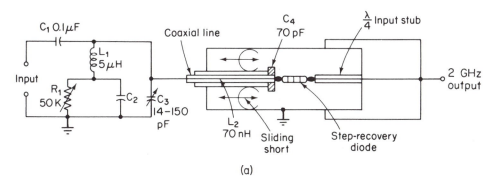

(a)

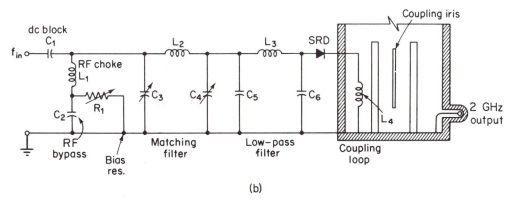

(b)

Figure 6-18   Frequency multiplier circuit employing a step-recovery diode: (a) assembly; (b) equivalent circuit.

a cavity which is tuned to the required harmonic and therefore discriminates against the unwanted frequencies. Theoretically, any harmonic could be selected, but in practice the range extends from the second to the thirtieth; this limitation is the result of the power efficiency

$$\frac{\text{harmonic output power}}{\text{signal input power}} \times 100\%$$

and the harmonic frequency. For example, a tripler with an output frequency of 1 GHz may have an efficiency of 85%; by comparison, if a fifth harmonic of 10 GHz is selected, the efficiency can be as low as 15%. Because of these considerations, step-recovery diodes are not normally used at frequencies above 20 GHz.

Comparison of Varactor and Step-Recovery Diodes

The following is a summary of the comparison between step-recovery and varactor diodes:

1. *Frequency range*. Step-recovery diodes can operate up to 20 GHz, but varactor diodes can be successfully used at frequencies exceeding 100 GHz.
2. *Microwave power output*. Step-recovery diodes are capable of power outputs that exceed 1 W at 10 GHz, 8 W at 3 GHz, and 50 W at 0.25 GHz. Varactor diodes have maximum power outputs of 8 W at 3 GHz, 250 mW at 50 GHz, and 30 mW at 100 GHz. Either step-recovery or varactor diodes may be paralleled to extend the power output capability.
3. *Power efficiency*. A step-recovery diode tripler stage at 1 GHz has an efficiency of only 75% compared with 85% for a varactor diode. However, for the higher harmonics the snap-off multipliers are much more efficient. With GaAs varactors it is possible to achieve efficiencies of 35% at 40 GHz and 15% at 100 GHz.
4. *Bandwidth*. For multipliers that are required to operate over a band of frequencies (from 90% to 110% of the center frequency) varactor diodes are superior.
5. *Harmonic content*. Compared with varicaps, step-recovery diodes can produce a far greater number of significant harmonics. If the final tuned circuit is eliminated from a step-recovery multiplier, the multiple harmonics can appear in the output of a so-called "comb" generator.

Frequency Multiplication

We have already discussed the mechanism by which a step-recovery diode creates harmonics; now we need to know how a varactor diode achieves the same result. In Figure 6–15b we observe that the characteristic of C_T versus V_R is curved so that both the diode's capacitance and its capacitive reactance are nonlinear. Consequently, when an ac voltage at a particular frequency is applied across a reverse-biased varactor diode, the resulting current waveform will be distorted and will therefore contain harmonics. Moreover, since this nonlinear device behaves primarily as a reactance, high values of efficiency are possible. The basic circuit for a varactor diode tripler stage is shown in Figure 6–19. The resistor R_1 forms part of a time-constant circuit to generate a reverse bias whose value is determined by the amplitude of the input signal. The input filter L_1C_1 is used to prevent unwanted frequencies from being applied to the diode, while the filter L_3C_3 eliminates unwanted harmonics from the output. The "idler" circuit L_2C_2 is used to eliminate the result of heterodyning between the input and output frequencies.

We have learned that step-recovery diodes are capable of high-power outputs at low frequencies (less than 20 GHz), while varactor diodes produce relatively low-power outputs at much higher frequencies. Since each diode stage represents a power loss, the initial crystal oscillator of a frequency multiplier chain (Figure 6–20a) must be operated with a high-power output at a comparatively low frequency; in our example the oscillator (which determines the ultimate frequency stability) contains lumped circuitry with a 20-MHz crystal operating at the seventh overtone. The power output of 40 W is derived from a dc supply which feeds only the oscillator stage. The following two stages contain step-recovery diodes whose frequency multiplication factors and efficiencies are, respectively, ×10, 20% and ×5, 25%. The final stages are both varactor triplers with efficiencies of 50% and 25%, so that the final output power is only 250 mW at a frequency of 63 GHz. The physical construction of a varactor diode multiplier chain is illustrated in Figure 6–20b, where the input frequency is 72 MHz and a frequency multiplication factor of 18 is obtained from two triplers and a doubler. This creates an output frequency of 1.296 GHz.

Modern techniques have tended to supersede the use of frequency multiplier chains. GaAs FETs are capable of higher-power outputs at frequencies up to 20 GHz, while IMPATT and Gunn diode oscillators are superior from 20 GHz to over 100 GHz. However, to achieve a high degree of frequency stability, such oscillators may be controlled by a phase-locked loop which incorporates a low-power frequency multiplier chain.

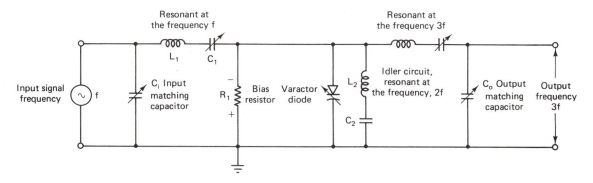

Figure 6–19   Tripler varactor diode multiplier circuit.

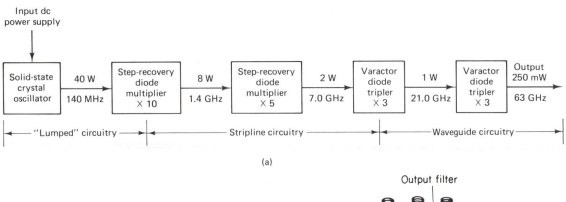

(a)

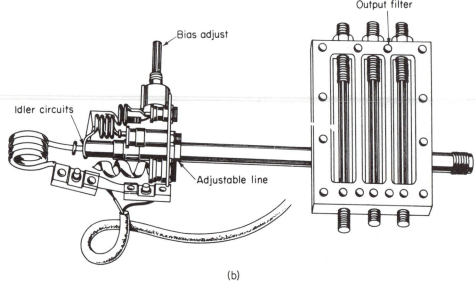

(b)

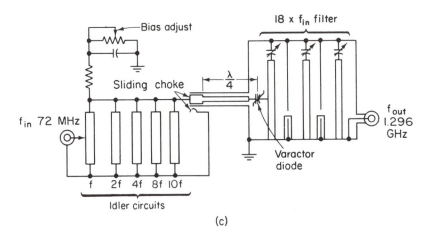

(c)

Figure 6-20 (a) Varactor diode/step-recovery diode frequency multiplier chain; (b) varactor diode multiplier assembly; (c) equivalent circuit.

6–6 PARAMETRIC AMPLIFIERS

A parametric device is one whose operation depends on the periodic variation of one of the device's parameters, such as a capacitance or an inductance. Although the use of low-noise parametric amplifiers occurred in the late 1950s, the idea behind them is quite old and was first proposed in the nineteenth century for mechanical systems. However, such amplifiers were not possible in the practical sense until suitable varactor diodes were developed.

Variable Inductance Devices

The nonlinear *B–H* characteristics of the ferrous metals, which are widely used for the cores of chokes and transformers, result in all iron-cored chokes being variable-inductance devices since the inductance of the choke varies with the current flowing through it. The production of harmonics in such devices is well known and the saturable reactor is widely used as a form of amplifier. As shown in Figure 6–21, a small dc current through an additional winding on the choke can control quite a large alternating current; this result can be used to create a form of amplifier. By using such a variable reactance a power gain is possible, but except with ferrite-cored reactors, the frequency of operation is severely limited by large eddy current and hysteresis losses in the core.

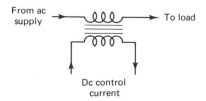

Figure **6–21** Principle of the saturable reactor amplifier.

Variable-Capacitance Devices

Variable-capacitance devices other than mechanically operated ones are a fairly new development arising from rapid advances in solid-state technology. It had been known for some time that the capacitance of a reverse-biased semiconductor diode varied slightly with changes in the bias, but the variation in the capacitance was small. The development of the modern epitaxial diode resulted in a solid-state device in which the variation in the capacitance was considerable. Such diodes have made possible the microwave parametric amplifiers to be described.

Parametric Amplification

In nearly all amplifying devices dc power is converted at a certain efficiency into ac power and the energy exchange takes place through the movement of charge carriers. In a parametric amplifier, the power source is an oscillator and the transfer of energy occurs by means of the changes in the variable parameter. As a result, the output signal power is at a different frequency from that of the power source. This assumes that the power supplied by the input signal is small compared with the power of the output signal.

While a linear amplifier contains signals only at the input frequency, the parametric amplifier always contains signals at frequencies other than that of the input signal. In general, a parametric amplifier contains at least three frequencies. The incoming signal is responsible for providing the *input* frequency, the ac power source is operated at the *pump* frequency and provides the power necessary to achieve amplification, and the amplified power output is either at the input frequency or at the socalled *idler* frequency, which is determined by the input and the pump frequencies. These three frequencies are related by

$$f_p = f_1 + f_2 \qquad (6\text{--}11)$$

where

f_1 = input frequency
f_p = pump frequency
f_2 = idler frequency

If the parametric amplifier is compared with a superheterodyne mixer (which may be thought of as a type of parametric amplifier with a loss rather than a gain), the input, pump, and idler frequencies correspond to the signal, the local oscillator, and the intermediate frequencies.

A simple qualitative explanation of the mechanism by which a parametric amplifier operates can be given by considering an elementary but rather special form in which only two frequencies are present. If the input and idler frequencies of a parametric amplifier are equal, the amplifier is said to be degenerate, and in order to satisfy the equation $f_p = f_1 + f_2$, the pump must operate at twice the signal frequency.

The circuit shown in Figure 6–22 is essentially a lossless resonant circuit in which the separation of the plates of the capacitor can be varied mechanically. If the circuit is lossless and is energized in some way, the voltage across the capacitor plates will be sinusoidal with a constant rms value. This assumes that the separation of the plates is not altered.

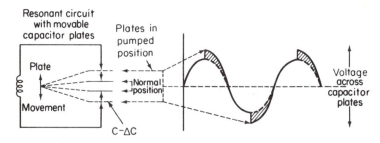

Figure 6–22 Lossless resonant circuit including a mechanically variable capacitor.

We will now consider that the capacitor plates are pulled apart at the peak of each half-cycle, as shown in Figure 6–23 and then pushed back again at the zero-voltage points which occur in the V_C (voltage across the capacitor) waveform. Since the plates are pulled apart while there is a voltage across the capacitor, the mechanical "pump" will have to do work by moving the plates against the attractive effect of the electrostatic field within the capacitor. This mechanical energy can only be converted into electrical energy, as there is no form of energy loss in an ideal capacitor; consequently the energy stored in the capacitor is increased.

At the instant the plates are pulled apart, the electrical charge stored, q, remains the same, but the capacitance C is reduced by increasing the plate separation. The capacitor's voltage V_C must therefore increase to satisfy the relationship $q = CV_C$. As the capacitor plates are always pushed back when the capacitor voltage is zero, no energy is transferred from the pump to the electrical circuit during this operation. If the capacitor is repeatedly "pumped" in this manner, there will be a continuous transfer of energy to the electrical circuit. The voltage across the capacitor and the stored energy in the resonant circuit will increase in steps as shown in Figure 6–23, and the energy at the pump repetition frequency will be converted to energy at the signal frequency, which is one-half of the pump frequency. The additional electrical energy stored is then obtained from the source of mechanical power which moves the plates.

An electrical circuit which is completely isolated from any other circuit is obviously not an amplifier, and to convert the device of Figure 6–22 into an amplifier, several alterations are necessary. To start with, the resonant circuit must possess

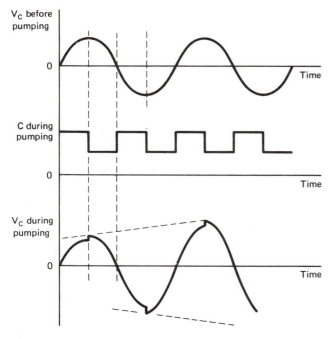

Figure 6–23 Basic principle of the parametric amplifier.

some losses, so that the initial signal will die away unless there is a steady input signal which is amplified to maintain a continuous output. Since the amplified signal is the same frequency as the input signal and appears across the same two terminals, nonreciprocal devices must be used to prevent the output power from returning to the input signal source.

In a practical circuit the mechanical pump is replaced by a varactor diode whose capacitance can be varied by applying a voltage across it. However, at first glance it would appear that the pump source must supply a square-wave voltage. Since it is very difficult indeed to supply a square wave at the high frequencies required, it is fortunate that the system will still amplify if a sinusoidal pump supply is used instead. This is possible since the strongest component in a square waveform is its fundamental sine wave. As discussed previously, the frequency of the sinusoidal pump must be twice that of the input signal and their waveforms must bear the correct phase relationship.

A possible form of microwave parametric degenerate amplifier is shown in Figure 6–24. The amplifier is said to be degenerate because the input and the idler frequencies are the same, and consequently the pump must act at twice the signal frequency in order to satisfy the equation $f_p = f_1 + f_2$. There are certain disadvantages to this arrangement:

1. It is difficult to obtain adequate power sources at microwave frequencies.
2. The pump and input signals must have the phase relationship which is illustrated in Figure 6–23. This requirement is not capable of being satisfied at microwave frequencies.

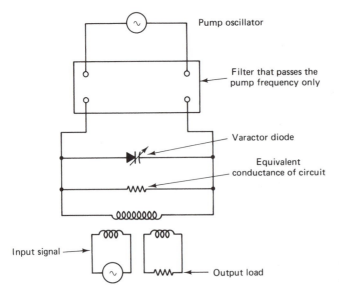

Figure 6–24   Basic principle of the degenerate parametric amplifier.

Three-Frequency Parametric Amplifiers

In this circuit the only frequency relationship that must be obeyed is

$$f_p = f_2 = f_1 + f_3$$

where

$$f_p = f_2 = \text{pump frequency}$$
$$f_1 = \text{signal frequency}$$
$$f_3 = \text{idler frequency}$$

Under these conditions the idler frequency may be greater or less than the input frequency. This device can therefore be an up converter or a down converter.

The basic arrangement of the amplifier consists of two resonant circuits, one resonant at the input signal frequency and the other at the idler frequency, with the two coupled together by a variable parameter element. In virtually all practical arrangements the pumping oscillator will also involve the use of a circuit resonant at the pump frequency so that there will be three resonant circuits in all. A possible schematic is shown in Figure 6–25.

It is sometimes convenient to represent the action of the variable reactance, pump, and idler circuits as a negative conductance. In the absence of any output load on the parametric amplifier, the system will be unstable if the total equivalent negative conductance of the idler circuit and the variable capacitance is greater than the positive conductance of the input circuit. The magnitude of the equivalent negative conductance is controlled by the size of the capacitance variation, which in turn is controlled by the pump power. When the device is used as an amplifier, a load resistance will be added to the idler circuit. This will cause the total conductance of the system to be positive again unless the pump power is increased to raise once more the magnitude of the capacitive variation.

It is apparent that one of the disadvantages of the parametric amplifier is its instability. For any given resistive load the pump power must be adjusted until the

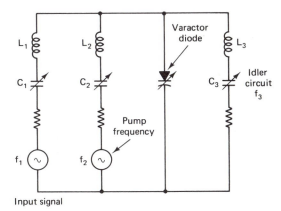

Figure 6–25 Nondegenerate three-frequency parametric amplifier.

total shunt conductance of the system is small but positive for any given signal. The input conductance and its related power are quite small, while the output conductance may be relatively large. The gain in the energy occurs because of the power extracted from the pump source by means of the variable parameter.

In the nondegenerate three-frequency amplifier, the signals that are increased, occur in the idler and input circuits, so that once again there must be a phase coherence between the frequencies. However, the output power in the idler circuit is not supplied by the input signal power alone; it is, in fact, generated by the combined action of the input signal and the pump. Consequently, the idler circuit always starts its oscillation at the correct phase for amplification since it extracts energy from the pump rather than forces energy back into it. Since this adjustment of the idler signal phase occurs automatically, phase coherence between the pump's output and the input signal is not necessary.

The difficulties arising from the direct connection between the input and the output signals can easily be overcome by using either an isolator, in which case the amplifier is known as a "transmission type" (Figure 6–26a), or a circulator, where the amplifier is termed a "reflector type" (Figure 6–26b).

In a parametric amplifier a Gunn diode oscillator is commonly used as the pump source and the varactor diode is manufactured from gallium arsenide. An example of a practical parametric amplifier is illustrated in Figure 6–27.

Practical Limitations of Parametric Amplifiers

Frequency.   The upper frequency limit is set by the frequency at which the variable capacitance can be pumped and the difficulty of obtaining a source power at the pump frequency. Practical amplifiers have been developed up to about 40 GHz, while experimental amplifiers have operated as high as 200 GHz.

The lower frequency limit is set by the isolators and circulators that have to be used. This limitation does not apply to up and down converters (where the input and the output frequencies are different) and such devices have been made to operate with quite low input frequencies.

Bandwidth.   The bandwidth of the simple parametric amplifier is usually quite small, as all three frequencies are tuned by resonant circuits or cavities. It is, however, possible to increase the bandwidth by stagger tuning so that bandwidths of the order of 25 MHz at 9 GHz are quite common; however, this is less than 0.25% of the center frequency. As in most amplifying devices, the "gain × bandwidth" product is a constant for a given circuit so that the bandwidth can always be increased at the expense of the gain. This principle is exploited in the form of a broadband parametric amplifier which can be designed with a bandwidth of more than 25% of the center frequency. The operation of this circuit is based on the voltage wave which travels along a matched artificial line (Section 2–5). To achieve amplification the line is formed from series lumped inductances and shunt varactor diodes (Figure

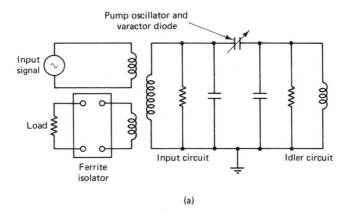

(a)

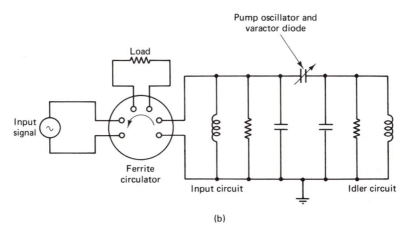

(b)

Figure 6–26 (a) Transmission-type and (b) reflector-type parametric amplifiers.

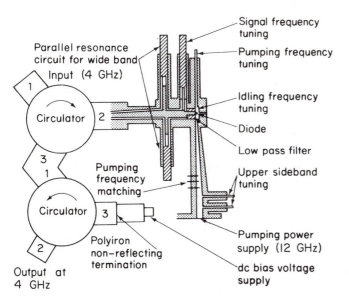

Figure 6–27 Practical 4-GHz parametric amplifier.

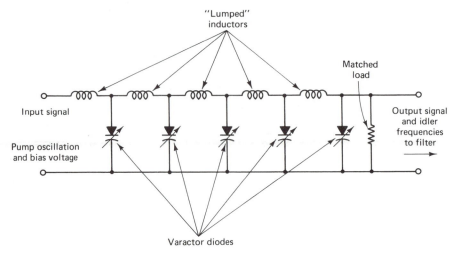

Figure 6-28   Traveling-wave parametric amplifier.

6-28). The input signal and pump frequencies are then applied at the beginning of the line, while outputs at the signal and idler frequencies appear at the line's termination. As the signal and the idler voltages travel down the line, they are amplified at the expense of the power drawn from the pump. It would appear that the longer the line, the greater would be the gain but, in fact, the losses on the line require the length to be optimized.

The terminating filter will accept the input frequency (f_s), idler (f_i), and pump (f_p) frequencies but must reject the frequency $f_s + f_p$. If the filter is correctly loaded at the desired output frequency, there will be no reflection back to the input end of the line, so that the input and output signals will be isolated and no circulator is therefore required.

Gain.   The limit of the gain is set by the stabilities of the pump source and the variable reactor element. If the amplifier is operated at high gain, instability is liable to occur. A gain on the order of 20 dB is a practical figure and this has been increased experimentally to 80 dB by using a form of regeneration.

Noise.   Since capacitance reactance is involved in the action of a parametric amplifier, such stages have a very low noise figure. The level of noise from most parametric amplifiers is so low that accurate measurements are very difficult to obtain. Most good parametric amplifiers have a measured noise figure between 1 and 2 dB. The small amount of noise that does exist in a parametric amplifier is due to the varactor diode's spreading resistance and the resistance associated with the isolator or the circulator. Since noise is a function of temperature, it may therefore be reduced by cooling the complete amplifier arrangement; the cooling is, of course, not essential but is an attempt to improve the amplifier's performance.

Various cooling methods can be used, including cryogenic equipment which is capable of lowering the temperature to a few degrees above absolute zero. However, a cryogenic system is large and expensive and does not provide the anticipated improvement in performance, possibly because the spreading resistance may increase as the temperature is lowered. The low noise figure is probably the greatest advantage of the parametric amplifier when used in space communication systems, radiotelescopes, and tropospheric scatter receivers.

6–7 TUNNEL DIODES

We have already seen that the action of the parametric amplifier can be explained in terms of negative conductance or its reciprocal, negative resistance. When the tetrode tube was developed in the 1920s, it was found to exhibit a *dynamic* negative resistance over part of its plate voltage versus plate current characteristic. In this case the term "dynamic negative resistance" means that over a (small) portion of the characteristic, a *decrease* of the plate current (ΔI_b) is accompanied by an *increase* in the plate voltage (ΔE_b). The value r of the dynamic negative resistance is then given by

$$r = -\frac{\Delta E_b}{\Delta I_b} \tag{6–12}$$

The "negative resistance" tetrode was connected across a tank circuit and the result was the so-called "dynatron" oscillator.

The tunnel diode was the first semiconductor device which was shown to possess a dynamic negative resistance region. It was developed by Leo Esaki in 1958 and was connected across a tank circuit to create an oscillator with a low-power output at very high frequencies. However, in the early days the generated frequency was found to be extremely unstable, so that little use was made of tunnel diodes. This problem was eventually overcome, so that tunnel diode amplifiers and oscillators are now commonly found.

Although the tunnel diode uses a *p-n* junction, its manufacture differs from that of the basic semiconductor diode in two important respects:

1. The level of doping extends from 100 times to several thousand times that of the conventional diode.
2. As the result of the high levels of doping, the depletion layer at the junction is extremely thin and is only 100 μm wide. This is about 1% of the normal thickness, and consequently the transit time for the device is extremely short, especially as gallium arsenide rather than silicon is used as the semiconductor material. Consequently, the tunnel diode has obvious applications to microwave circuitry and high-speed switching.

Tunnel Diode Characteristic

For appreciable current to flow in a conventional diode, it must be provided with a forward bias (several hundred millivolts) which exceeds the barrier potential. This allows the electrons (majority charge carriers) in the *n* region to possess the required energy to cross into the *p* region. However, with the tunnel diode a small forward voltage of only tens of millivolts will provide the electrons with sufficient energy to "tunnel" their way through the extremely thin depletion layer. These majority charge carriers literally "punch" through with velocities which are much higher than those associated with ordinary diodes and in the process the electrons lose virtually none of their energy. In the forward voltage versus forward current characteristic of Figure 6–29a we are describing the action of the tunnel diode between points O and *P*. At point *P* (where the peak current I_p occurs) the energy levels of the electrons in the *n* region are the same as the vacant energy-level states of the valence electrons in the *p* region; consequently, the maximum number of electrons can tunnel through the depletion layer.

When the forward bias is increased beyond point *P*, the energy-level state of the valence electrons is further reduced, so that the empty energy levels in the *p* region drop below the energy levels of the electrons in the *n* region. The action of the forbidden zone then comes into play so that between points *P* and *V* there is a continuous reduction in the number of electrons capable of tunneling through the depletion region. The diode current then falls to a minimum at point *V* (where the valley current I_v occurs).

Between points *P* and *V* the tunnel diode exhibits its property of dynamic negative resistance since as the forward voltage is increased over this section, the forward current is reduced.

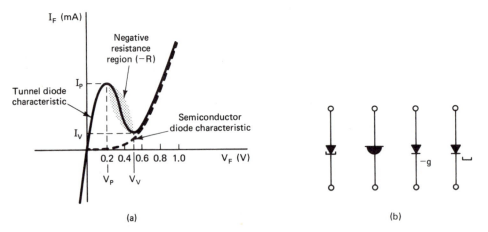

(a)   (b)

Figure 6–29   (a) Tunnel and semiconductor diode characteristics; (b) tunnel diode symbols.

Beyond point V the tunneling action ceases entirely and ordinary diode current occurs as the result of the forward bias overcoming the potential barrier. For comparison purposes the characteristics of both the tunnel diode and the conventional diode are illustrated in Figure 6-29a, while common tunnel diode symbols are shown in Figure 6-29b. Notice that the tunnel diode's reverse current is much higher due to the heavy doping; however, this has little significance since the tunnel diode is not normally reversed biased.

The construction of a tunnel diode is depicted in Figure 6-30 and its typical specifications are:

Maximum forward current $I_{F\,(\text{max})}$	5 mA
Maximum forward voltage $V_{F\,(\text{max})}$	500 mV
Maximum reverse current $I_{R\,(\text{max})}$	10 mA
Maximum reverse voltage $V_{R\,(\text{max})}$	30 mV
Peak current I_P	1.0 mA
Peak voltage V_P	60 mV
Valley current I_V	0.1 mA
Valley voltage V_V	350 mV

Note that the value of the ratio $I_P : I_V = 1.0$ mA$:0.1$ mA $= 10:1$. This is typical of germanium tunnel diodes, while the ratio is nearer to $20:1$ for GaAs; a high value of $I_P : I_V$ is important for computer operations.

The figures immediately indicate that the tunnel diode is a low-power device. To operate over the negative resistance region the required value for the forward bias is $(350 - 60)/2 = 145$ mV, while the highest possible voltage and current variations are $350 - 60 = 290$ mV (peak to peak) and $1.0 - 0.1 = 0.9$ mA (peak to peak), respectively. This corresponds to an rms power of $(290 \times 0.9)/8 = 33$ μW. In addition, the maximum forward power rating is only 500 mV $\times 5$ mA $= 2.5$ mW.

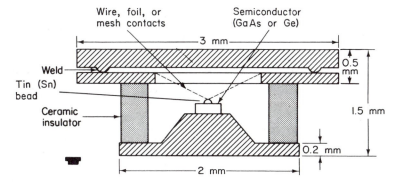

Figure 6-30   Pile-type tunnel diode construction.

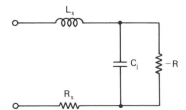

Figure 6–31 Equivalent circuit of a tunnel diode.

An approximation to the value of the negative resistance R is given by

$$R = -\frac{\Delta V_F}{\Delta I_F} = -\frac{V_V - V_P}{I_P - I_v} = -\frac{(350 - 60)\text{ mV}}{(1.0 - 0.1)\text{ mA}}$$

$$\approx -300\ \Omega$$

This is somewhat high; near the center portion of the negative resistance characteristic, the value of R is closer to $-100\ \Omega$. The full equivalent circuit of a tunnel diode is shown in Figure 6–31. I_s is primarily the stray lead inductance, whose value is of the order of nanohenrys. The value of R_s (typically a few ohms) is the result of combining the bulk resistance of the device, the resistance of the connecting leads with their skin effect, and the contact resistance between the leads and the semiconductor material. The junction diffusion capacitance C_j normally has a value between 1 and 10 pF.

Tunnel Diode Applications

As stated previously, the tunnel diode is capable of providing either oscillation or amplification at microwave frequencies. In either application the diode must be loosely coupled to its tank circuit; otherwise, instability will occur. The necessary power for amplification or oscillation is then derived from the dc voltage which biases the diode near the center of its negative resistance characteristic.

    Tunnel diode oscillator.   The basic circuit for a tunnel diode oscillator is shown in Figure 6–32. The capacitive divider C_1, C_2 is included to control the degree of

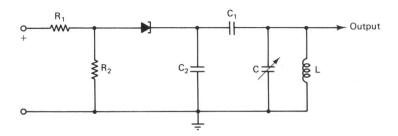

Figure 6–32 Basic circuit of the tunnel diode oscillator.

coupling between the tank circuit and the diode. When power is first applied, there is a brief surge of current through C and C_1, through the junction capacitance of the tunnel diode, and through R_1 to the positive terminal of the source. This surge provides sufficient energy to shock the tank circuit into oscillation. The level of the surge current is held to an acceptable value by the limiting action of the dropping resistor R_1. The voltage divider action of R_1 and R_2 now sets the dc bias for the tunnel diode at point P (Figure 6–29a), which allows the diode to operate on the linear portion of its negative resistance slope.

The voltage applied across the tunnel diode is determined not only by the dc potential developed across R_2, but also by the varying voltage across the resonant tank circuit. The dc bias developed across R_2 would (by itself) operate the diode at point P on the characteristic curve. Voltage variations (positive and negative half-cycles) of the tank circuit would then shift the operation of the diode back and forth along the negative resistance portion of the characteristic. Remember that the tank circuit has its own positive resistance, which naturally creates energy losses. This positive resistance in the tank circuit absorbs power and would eventually damp out the initial oscillation. However, if the negative resistance of the tunnel diode is equal to or greater than the resistance of the tank circuit, the diode will act as a switch which will resupply energy to the tank circuit, so that a continuous oscillation can be sustained.

During the positive half-cycle (with respect to ground) of the tank circuit's operation, the forward bias on the tunnel diode is reduced so that the diode current increases. This resupplies the tank circuit with additional energy to overcome the natural losses. During the negative half-cycle of the tank circuit's operation, the forward bias is being increased, but the current will correspondingly decrease as the valley voltage V_V is approached. Little current is therefore provided to draw energy from the tank circuit. It follows that more energy is supplied to the tank circuit than is taken from it and provided that the energy gain is sufficient, a continuous oscillation will be sustained up to frequencies of about 100 GHz.

The tunnel diode is a nonlinear device and can therefore be used for frequency conversion since it can behave as a combination of oscillator and mixer stage.

Tunnel diode amplifier (TDA).   A tunnel diode can be connected as an amplifier either in parallel or in series with a resistive load. These circuits are illustrated in Figure 6–33, but in both cases the diode's loss resistance and its stray inductance have been omitted.

In the parallel arrangement the power gain G_p of the amplifier is given by

$$G_p = \frac{R}{R - R_L} \tag{6–13}$$

As the negative resistance of the diode approaches the value of the load resistance, the power gain increases. In the extreme case where $R = R_L$, the gain is infinite and the circuit will oscillate.

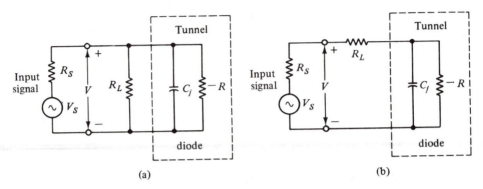

Figure 6–33 Tunnel diode amplifier equivalent circuits for (a) parallel, and (b) series loading.

In the series arrangement the power gain G_p is

$$G_p = \frac{R_L}{R_L - R} \tag{6–14}$$

Provided that $R_L > R$, the circuit will remain stable with a power gain of greater than unity.

    To provide isolation between the source and the load the tunnel diode amplifier can be connected to a circulator as shown in Figure 6–34. Provided that the output port is correctly terminated and the characteristic impedance of the circulator is greater than the diode's negative resistance, the amplifier can provide the required power gain.

    In general, tunnel diode amplifiers require a very simple dc power supply, are capable of broadband operation, and have low noise figures (less than 5 dB at 10

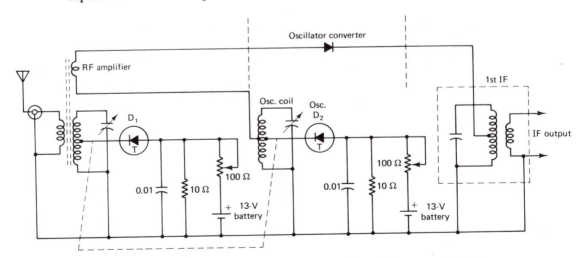

Figure 6–34 Tunnel diode microwave amplifier and frequency converter.

GHz) due to their small current levels. A schematic of a tunnel diode microwave amplifier is illustrated in Figure 6–34. Gallium arsenide FET amplifiers tend to be preferred up to 20 GHz, but tunnel diodes have a superior performance between 20 GHz and 50 GHz. Finally, tunnel diodes are immune to the background radiation encountered in the solar system and are therefore suitable for space communications.

6-8 TRANSFERRED ELECTRON DEVICE: THE GUNN DIODE

The Gunn diode is a solid-state (gallium arsenide crystal) bulk effect source of microwave energy (Figure 6–35). The discovery that microwaves could be generated by applying a steady dc voltage across a chip of a n-type gallium arsenide crystal was made in 1963 by J. B. Gunn. The operation of this crystal device results from the excitation of electrons into energy states higher than those in which they normally occur. Notice, however, that this device is not really a diode since no junction is involved and it is only for convenience that we refer to the positive end as the anode and the negative end as the cathode.

In a gallium arsenide semiconductor there exist empty valence bands, which are higher than those occupied by the electrons. These higher valence bands have the property that their electrons are less mobile under the influence of an electric field than when they exist normally in a lower valence band.

To simplify the explanation of this effect, assume that the electrons in the higher valence band have essentially no mobility. If an electric field is applied to the gallium arsenide semiconductor, the current will increase with a rise in the voltage, provided that the voltage is at a low level. However, if the voltage is made higher enough with an electric field intensity of more than 300 kV/m, it is possible to excite electrons from their initial band to the higher band, where they become virtually immobile. If the rate at which electrons are removed is high enough, the current will decrease

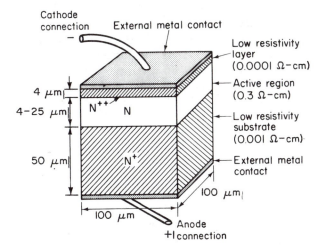

Figure 6–35   Composition of a Gunn diode.

even though the voltage is being increased. This is an equivalent negative resistance effect which can therefore form the basis for an oscillator circuit.

So far it is clear that:

1. The Gunn diode is not a diode in the normal sense since no junction is involved and the device is composed primarily of n-type GaAs bulk material (it should be noted that p-type material cannot be used for this purpose). However, since a dc voltage is applied across the device, we may refer to its positive end as the anode and its negative end as the cathode.

2. The principle of the Gunn diode depends on the transfer of the electrons to the higher energy level, owing to the application of a very strong electric field intensity. For this reason the diode is referred to as a transferred electron device (TED).

Formation of Domains

If a voltage is applied across an unevenly doped n-type gallium arsenide crystal, the crystal will be divided into regions or domains with electric fields of different intensities. In particular, small domains can form within which the field will be very strong, while in the surrounding crystal material the electric field will be comparatively weak.

It is not difficult to see that such a domain is unstable. For example, we will assume that there is a sudden increase in the electron density at some point in the crystal which tends to reduce the electric field to the left of the disturbance while increasing the electric field to the right. In a negative resistance material the decreasing field to the left of the disturbance will cause an increase in the current flowing into the disturbed region, while the increase in the field to the right will tend to lower the current outside this region. This current pattern will have the effect of building up the charge disturbance even more; the situation will then become unstable and will result in a redistribution of the electric field within the crystal.

The domains formed in the gallium arsenide crystal are not stationary since the electric field acting on the electron charge will cause the domain to move across the crystal; this is illustrated in Figure 6–36. The domain will travel across the crystal from one electrode to the other, and as it disappears at the anode, a new domain will form near the cathode.

The Gunn Oscillator

The Gunn oscillator will have a frequency inversely proportional to the time required for the domain to cross the crystal. This time is proportional to the length of the crystal and depends to some degree on the applied voltage. Each domain results in a pulse of current at the output, so that the Gunn oscillator produces a microwave frequency which is determined, for the most part, by the physical length of the chip.

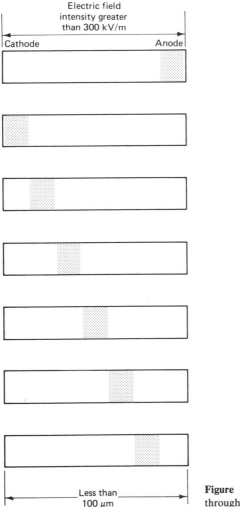

Figure **6–36** Movement of a domain through a gallium arsenide chip.

It is possible for the Gunn diode to oscillate by using a simple resistor as the load. However, the efficiency of such an arrangement is only a few percent. Preferably the diode should be placed in a resonant cavity, which improves the efficiency and allows some variation in the operating frequency. Such an oscillator (Figure 6–37) is capable of delivering power outputs of a few watts at 30 to 40 GHz (continuous operation) and up to 200 W in pulsed operation. The power output capability of this device is limited by the difficulty of removing heat from the small chip. Examples of commercial Gunn oscillators are illustrated in Figure 6–38.

The advantages of the Gunn oscillator are its small size, ruggedness, and low cost of manufacture.

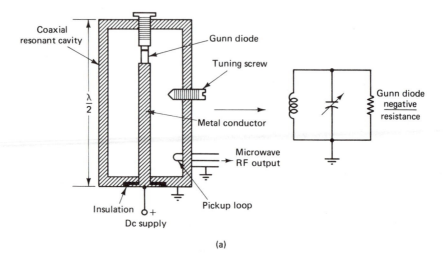

(a)

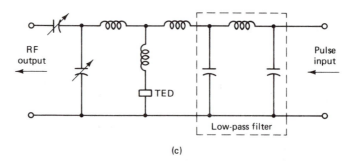

(b)

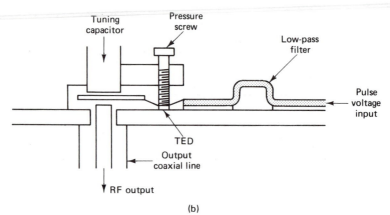

(c)

Figure 6-37   (a) Resonant cavity Gunn diode oscillator; (b) coaxial line Gunn diode oscillator; (c) equivalent circuit for part (b).

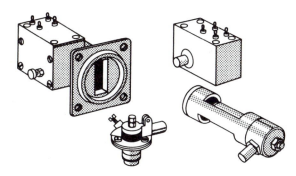

Figure 6–38 Commercial Gunn diode oscillators. (Courtesy of Watkins-Johnson Company.)

6–9 ACTIVE MICROWAVE DIODES

Referring back to the tunnel diode, the dc bias point on its characteristic was ideally positioned at the center of the negative resistance region. When the ac voltage from the resonant circuit was subsequently applied across the diode, the alternating component of the diode current was 180° out of phase with the ac voltage. Consequently, any device with this phase relationship automatically exhibits the property of negative resistance and may be used for oscillation and amplification.

The avalanche diode was first put forward as a theoretical consideration by W. T. Read in 1958. It was proposed that such a diode could produce a negative resistance effect at microwave frequencies by using carrier impact ionization and drift in the high-field intensity region of a reverse-biased semiconductor junction. Since that time three distinct versions of this device have been studied. In chronological order the first was the IMPATT diode, originally developed by R. L. Johnston and others in 1965; the acronym IMPATT stands for IMPact ionization Avalanche Transit Time effect. Such a diode, when manufactured from silicon, has a higher efficiency

$$\frac{\text{RF power output}}{\text{dc power input}} \times 100\%$$

of between 3% (CW) and 60% (pulsed) and can operate at frequencies up to and beyond 100 GHz. In 1967, H. J. Prager operated an avalanche diode in its TRAPATT (TRApped Plasma Avalanche Triggered Transit) mode. This version is particularly suitable for the lower microwave frequencies from 1 to 3 GHz and can generate pulsed powers of several hundred watts with efficiencies from 20 to 60%. In 1971, the BARITT (BARrier Injected Transit Time) diode was discovered by D. J. Coleman. In this device the charge carriers which traverse the drift region are produced by minority carrier injection from forward-biased junctions, as opposed to being extracted from the plasma of an avalanche region. BARITT diodes have low-noise figures of less than 15 dB, but their output powers and bandwidths are relatively small. We will start by examining the basic principles of the theoretical Read diode and then study the IMPATT, TRAPATT, and BARITT variations.

The Read Diode

This diode has four regions which together form an n^+-p-i-p^+ structure and are manufactured from silicon. The positive signs indicate a high level of doping and the i stands for intrinsic or pure (undoped) silicon. Essentially, the four regions are:

1. The avalanche or high-field region, which is in the vicinity of the n^+-p junction. The production of hole–electron pairs occurs within this region.
2. The space-charge region, which consists of the thin p region and the intrinsic region.
3. The drift region, which is involved with the intrinsic material alone. The generated holes must drift through this region toward the p^+ contact.
4. The second drift region, which is associated with the heavily doped p^+ section.

These regions are illustrated in Figure 6–39a.

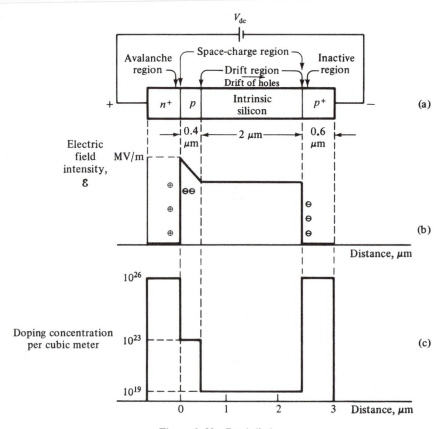

Figure 6–39 Read diode.

To form an oscillator the Read diode is reverse biased and mounted in a resonant cavity. The diode then presents capacitive reactance which is combined with the cavity's inductive reactance to create a resonant circuit. The negative resistance of the diode allows a continuous oscillation to be sustained by converting a portion of the dc power input associated with the bias supply into a microwave power output from the cavity.

We will now discuss the mechanism through which the avalanche diode can exhibit its negative resistance property.

Principle of Avalanche Breakdown

When the reverse bias voltage greatly exceeds the breakdown voltage, a maximum field intensity of several megavolts per meter occurs at the n^+-p junction (Figure 6-39b). Those holes which are moving in these high-field regions near the n^+-p junction have sufficient energy to knock valence electrons into the conduction band. This action creates electron–hole pairs whose rate of production is referred to as avalanche multiplication and is determined by the strength of the electric field intensity. By using the proper doping profile (Figure 6-39c), the field intensity will have a sharp peak, so that the avalanche multiplication occurs only in the immediate vicinity of the n^+-p junction. The conduction electrons then move into the n^+ region and their associated holes drift through the space-charge region to the p^+ region. The drift velocity v_d of the holes is of the order of 10^5 m/s, while the electric field intensity in the space-charge region is no more than 0.5 MV/m. If the length of the intrinsic region, l, is 2 μm, the drift time through this region is

$$\frac{l}{v_d} = \frac{2 \times 10^{-6}}{10^5} \, \text{s} = 2 \times 10^{-5} \, \mu\text{s}$$

Operating Conditions in the Avalanche Diode Oscillator

Assume that the Read diode is placed in the cavity and the circuit is oscillating with a continuous microwave output. The diode is then subjected to an alternating voltage as well as its dc reverse bias (Figure 6-40). At the peak of the "positive" half-cycle the combination of the dc and ac field intensities is well above the breakdown point, so that avalanche multiplication occurs. However, since the avalanche effect involves a chain reaction, the growth of the resulting charge or hole current I_h is not instantaneous but exponential, so that a certain time elapses before the current reaches its maximum value. By arranging that this time interval, $T/4$, is equal to one quarter-period of the microwave voltage, the peak of the current lags the peak of the voltage by 90° (one-quarter of a cycle). Subsequently, the combination of the dc and ac fields falls below the breakdown point during the negative half-cycle and the charge current decays exponentially to a small value in a further $T/4$ time interval. Concurrently, the generated holes are injected into the drift region and move toward the p^+ region which is connected to the negative terminal. As these injected holes move across the drift space, an electron flow current I_e is induced into the external circuit (Figure 6–40c) and dc power is drawn from the reverse bias supply.

It is clear that the induced current in the external circuit must be equal to the average current which occurs in the space-charge region. When the pulse of hole current I_h is rapidly created at the n^+-p junction and reaches its peak, the constant current I_e flows in the external circuit and continues for a time $T/2$ while the holes are moving across the space-charge region. At the midtime of this condition (point Q), the external current I_e has been delayed by a further time of $T/4$, so that there is another 90° shift relative to the I_h peak. Consequently, there is a total phase shift of $90° + 90° = 180°$ between the external current and the alternating

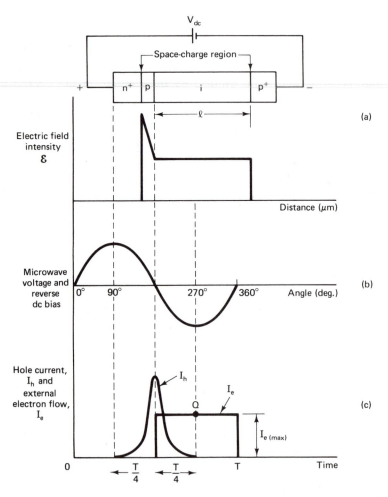

Figure 6–40 Phase relationship between the microwave voltage and the external electron flow associated with a Read diode. (Courtesy of AT&T Bell Laboratories.)

microwave voltage; the result is a certain value of negative resistance which will enable a continuous oscillation to be sustained.

Since the drift time $(2 \times 10^{-5}$ μs) in the intrinsic region is one half-period of the microwave voltage, $T/2 = 2 \times 10^{-5}$ μs and the cavity should be resonant at the generated frequency, $f = 1/T = 1/(4 \times 10^{-5} \times 10^{-6}) = 2.5 \times 10^{10}$ Hz $= 25$ GHz. Note that there is a further drift of the charge carriers in the p^+ layer, so that the Read device has a double-drift region, which results in the most efficient form of avalanche diode.

The external current I_e is in the form of an approximate square wave with top and bottom levels of $I_{e\,(max)}$ and nearly zero. The average current drawn from the reverse bias supply is $I_{e\,(max)}/2$ and the fundamental current component at the microwave frequency has a peak value of $2I_{e\,(max)}/\pi$. If the peak value of the microwave voltage is V, the alternating power delivered to the cavity is

$$\frac{V}{\sqrt{2}} \times \frac{2I_{e\,(max)}}{\pi} = \frac{\sqrt{2}}{\pi} VI_{e\,(max)} \qquad \text{watts}$$

The Q factor of a circuit is defined as the ratio $2\pi \times$ peak energy stored during the cycle : average power dissipated over the cycle. Since the Read diode delivers energy to the cavity, its effective Q is negative and this value must balance with the cavity's positive Q to maintain a continuous oscillation.

IMPATT Diodes

The Read device with an n^+-p-i-p^+ (or p^+-n-i-n^+) structure is the most efficient form of IMPATT diode since its action relies on the double-drift (i and p^+) region. We have learned that the negative resistance effect is due to:

1. The impact multiplication avalanche effect, which causes the hole carrier current to lag the microwave output voltage by 90°.
2. The effect of transit time through the drift region; this results in the external current lagging the microwave voltage by a further 90°.

In fact, many forms of p-n device may be made to behave as IMPATT diodes provided that they contain a very heavily doped region followed by a drift region where no avalanching occurs. Some practical examples are the one-sided abrupt p-n junction, the linearly graded p-n junction and the p^+-i-n^+ diode, which are all illustrated in Figure 6–41 together with their doping profiles. For each of these diodes the basic explanation is similar to that for the Read device.

The fabrication of the one-sided abrupt junction diode involves the formation of an n-type expitaxial layer over the n^+ substrate. On top of this is diffused a thin layer of p-type impurity creating an avalanche zone next to the p^+ diffused region. By contrast, the Read-type IMPATT diode was initially more difficult to manufacture. For the p^+-p-n-n^+ version, the problems have been overcome by growing an n layer

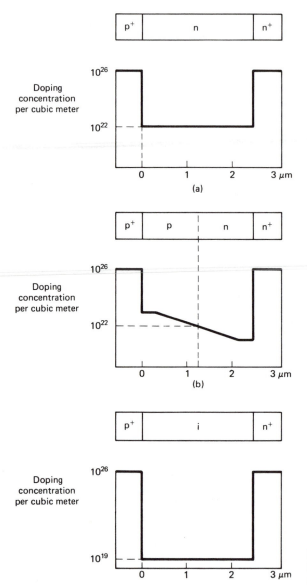

Figure 6–41 Three types of silicon IM-PATT diode: (a) abrupt junction; (b) linearly graded junction; (c) $p^+ - i - n^+$. (Courtesy of AT&T Bell Laboratories.)

on an n^+ substrate. Subsequently, the p layer is grown epitaxially, and ultimately the p^+ layer is formed by diffusion.

The unit construction of an IMPATT diode is shown in Figure 6–42. Such diodes are at present the most powerful semiconductor microwave sources and can be manufactured from germanium, silicon, gallium arsenide, or indium phosphide. Theoretically, GaAs should provide the greatest efficiency, the highest operating fre-

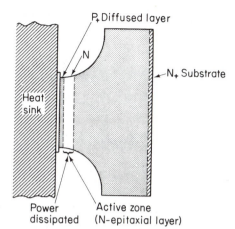

Figure 6–42 Construction of $p^+ - n - n^+$ IMPATT diode.

quency, and the lowest noise figure. However, GaAs is more difficult in the fabrication process and is more expensive than silicon.

Commercial IMPATT diodes (Figure 6–43) have a frequency range from 425 MHz to over 107 GHz and an efficiency which extends from 1% above 100 GHz to 60% below 1 GHz. The microwave power outputs lie between about 1 W (CW) and over 400 W (pulsed).

The major disadvantage of the IMPATT diode is its high noise figure, which is due primarily to the avalanche process and to the high level of operating current.

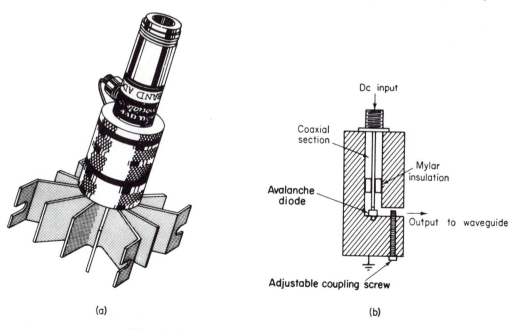

(a)   (b)

Figure 6–43 Commercial IMPATT diode. (Courtesy M/A-Com, Inc.)

A typical noise figure is of the order of 30 dB, which is significantly worse than the figures for Gunn diodes.

The equivalent circuit of an IMPATT diode may be reduced to a negative resistance of about 10 Ω in parallel with a capacitance of the order of 1 pF. The single-stage IMPATT amplifier has a gain of 5 to 10 dB with a bandwidth which is approximately one-tenth of the center frequency. Operating frequencies are possible to 100 GHz, but high noise figures are still a problem.

TRAPATT Diodes

Compared with the IMPATT diode, the TRAPATT (TRApped Plasma Avalanche Triggered Transit) diode is a microwave source with a higher efficiency, but its working frequency is limited to below 10 GHz and requires a greater voltage swing for its operation as an oscillator or as an amplifier. The typical TRAPATT diode has a p^+-n-n^+ (or n^+-p-p^+) structure (Figure 6–44) and is manufactured from silicon. For high peak power outputs the width of the p-type depletion layer is about 8 μm and its doping allows the diode to be fully "punched through" when breakdown occurs; in other words, the depletion layer's electric field intensity just before breakdown greatly exceeds the level associated with the drift velocity.

During the operation of the device as an oscillator, the TRAPATT diode is mounted inside a coaxial cavity at a position where there is the maximum RF voltage

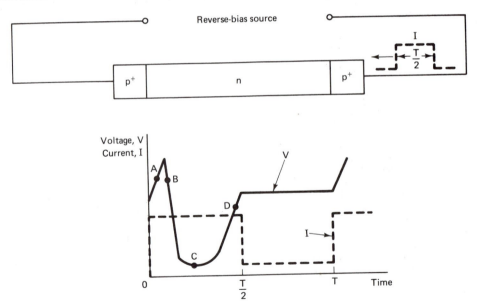

Figure 6–44 Action of the TRAPATT diode: A, charging; B, formation of plasma; C, extraction of plasma; (d) charging. (A. S. Clorfeine, R. J. Ikola, and L. S. Napoli, *RCA Review*, Vol. 30, Sept. 1969, pp. 397–421; reprinted by permission of RCA Laboratory.)

swing. It is then assumed that the reverse-biased diode starts off as an IMPATT device but when the oscillation builds up to a certain level, the circuit automatically switches to the TRAPATT mode. The total voltage (dc bias together with the RF swing) is then much higher than the IMPATT value at the threshold of breakdown. When avalanche occurs under these conditions, it generates a plasma of holes and electrons. This plasma is in a form of a neutral gas in which the electrons have been stripped away from their atoms; as an example, the ionosphere is a plasma which is created by ultraviolet radiation from the sun. The presence of the plasma creates across the junction a high potential difference which opposes the reverse-bias voltage. This reduced voltage traps the plasma of holes and electrons, but the external current is now flowing so that as the voltage rises, the plasma is ultimately released and the resulting current pulse moves across the drift region; however, the associated drift time is comparatively long because of the reduced voltage. The total transit time is composed primarily of the delay in releasing the plasma and the drift interval. Compared with the IMPATT diode, the TRAPATT diode's transit time is much greater, so that its operating frequency is only on the order of a few gigahertz. However, since the current pulse is associated with the relatively low voltage, the amount of power dissipation is low and the efficiency is correspondingly high. This is comparable with the high efficiency obtained when operating under class C conditions. Moreover, since the high current coincides in time with a low voltage, and vice versa, the device exhibits the required property of negative resistance.

Because of the very high power densities (between 10 and 10^2 W/m^2) that occur with the device, the TRAPATT diode is not suitable for continuous operation, but with a duty cycle of 0.001, it is possible to obtain peak powers of 250 W at 3 GHz (efficiency 35%) and 550 W at 1 GHz (efficiency 60%). A major disadvantage of the TRAPATT diode is a high-noise figure of about 60 dB, which severely limits its use as an amplifier. In addition, the current pulse is of extremely short duration, so that a TRAPATT amplifier generates strong harmonics which must be largely eliminated when operating at the fundamental frequency.

BARITT Diodes

The BARITT (BARrier Injected Transit Time) diode is the latest addition to the list of active microwave diodes. Although these devices have drift times similar to those of IMPATT diodes, the holes or electrons which traverse the drift regions are produced by injecting *minority* charge carriers from a *forward*-biased junction and not by an avalanche effect due to *reverse* bias.

Several different structures (*p-n-p*, *p-n-i-p*, *p-n*-metal, metal-*n*-metal) have been used to create BARITT diodes. In the *p-n-i-p* version, a forward-biased *p-n* junction releases holes into the intrinsic region. These holes drift through this region with the saturation velocity and ultimately arrive at the *p* contact. The external current allows the diode to present its negative resistance during one half-cycle of the microwave voltage; this is similar to the Read diode.

Let us examine in detail the action of a p^+-n-p^+ BARITT diode (Figure 6–

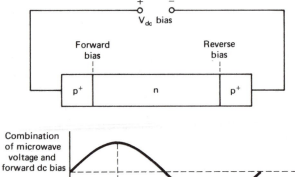

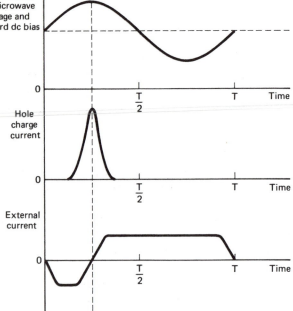

Figure 6–45 Action of the BARITT diode.

45). This device is essentially two junctions which are set back to back. The left-hand junction is forward biased while the right-hand junction is reversed biased. Provided that the amount of reverse bias is below the "punch-through" voltage, the amount of current represents only the normal leakage associated with a reverse-biased junction. However, if the reverse bias is raised to the point where the punch-through mode occurs, the depletion layer will extend across the entire region until it reaches the forward-biased junction. When this occurs, the hole charge carriers at the forward-biased junction will be injected into the *n* region so that the charge current will increase rapidly (Figure 6–45). For the device to operate correctly, the electric field intensity must be sufficient to cause the holes to drift across the *n* region with their saturation velocity but must also not exceed the level that would result in an avalanche effect.

Let us assume that a BARITT diode is initially biased below the voltage required for punch-through to occur. The diode is mounted in a cavity so that when the microwave voltage is also applied across the diode, the total voltage will exceed the punch-through potential on the positive half-cycle and a sharp pulse of hole current is created. During the drift time a square wave of external current will deliver energy from the dc bias source to the cavity so that a continuous oscillation will be maintained.

Because of their long transit time, BARITT diodes are primarily suited for the C band. However, they can operate over the frequency range 4 to 8 GHz but with narrow bandwidths, low efficiencies (about 2%), and microwave power outputs of only a few milliwatts. They may therefore be used as the local oscillator in microwave receivers.

6–10 PASSIVE MICROWAVE DIODES

In Section 6–9 we discussed the "active" diodes, which were capable of producing oscillation and amplification. We are now turning our attention to the "passive" diodes, which perform such functions as detection, mixing, modulation, and switching.

PIN Diodes

A PIN diode is basically a microwave device with three regions (Figure 6–46). The center layer consists of n-type silicon which is so lightly doped that it is regarded as intrinsic material. On one side of the intrinsic region is a narrow (heavily doped) p^+ region while on the other side is an equally narrow n^+ region; the combination of p^+-i-n^+ regions then gives the diode its name. Although gallium arsenide is used

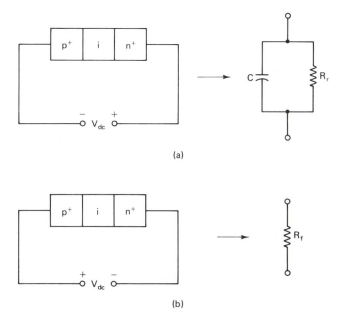

(a)

(b)

Figure 6–46 PIN diodes and equivalent circuits: (a) reverse biased; (b) forward biased.

in the manufacture of PIN diodes, it presents more fabrication problems than silicon and has a lower power capability. In the construction process the intrinsic region is grown expitaxially on an n-type substrate and the p region is obtained by diffusion.

When the PIN diode is reversed biased, its equivalent circuit at microwave frequencies (Figure 6–46) consists primarily of a small capacitance C, in parallel with a high reverse resistance R_b, whose value is typically a few kilohms. However, under forward-bias conditions, avalanche effects associated with the p^+ and n^+ regions causes large numbers of holes and electrons to move into the intrinsic layer. The equivalent resistance R_f of the device then falls to a low value of a few ohms.

Assume that a PIN diode is mounted across a waveguide or a coaxial cable. When reverse biased the diode's high resistance will have little effect on the incident power, so that virtually no reflection occurs. The diode is now behaving as a switch in the "power on" condition. However when the diode is forward biased, its low resistance creates a severe mismatch and virtually all of the incident power is reflected. This "power off" condition may be further improved by connecting a number of PIN diodes in parallel to achieve a lower forward resistance and a higher power capability.

If the PIN diode's bias is abruptly changed back and forth, the microwave power will be amplitude modulated by a pulse or square wave. For example, many experiments described in Chapter 8 require that the microwave power be amplitude modulated by a 1-kHz square wave. This could be done by connecting a PIN diode across the output of a Gunn oscillator; such a diode would then be biased by a 1-kHz square-wave voltage of sufficient amplitude. The advantage of this type of modulation is that the p^+-i junction neither rectifies nor generates harmonics.

Because of the short transit time associated with the intrinsic region, the PIN diode's switching time is of the order of a few nanoseconds. Operating frequencies extend to 100 GHz and a single diode may handle a peak power of 25 kW with a duty factor of 0.001. Greater peak powers are possible if a number of PIN diodes are paralleled.

Schottky Hot-Carrier Barrier Diode

In our discussion of the MESFET transistor we encountered the Schottky barrier gate in which a potential barrier was created in a metal–semiconductor interface by establishing stable charges in a doped material without the need for a chemical junction. The arrangement is therefore similar to the original microwave detector, which was the point contact diode. The only difference is that the metal interface is a surface rather than a point. This has the twin advantages of less noise and lower forward resistance, but the disadvantage of a greater shunt capacitance. Consequently, point-contact diodes with their lower capacitances are still used at frequencies of 150 GHz and above.

When forward bias is applied to a Schottky diode, the potential barrier is decreased at the metal-semiconductor junction, so that the majority charge carriers (electrons) are injected into the metal with energy levels which are much higher than those of normal free electrons. Consequently, we refer to these high-energy electrons as "hot carriers."

Once inside the metal, the majority hot carriers lose their excess energy very rapidly, with an average lifetime of only a few picoseconds. At the same time there is no flow of minority carriers in the reverse direction. Consequently, under reverse-bias conditions there is a very rapid response because the energy stored at the interface is low. Furthermore, since there is virtually no current from the metal to the semiconductor with the reverse bias, there is no delay time associated with electron–hole recombination (such as occurs with the junction diode).

Because of their low noise levels and fast response, Schottky diodes are commonly used in microwave mixer stages, where they provide high efficiencies for frequency conversion. Figure 6–47 illustrates the schematic layout and physical assembly for a broadband Schottky diode balanced mixer stage in a microwave receiver.

In the manufacture of Schottky diodes we use n-type epitaxial material which

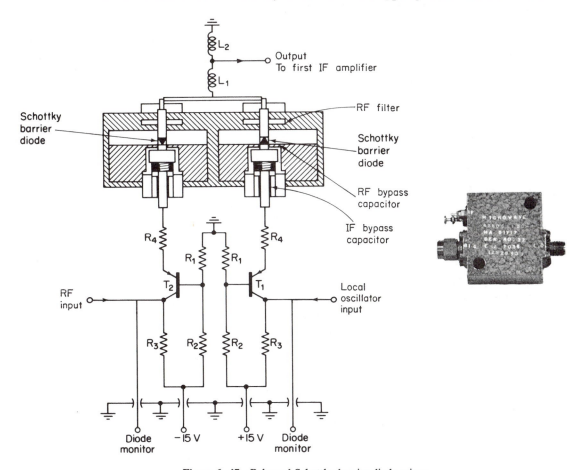

Figure 6–47   Balanced Schottky barrier diode mixer.

is fabricated from either gallium arsenide or silicon. As stated previously, GaAs has less noise and is capable of operating at higher frequencies, while silicon is easier to fabricate and is used primarily at frequencies below 10 GHz. With *n*-type silicon good barriers are formed with gold or platinum contacts.

The use of GaAs Schottky barrier diodes as mixers and detectors extends as high as 100 GHz; at these frequencies a typical noise figure is of the order of 15 dB. At lower frequencies in the S-band the noise figure drops to about 5 dB.

Backward Diodes

You will recall that the tunnel diode (Section 6–7) consisted of heavily doped *p* and *n* regions which were fabricated from gallium arsenide. The result was that the tunneling occurred when the diode was forward biased and that a large current flowed under reverse-biased conditions. Moreover, the tunnel diode had a very narrow depletion region so that it could function at high operating frequencies.

The backward or back diode has similar properties to the tunnel diode *except* that the manufacturing process is changed to prevent the effect of tunneling in the forward-biased condition. Therefore, there will be neither peak nor valley voltages and the negative resistance region will be eliminated. In fact, the normal forward current will not start to flow until the forward voltage exceeds about 900 mV. On the other hand, a high current will flow for very low reverse voltages, due to the very heavy doping. This is the exact opposite of the manner in which a normal junction diode operates, which explains the use of the term "backward" or "back" diode.

For comparison purposes the characteristics of the point contact and backward diodes are illustrated in Figure 6–48. The main advantage of the back diode is its

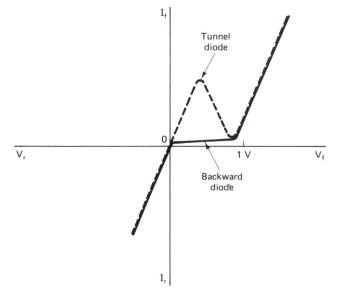

Figure 6–48 Comparison of the characteristics of backward and tunnel diodes.

considerably higher current ratio (reverse current : forward current) and its lower noise level. However, its signal-handling capacity is much less because of the large current which flows when the forward voltage exceeds 900 mV. The backward diode is therefore a low-power device of high sensitivity and its principal use is as a small-signal detector. It is also used widely as the mixer device in a microwave receiver, in which the required local oscillator input is appreciably less than if a point contact diode were used. Coaxial versions of the backward diode are illustrated in Figure 6–49.

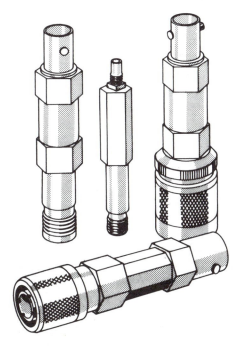

Figure 6-49 Commercial coaxial backward diodes. (Courtesy M/A-Com, Inc.)

6-11 MASERS

Certain substances, when irradiated by electromagnetic energy at microwave frequencies, are capable of partially absorbing this energy. It commonly occurs that the absorption takes place at very sharply defined frequencies and that the substance therefore acts as a resonant circuit by storing the electromagnetic energy. The manner in which certain substances can be employed to produce amplification at microwave frequencies is the subject of this discussion.

The term "maser" is compounded from the initial letters of the phrase, Microwave Amplification by Stimulated Emission of Radiation. The operation of this device can only be explained in terms of the properties of atomic and molecular particles and their interaction with electromagnetic waves.

The related properties and interactions are:

1. Atomic and molecular particles tend to exist naturally at discrete energy levels.
2. These particles can jump from one energy level to another under the influence of an electromagnetic wave with the correct frequency. In the process the particles give up to the wave, or absorb from it, a discrete quantity of energy; this depends on whether the particles drop to a lower level or are lifted to a higher level.

In the practical maser it is arranged that a large number of particles exist at a high energy level. These are then exposed to the stimulating influence of an electromagnetic wave, so that they drop to a lower energy level and give up energy to the wave, which is therefore amplified. To achieve this, the gap between the energy levels of the particles must be related to the frequency of the wave by the equation

$$\text{change in energy level} = h \times f \quad \text{J} \tag{6–15}$$

where

$$h, \text{Planck's constant} = 6.6 \times 10^{-34} \text{ J-s}$$
$$f = \text{frequency (Hz)}$$

The simplest form of maser was the original ammonia gas maser. However, modern amplifiers use the energy levels of solid-state substances to achieve similar results and consequently avoid the necessity for a vacuum. The problem is to find substances with their energy levels separated by the correct amount so as to provide those transitions which exist in the microwave region.

If a maser is to operate properly, the energy levels occupied by the particles must be separated by those amounts which are proportional to the microwave frequencies. In addition, favorable conditions must exist for the particles to interact with the microwave signals so that amplification is achieved. Such conditions do not occur naturally.

Paramagnetic ions provide the energy levels which are suitable for use in solid-state masers. These ions are included as trace impurities in nonmagnetic host crystals. The ion used most successfully is from the element chromium, which is diluted in a ruby gemstone; this results in the pink ruby crystal.

The values of the energy levels are determined by the strength of the dc magnetic field which is applied to the crystal. As a result, the basic electron energy level is split into sublevels (Figure 6–50). For the ruby crystal, four such levels exist and the energy spacing of these four states corresponds to emission and absorption frequencies in the microwave band. Transition between any pair of these states can be achieved by applying an RF magnetic field at a frequency corresponding to the quotient of the difference in energy between the two states and Planck's constant. The equation is

$$f = \frac{\text{difference in energy-level values}}{h} \tag{6–16}$$

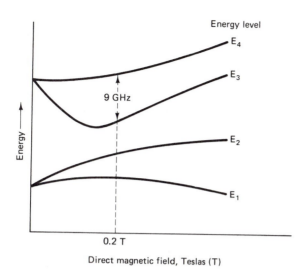

Figure 6-50 Effect of a direct magnetic field on the energy level distribution.

where f is the frequency of the RF magnetic field and h is Planck's constant. As illustrated in Figure 6-50, the spacing between the energy levels varies with the applied magnetic field so that some external control of the operating frequency is possible.

Three-Level Masers

The number of electrons, atoms, or molecules in a particular energy state falls off exponentially as the energy of the states is increased. For any three levels the number of electrons in each state (the state population) might be as shown in Figure 6-51. If the state energies are E_1, E_2, and E_3 as shown, the substances will be capable of absorbing radiation at any of the frequencies f_{13}, f_{23}, and so on, where f_{13} is defined by

$$\text{frequency}\, f_{13} = \frac{E_3 - E_1}{h} \qquad (6\text{-}17)$$

If the crystal is irradiated with radiation at a frequency f_{13}, a certain amount of the energy will be absorbed and some electrons will be elevated from state 1 to state 3 in the process; at the same time induced emission from state 3 to state 1 will also take place. Provided that the interchange between the two levels takes place quickly enough, the result is a condition of saturation in which the populations of states 1 and 3 become equal; this is referred to as *population inversion*. Then there will be more electrons in state 3 than in state 2 and a possible use of the medium becomes apparent. If a signal now irradiates the system at a frequency f_{23}, stimulated radiation occurs as the excess population of state 3 falls back to state 2; there will, of course, be some elevation of electrons from state 2 to state 3 by the signal, but since there are always more electrons in state 3 than in state 2, the result will be

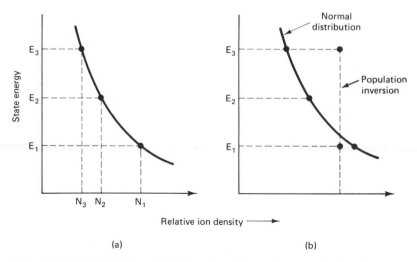

Figure 6–51   Principle of population inversion: (a) normal state population; (b) saturation-state population for a three-level maser pumped at the frequency $f_{13} = (E_3 - E_1)/h$ hertz.

one of enhanced radiation rather than absorption. The population of state 3 is maintained by the frequency f_{13}, which continues to "pump" electrons up from state 1. This is a disadvantage of the maser amplifier in that it requires a pump source which operates at a frequency greater than that of the signal frequency.

Three-Level Cavity Masers

A typical arrangement of the three-level cavity maser is shown in Figure 6–52. A paramagnetic crystal is located at the bottom of a cavity resonator which is designed so that it can resonate simultaneously at frequencies f_{13} and f_{23}.

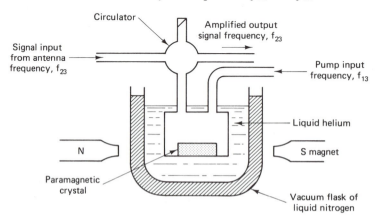

Figure 6–52   Three-level cavity maser.

An externally applied magnetic field, in which the crystal is situated, allows some adjustment of the energy level spacing and is therefore a factor in controlling the values of the frequencies f_{13} and f_{23}. The pump frequency power is fed into the cavity at the frequency f_{13} from an external oscillator which is capable of producing saturation. The input signal from the antenna at the frequency f_{23} travels through a circulator into the cavity and, after being amplified by the stimulated radiation, returns through the circulator and passes to the next amplifier, which might be, for example, a traveling-wave tube.

This maser amplifier will produce a stable gain of up to 20 dB at frequencies of about 3 GHz. The pump power is about 10 mW and the bandwidth is of the order of 20 MHz, although this is a function of the pump power. The major advantage of such amplifiers is their very low noise figures, which are obtained by operating the maser at the temperature of liquid helium. Naturally, this is very difficult to do in practice, and the necessity for this degree of cooling precludes their use in anything but large static equipment. Their main use is in satellite communication systems, but they have, to a certain extent, been superseded by the improvements which have recently taken place in the performance of parametric amplifiers.

Traveling-Wave Masers

In the cavity maser, a weak input signal is fed into the cavity and creates a relatively strong field (because of the magnifying effect of the cavity) in the region where the crystal is situated. The amplifying effect is therefore achieved by allowing a relatively strong field to interact with a small volume of crystal in which there is a high-energy storage. An alternative method of achieving the same purpose would be to cause a weaker field to interact over a larger volume of crystal in which there is a lower-energy storage.

This principle is used in the traveling-wave maser, which requires a slow-wave structure with a group velocity of about 1% of the free-space velocity. A suitable circuit is provided by a capacitively loaded comb structure which is illustrated in Figure 6–53. A finger spacing of 2 mm would then correspond to a bandwidth of 750 MHz. This is the major advantage of the traveling-wave maser compared with the cavity maser. However, the traveling-wave type must still be cooled by immersing it in liquid helium.

6–12 LASERS

The word "laser" is an acronym for Light Amplification by Stimulated Emission of Radiation. The first (three-level) ruby laser was developed by T. H. Maiman in 1960 and was followed by A. Javan's gas (helium-neon) laser in 1961. Both lasers and masers use the conversion of atomic or molecular energy to produce electromagnetic radiation by the process of stimulated emission. If the wavelength of the emitted radiation is of the order of 1 cm (30 GHz), the device is a maser. By contrast, if

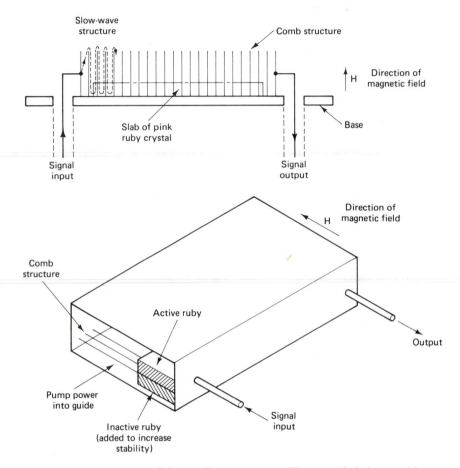

Figure 6–53 Principle of the traveling-wave maser. (The assembly is immersed in liquid helium.)

the radiated EM energy lies in the infrared or light regions where the frequencies exceed 100 THz and the wavelengths are of the order of microns, the device is referred to as an optical maser or laser; however, there is no sharp dividing line between the regions in which we use the terms "laser" and "maser."

In 1962, W. P. Dumke demonstrated that the action of a laser occurred in a *p-n* junction of gallium arsenide. When this junction was forward biased and cooled by liquid nitrogen, the output radiation pulse possessed a frequency of approximately 357 THz. Since that time other laser materials have been discovered and the range of the wavelengths now extends from the middle of the infrared region out to the ultraviolet band.

From this introduction it is clear that we need to discuss the features of the ruby, gas, and semiconductor lasers. However, we should be aware that there are also liquid (dye) and chemical lasers.

Ruby Lasers

The original laser consisted of a circular ruby rod whose ends are mirrored. This rod was surrounded by a xenon flash-tube pump which was connected to a bank of capacitors. When these capacitors were discharged, the xenon tube emitted an extremely intense flash of light with a duration of several milliseconds and a frequency band containing a component of 545 THz. A photograph of Maiman's original laser is shown in Figure 6–54a and Figure 6–54b illustrates its simplified equivalent diagram.

The basic operation of the ruby laser is similar to that of the three-cavity maser described in Section 6–10. One difference is that no direct magnetic field is needed to modify the energy states because they are already at the required levels for laser

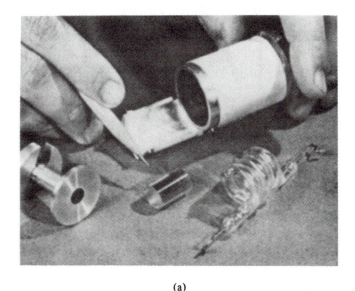

(a)

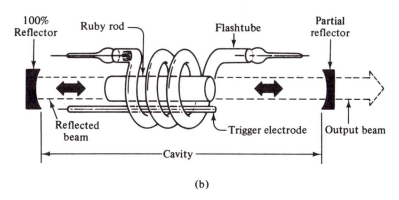

(b)

Figure 6–54   (a) Maiman's original ruby laser; (b) simplified equivalent diagram.

action. Moreover, the laser is primarily an oscillator device, while the maser is used as an amplifier.

When the xenon flashlight pump energizes the ruby rod, chromium ions are raised from energy state E_1 to the level of E_3. Many of these ions then fall back to energy level E_2, but this transition process involves heat and no laser radiation. However, when the E_2 state decays by stimulated emission there is an intense pulse of radiation with a frequency of 432 THz in the red region of visible light; the duration of this pulse is of the order of microseconds or nanoseconds. E_2 is referred to as the metastable state because its lifetime is relatively long (milliseconds).

The length of the circular ruby rod is about 1 cm and its ends are coated with silver to provide the effect of two parallel mirrors. One mirror is fully coated to give total reflection, while the other is partially silvered to allow the emission of the laser's output pulse. For correct operation the distance between the mirrors must be a whole number of wavelengths corresponding to the output frequency. The combination of the ruby rod and the mirrors then forms a so-called Fabry–Perot resonator. Photons travel longitudinally along the axis of the ruby rod (although a number escape from the cylindrical sidewall) and are reflected at the ends to create further stimulated emission. The xenon pump then sustains the population inversion so that the continuous photon buildup results in the generation of the very intense output pulse with a power of several megawatts. After the laser's output has emerged from the partially silvered mirror, the ruby rod is back to its original state and awaits the arrival of the next pulse from the pump.

Although the ruby laser is commonly operated at room temperature, its performance can be improved by cooling the rod to a constant temperature; this overcomes the effect of the heat dissipated in the rod as part of the energy delivered from the pumping source.

The output laser pulse has two important properties. First, the pulse is monochromatic, which means that only a single frequency is involved. In fact, this is not quite true, but the frequency spread is only 1 GHz in a center frequency of about 430 THz. Second, the light pulse leaving the ruby crystal is extremely narrow, with a divergence typically less than 0.1°. For example, it could be arranged that a beam with a diameter of 1 cm at the earth's surface would have a diameter of only about 300 m on reaching the moon; this would be achieved without the necessity for an antenna system. In addition, the light from a laser can be polarized with its electric and magnetic fields in time phase. These characteristics refer to the degree of *coherence* of the laser's output. To summarize, the high levels of directivity (otherwise referred to as collimation), intensity and monochromaticity are unique to the laser and are found in no other radiation source.

A commercial pulsed ruby laser has an efficiency of only 1% and an average power range of 1 to 20 W; the peak power is of the order of megawatts. The pulse, whose frequency is 432 THz, has a repetition rate of one per second and a duration between 0.3 and 6 μs.

A more suitable crystalline substance for continuous (CW) operation of a laser is yttrium-aluminum-garnet or YAG (*not* to be confused with the YIG ferrite material of Section 4–12). To create the most common four-level type of solid-state laser, the YAG lattice is doped with atoms of the rare earth element neodymium. In the four-level system the energy levels are E_1 (ground state), E_2 (laser terminal state), E_3 (metastable state), and E_4 (highest excitation state). The pump then excites electrons from E_1 to E_4 but the output occurs as the electrons fall back from E_3 to E_2 so that the laser frequency is f_{23}. The result is a better performance (compared with the three-level system) with less optical pumping required, so that there is a reduction in the waste of energy.

Q-switching (Q-spoiling).   From the process so far described for the three-level system, it is clear that the degree of population inversion is limited since as the population of the metastable state E_2 reaches the amount required for stimulated emission, its energy is quickly released in the laser pulse. The peak power of the pulse could be increased if the stimulated emission were delayed until a greater population inversion was achieved. The process involved is called Q-switching or Q-spoiling, where Q is the quality factor of the resonant cavity. One method is to leave the front face of the ruby rod partially silvered while the rear face is totally unsilvered. The back reflector is then in the form of an external mirror which is rotated at high speed (Figure 6–55).

When the plane of the mirror is exactly perpendicular to the axis of the laser, the cavity is resonant but as soon as the mirror swings away from this position, the cavity is no longer resonant (the Q factor is "spoiled"). Consequently, there is no buildup of photons by multiple reflections and no laser action occurs. However, as the mirror is rotating, the flash from the xenon pump causes a large buildup in the inversion population. When the mirror returns to the position where the photons

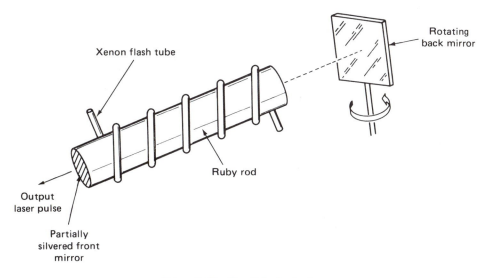

Xenon flash tube

Rotating back mirror

Ruby rod

Output laser pulse

Partially silvered front mirror

Figure 6–55 Q-switched ruby laser.

are reflected back into the rod, stimulated emission occurs and the energy of the highly populated E_2 level is released in one intense laser pulse. This arrangement is called a "Q-switched" or "giant pulse" laser. As an example, if the energy in the pulse is 10 J and its duration is 0.1 μs, the pulse power is 10 J/(0.1 $\times$ 10^{-6}) s = 10^8 W = 100 MW.

A second method of Q-spoiling occurs in the Kerr cell, where a transverse electric field controls the transmission of light through a material in the laser path. If we control the polarization of the light transmitted through the cell, the Q of the laser's cavity can be lowered to the point where the cell is only pulsed by the "on" condition of the applied electric field. In yet another method a container of a bleachable dye is interposed between one end of the rod and its associated mirror. The dye solution remains opaque until the population inversion reaches a level where the spontaneous emission is sufficiently intense to optically bleach the dye. This results in a clear path to the mirror so that the resonant cavity is then established.

Gas Lasers

Solid-state lasers tend to be inefficient because the creation of the population inversion requires a large amount of optical energy, and unfortunately most of the flashlight's spectrum is not of the correct frequency for the desired pump action. A large amount of the pump's energy is therefore wasted as heat, and this requires that the laser be pulsed so that some degree of cooling can occur between the pulses.

Many laser applications require steady [continuous wave (CW)] operation, and this is possible with a gas laser which is continuously pumped from an RF source (Figure 6–56). The RF current traveling through the gas produces some ions and free electrons. The ions and electrons are accelerated by the electric field and excite other gas atoms by collision; the result is a plasma of free electrons and gas ions. The excited atoms can then decay to lower energy states by spontaneous emission, collisions between atoms, and collisions between electrons and excited atoms.

The most common type of gas laser uses a mixture of helium and neon gases at a pressure of 0.3 mm of mercury. The container is made of glass or quartz to which metal contact rings are attached. These rings are connected to the RF pumping source, whose required frequency is approximately 28 MHz. It should be emphasized that the creation of a population inversion in a gas laser is much more complex than in a solid-state laser because of the numerous factors involved. Consequently, there are a number of possible energy transitions which can create stimulated emission. However, the He-Ne laser is commonly used to create a continuous output at a wavelength of 0.6328 μm. This corresponds to a frequency of 472.5 THz and lies in the red region of the visible spectrum. However, the strongest possible emission occurs at the wavelength of 1.1 μm, which is equivalent to a frequency of 273 THz in the infrared region.

To create the resonant cavity, we need a mirror at each end of the plasma tube. It is essential that the mirrors be precisely aligned and therefore they are mounted externally to facilitate their adjustment (Figure 6–56). However, we must then arrange

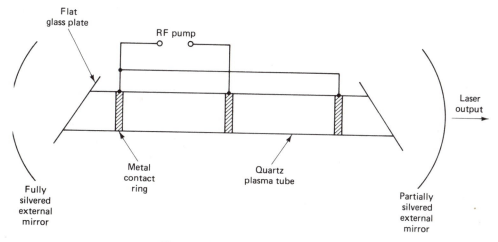

Figure 6-56 Structure of a gas laser.

that the light passes out from the tube with minimum reflection. This is achieved by terminating the tube at either end with an optically flat glass plate which is inclined to the tube axis at the so-called Brewster angle; the reflection from the glass plate back into the tube is then zero provided that the light is correctly polarized.

To improve the stability of the resonant cavity, it is common practice to use spherical rather than plane mirrors; one of the mirrors is then partially silvered to allow the passage of the laser's output, which has less divergence than the output of the solid-state laser.

Although the He-Ne laser produces a CW output, its efficiency is typically less than 1%. It follows that for a 20-mW CW output, the input power from the pump would be several watts. By contrast, a carbon dioxide laser is capable of producing a few kilowatts of infrared CW power at a wavelength of 10.6 μm (28.3 THz) and with an efficiency of about 25%.

Semiconductor Junction Lasers

In 1962, W. P. Dumke demonstrated that laser action was possible in a p^+-n^+ junction which was manufactured from gallium arsenide. However, this type of laser differs from the solid ruby and gas lasers in a number of important features:

1. The thickness of the junction laser may only be about 25 μm and its total volume can be less than 1 mm³.
2. The population inversion occurs in a very thin region at the junction.
3. The required pump is very simple and is in the form of a forward dc bias across the junction.
4. The transition of the charge carriers is between the conduction and valence energy *bands* rather than particular energy levels.

5. The laser characteristics are influenced by junction features such as the degree of doping.

The action of the semiconductor laser depends on pumping electrons and holes into the p^+-n^+ junction; for this reason the device is sometimes referred to as an injection laser, whose structure is illustrated in Figure 6–57. Basically, we have a single crystal of gallium arsenide with a *p-n* junction formed by the standard diffusion method using heavily doped zinc to create the p^+ region. The end faces form the Fabry–Perot resonator and are parallel; one face is totally reflective, while the other is semireflective to allow a passage for the laser's output. By contrast, the side faces are roughened to suppress all unwanted modes of propagation.

In the fabrication process we start with an n^+-type gallium arsenide wafer on which a p^+-type laser is grown epitaxially. After wrapping, the top and bottom surfaces are metalized to allow contact with the forward bias. The wafer is then sliced into slivers, and finally the reflective surfaces are evaporated on to opposite faces of one sliver. It should be pointed out that although the *p* and *n* regions are heavily doped, the degree of doping is less than that of a tunnel diode, so that the characteristic contains no negative resistance section. The sequence of fabrication is illustrated in Figure 6–58; this does not include the heat sink that may be necessary to allow for the large forward current of several amperes.

The operation of the semiconductor junction laser depends on the action of the resonant cavity and on creating a population inversion at the junction. The energy levels involved are those of the conduction and valence bands and suitable energy differences are only found in a semiconductor such as gallium arsenide, not in germanium or silicon. Under the condition of zero bias, the junction is depleted of charge carriers, but when a sufficiently high forward bias is applied, the potential barrier is overcome. Due to the heavy doping, large numbers of holes and electrons will be injected into the layer at the junction. As a result, this layer contains high concentrations of electrons and holes in the conduction and valence bands. These changes in the concentrations of the two energy bands represent the necessary population inver-

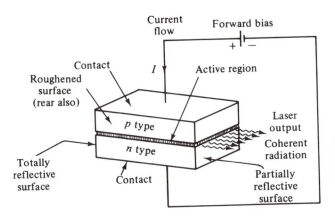

Figure 6–57 Principle of the semiconductor junction laser. (Eleccion, Marce, "The family of lasers: A survey." *IEEE Spectrum*, 9, No. 4, 23–40, March 1972. © 1972 IEEE.)

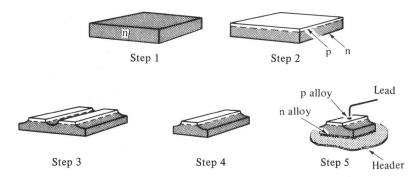

Figure 6-58 Fabrication of the semiconductor junction laser. Step 1, sample of *n*-type material; step 2, addition of diffused *p* layer; step 3, isolation of junctions by cutting or etching; step 4, individual junction to be cut or cleaved into devices; step 5, mounted laser structure.

sion and the layer at the junction is then called the inversion region. Because it is easier to inject electrons rather than holes, most of the inversion region lies on the *p* side of the junction. Possible transitions can then occur between the electrons at the bottom of the conduction band (where the electron concentration is greatest) and the top of the valence band (where the hole concentration is least).

Notice that the electron–hole recombination is taking place directly so that photons are emitted to create the laser action. The percentage of the original injected electron–hole pairs which result in emitted photons is about 70% and the laser has an overall efficiency of 30%. By contrast, most electron–hole recombinations take place indirectly in silicon and germanium so that in their cases little radiation would result.

The transition that occurs in the GaAs laser corresponds to frequencies in the infrared region. The precise wavelength depends on the chemical composition of the gallium arsenide material and lies in the range 0.75 to 0.90 μm (400 to 333 THz).

Commercial semiconductor junction lasers have emitting junction widths between 75 μm and 1.5 mm and can produce pulse power outputs between 2 and 70 W, with corresponding driving currents of 10 to 250 A at room temperature. The pulses have durations of the order of hundreds of nanoseconds with repetition rates between 10 and 20 kHz; the average power output then lies in the milliwatt range. For CW operation a power output of 1 W is possible and this is far more than is necessary to drive a fiber-optic system (Section 3–7).

Heterojunction Lasers

The semiconductor laser so far described contains only a single junction in one type of material; this limits the efficiency and makes it difficult to operate the device at room temperature. However, if multiple layers are used to form a *heterojunction* structure, the laser can then be used for CW operation as required by communication systems. The secret is to confine the injected charge carriers to a very narrow region so that the population inversion can occur with lower levels of the forward current.

One successful method is to confine the very thin (less than 1 μm) laser active p-type GaAs region between two AlGaAs layers which are epitaxially grown on the p region (Figure 6–59). Each AlGaAs layer then forms a boundary which injection electrons cannot cross because of the potential barriers existing at the GaAs/AlGaAs junctions. Not only does this double heterojunction confine the injected carriers to the active region, but the changes in the refractive indices at the boundaries confine the laser output to a very narrow strip whose thickness is about 0.2 μm. This form of laser is then very suitable for fiber-optic communication systems, which presently operate at a wavelength of 0.85 μm. However, frequencies corresponding to wavelengths of 1.30 or 1.55 μm suffer appreciably less attenuation in optical fibers, and these frequencies can be generated by heterojunction InGaAsP lasers.

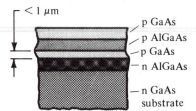

Figure 6–59 Heterojunction laser structure.

Laser Applications

Present and future laser applications may be summarized as follows.

Holography. Laser beams are used to photograph three-dimensional objects by storing both the amplitude and the phase information on a hologram (film). The hologram is then illuminated by a laser beam so that the object is recreated in three dimensions.

Communications. One of the main properties of a laser beam is its time coherence or monochromaticity. The laser beam is therefore ideal for communications and is used as a transmitter for optical fiber systems. In addition, the beam's high frequency allows an enormous bandwidth so that many channels of information can be carried by a single beam. The beam's space coherence or directionality fulfills one of the main requirements for communications in space. However, a commercial laser system for voice and television is not practical on the earth because of the difficulties associated with atmospheric absorption.

Microelectronics. If a laser beam is focused by an optical lens, the energy density of the image is capable of heating, melting, or vaporizing small areas of a variety of materials. For example, in a microwave integrated circuit a laser beam can cut a path a few microns wide and can also be used to weld connections in various parts of the circuit.

Security.   An infrared low-power laser will detect intruders when its invisible beam is broken.

Surgery.   Low-power lasers are used in eye surgery for restoration of detached retinas and other applications.

Rangefinders.   A typical airborne laser rangefinder is used to obtain the profile of the terrain. The aircraft travels in level flight and its altitude above ground level is automatically measured 1000 times per second. The transmitter is essentially a laser/lens array, while the receiver has a lens system, an optical bandpass filter, and a photomultiplier. A range finder of this type was used to land an Apollo spacecraft on the moon. By leaving behind a special reflector it was possible to use a laser technique to measure the distance between the earth and the moon, some 243,000 miles.

Atmospheric pollutants.   Infrared lasers interact with resonances associated with the atoms of pollutants so that specific frequencies are absorbed; in this way the presence and nature of pollutants can be detected.

Radar.   A scanning laser radar system has been developed to ascertain the position and orientation between two spacecraft. Compared with a conventional microwave radar the beam has higher directivity and is extremely narrow. In addition, the required antenna system is very small.

Fusion (the fuel of the future).   A gas dynamic laser uses a supersonic flow of gas to remove waste energy, while a high temperature creates the laser action. This opens the door to energy lasers, which may be used in developing fusion techniques.

Television.   A color television display may in the future be created by using three primary-color laser beams. An argon gas laser would produce the blue and green beams, while the red beam would be obtained from a krypton laser. After modulating the beams and focusing them on to a large screen, the resulting image would be far clearer and more intense than any display which is commercially available at the present time. Another consumer application of the laser lies in the compact disc (record) player.

Meteorology.   Existing methods of detecting the presence of fog banks are unreliable if the fog bank is patchy and therefore not uniform. By contrast, the beam from a laser system would be much less subject to scattering, and consequently its detection capability would be far greater.

"Star wars."   High-energy laser systems are presently under development and will possibly be mounted on orbiting space stations. The laser beams would then be capable of vaporizing hostile satellites and missiles. Alternatively, the laser beams could be operated from earth stations and then focused by mirrors mounted in space.

PROBLEMS

BASIC PROBLEMS

6–1. With the aid of a diagram, detail the construction of a MESFET. What is the principle of the Schottky barrier gate? Sketch some typical drain characteristics for a MESFET.

6–2. Discuss the differences in terms of construction and performance between monolithic and hybrid microwave integrated circuits.

6–3. Describe the construction of a microwave varactor diode and draw a schematic of its equivalent circuit. Sketch a typical characteristic of the varactor diode's capacitance versus reverse voltage.

6–4. Discuss the basic principles of a parametric amplifier in terms of a capacitor that is mechanically pumped. Describe the operation of a degenerate parametric amplifier.

6–5. What factors determine the noise performance of parametric amplifiers? List their advantages and applications.

6–6. What are the basic operating principles of an IMPATT diode? Describe its manufacture and list its applications.

6–7. Describe the operation of a ruby laser and list its main parameters. What is the purpose of Q-switching such a laser?

6–8. How does the operation of a gas laser differ from that of a ruby laser? Compare the applications of these two types of laser.

6–9. What do the letters PIN (diode) stand for? Sketch the construction of such a diode and explain its operation. What are the possible applications of a PIN diode?

6–10. Describe the construction of a Schottky barrier diode and explain its operation. What are the applications of such a diode?

ADVANCED PROBLEMS

6–11. Describe the construction of a microwave bipolar transistor. Discuss the microwave bipolar transistor's parameters in relation to frequency range, power gain, power output capability, and noise figure.

6–12. Draw a block diagram of a frequency multiplier chain which includes both snap-recovery and varactor diodes. If the final microwave power output lies in the C band, indicate at each stage the power and frequency levels.

6–13. Draw the schematic of a three-frequency nondegenerate parametric amplifier and discuss the circuit's operation.

6-14. What is meant by a tunnel diode's negative resistance? Draw the voltage versus current characteristic of a tunnel diode and use it to explain the operation of a tunnel diode oscillator.

6-15. Discuss the Gunn diode in relation to its negative resistance property. Sketch the diagram of a Gunn diode oscillator with its resonant cavity. List the performance parameters of such an oscillator.

6-16. Discuss the basic operation of a TRAPATT diode. Compare the parameters of IMPATT and TRAPATT diodes.

6-17. Discuss the basic operating principle of a BARITT diode. What are its distinguishing features compared with other avalanche diodes?

6-18. Explain the basic operation of the maser device with particular reference to the effect of "population inversion." How does the maser radiate energy received from the pump source?

6-19. Draw a sketch of a semiconductor laser, discuss its construction, and describe its operation. What are the applications of such a laser?

6-20. Compare the voltage versus current characteristic of a backward diode with that of a tunnel diode. Is the backward diode an active or a passive device? What are the applications of the backward diode?

CHAPTER 7 _____

DIPOLES
AND
MICROWAVE
ANTENNAS

7-1 INTRODUCTION

An antenna is defined as an efficient radiator of electromagnetic energy (radio waves) into free space. The same principles apply to both transmitting and receiving antennas, so that we refer to these antennas as being reciprocal; however, the RF power levels for the two antennas are completely different. The purpose of a transmitting antenna is to radiate as much RF power (watts or kilowatts) as possible, either in all directions (omnidirectional antenna) or in a specified direction (directional antenna). By contrast, the receiving antenna is used to intercept an RF signal voltage which is sufficiently large compared to the noise existing within the receiver's bandwidth; at the receiving antenna the power level is only of the order of microwatts or picowatts.

The transmitting antenna itself is some form of metal structure which is capable of radiating electromagnetic energy into free space with an acceptable level of efficiency. We will start by deriving the basic dipole antenna and use it to introduce the meaning of such terms as radiation resistance, bandwidth, beamwidth, and antenna gain. This will automatically lead to an explanation of microwave antennas with their narrow beamwidths and high gains.

The principles of the various antenna systems are discussed in the following topics:

7-2. Hertz Dipoles
7-3. Radiated Electromagnetic Waves
7-4. Radiation Patterns
7-5. Parasitic Elements: Folded Dipole and Yagi Antennas

7–2 HERTZ DIPOLES

We already know that the main disadvantage of a twin-wire transmission line is its radiation loss. However, provided that the separation between the leads is short compared with the wavelength, the radiation loss is small since the fields associated with the two conductors will tend to cancel out. Referring to Figure 7–1a, the equal currents i which exist over the last resonant quarter-wave length of an open-circuited transmission line will instantaneously flow in opposite directions so that their resultant magnetic field is weak. However, if each $\lambda/4$ conductor is twisted back through $90°$ (Figure 7–1b), the currents i are now instantaneously in the same direction, so that the surrounding magnetic field is strong. In addition, the conductors carry a standing-wave voltage distribution with its associated electric field. The standing-wave voltage and current distributions over the complete half-wavelength are shown in Figure 7–1c. At the center feedpoint, the effective voltage is at its minimum level while the effective current value has its maximum level; the effective voltage distribution is drawn on opposite sides of the two sections to indicate that these sections instantaneously carry opposite polarities. In other words, when a particular point in the top section carries a positive voltage with respect to ground, the corresponding point in the bottom section has a negative voltage. The distributed inductance and capacitance (Figure 7–1d) together form the equivalent of a series resonant LC circuit. Notice, however, that the distributed capacitance exists between the two quarter-wave sections and that ground is not involved in this distribution.

By bending the two $\lambda/4$ sections outward by $90°$, we have formed the half-wave ($\lambda/2$) dipole or Hertz (Heinrich Hertz, 1857–1894) antenna. This antenna will be resonant at the frequency to which it is cut. We already know that the *electrical* wavelength in free space is given by

$$\text{electrical wavelength} = \frac{300}{f} \text{ m}$$
$$= \frac{984}{f} \text{ ft} \tag{7–1}$$

This leads to

$$\text{electrical half-wavelength} = \frac{492}{f} \text{ ft} \tag{7–2}$$

where f is the frequency in megahertz (MHz).

The voltage and current waves on an antenna travel at a speed which is typically

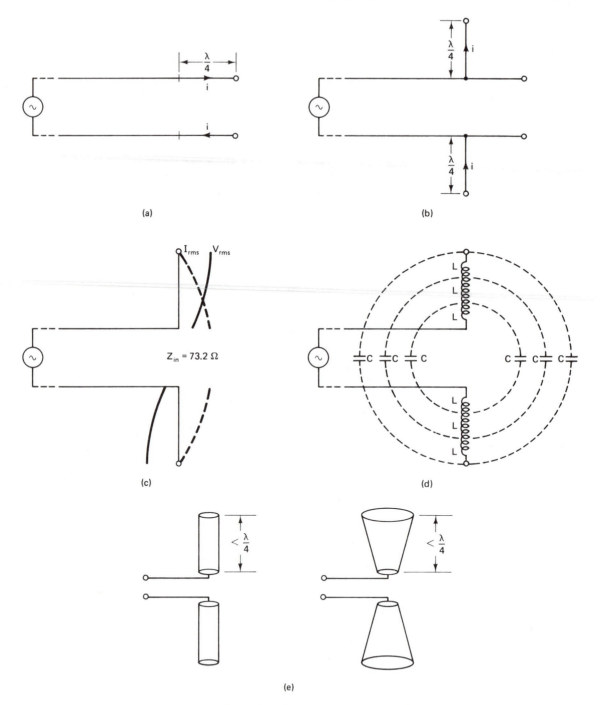

Figure 7–1 Principle of the hertz dipole antenna.

5% slower than the velocity of light. Therefore, the *physical* half-wavelength to which the Hertz antenna should be cut is shorter than the electrical half-wavelength:

$$\text{physical half-wavelength} = \frac{468}{f\,(\text{MHz})}\,\text{ft} \qquad (7\text{--}3)$$

For example, at a frequency of 100 MHz, which lies within the FM commercial broadcast band of 88 to 108 MHz, the length required for the Hertz antenna is $468/100 = 4.68$ ft. Such an antenna would be made from two thin conducting rods, each 2.34 ft long, and positioned remote from ground. At the frequency of 100 MHz, the antenna behaves as a series resonant circuit with a Q of approximately 10. Therefore, the thin dipole is capable of operating effectively within a narrow range which is centered on the resonant frequency. However, we must remember that if the operating frequency is below the resonant frequency, the antenna will appear to be short and will behave capacitively. Similarly, at frequencies above resonance, the antenna will be too long and will be inductive.

If it is required to operate a dipole over a wide range of frequencies, it is necessary to "broad-band" the antenna by lowering its Q without changing its resonant frequency. As an example, if we want to operate a dipole satisfactorily over the range of 125 to 175 MHz, the antenna should be cut to the midfrequency of 150 MHz and should possess a Q of $150/(175 - 125) = 3$. Since the resonant frequency of a series LC circuit is given by $1/2\pi\sqrt{LC}$ while $Q = (1/R)\sqrt{L/C}$, we must lower the distributed inductance L and increase the capacitance C. The solution is to shorten the antenna while increasing its surface area. This gives rise to such "broad-band" shapes as the cylindrical and biconical dipoles (Figure 7-1e).

When the Hertz dipole is resonant and the RF power is applied at the center of the antenna, the input impedance at the feedpoint is a low resistance which may be shown mathematically to have a value of 73.2 Ω (for this reason we often speak of a "70-Ω dipole"). This is referred to as the radiation resistance of the dipole and is the ohmic load which the half-wave antenna represents at resonance; to achieve a matched condition the dipole should be fed with a 70-Ω line. These results are based on the assumption of an ideal antenna; the losses associated with a practical dipole are discussed in Section 7-4.

The RF power P at the feedpoint is given by

$$P = I_A^2 \times R_A \qquad (7\text{--}4)$$

where

I_A = effective RF current at the center of the antenna (A)
R_A = antenna resistance (Ω)

As we move from the center of the antenna toward the ends (Figure 7-2a), the impedance increases from approximately 35 Ω (balanced either side with respect to ground) to about 2500 Ω (not infinity because of the "end" capacitance effects). It is therefore possible to select points on the antenna where the impedance can be

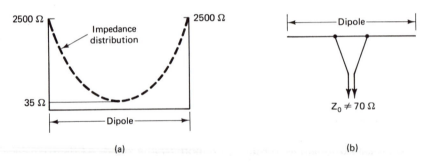

Figure 7–2 Impedance distribution along a half-wave dipole.

matched by a gradual taper to a line whose Z_0 is not 70 Ω; such an arrangement is known as a delta feed (Figure 7–2b).

> **Example 7–1**   Calculate the electrical length in centimeters of a Hertz dipole which is resonant at a frequency of 500 MHz.
>
> *Solution*
>
> $$\text{Electrical wavelength} = \frac{300}{500} \tag{7–1}$$
> $$= 0.6 \text{ m} = 60 \text{ cm}$$
>
> $$\text{Electrical length of the dipole} = \frac{60}{2} = \textbf{30 cm}$$

> **Example 7–2**   The physical length of a Hertz dipole is 2.4 ft. What is the resonant frequency of the dipole?
>
> *Solution*
>
> $$\text{Resonant frequency of the dipole} = \frac{468}{2.4} \text{ MHz} \tag{7–3}$$
> $$= \textbf{195 MHz}$$

> **Example 7–3**   The effective current at the center feedpoint of a resonant dipole is 1.3 A. Calculate the amount of RF power delivered to the dipole.
>
> *Solution*
>
> $$\text{RF power to the dipole} = (1.3 \text{ A})^2 \times 73.2 \ \Omega \tag{7–4}$$
> $$= \textbf{124 W}$$

7–3 RADIATED ELECTROMAGNETIC WAVES

We have established that there are electric and magnetic fields in the vicinity of the antenna (Figure 7–3a). Since the instantaneous voltage and current on the antenna are 90° out of phase, the same phase relationship applies to the E and H fields (Figures 7–3b). These fields are continuously expanding out from the antenna and

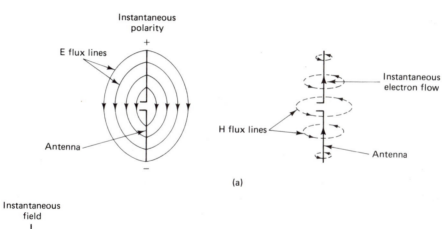

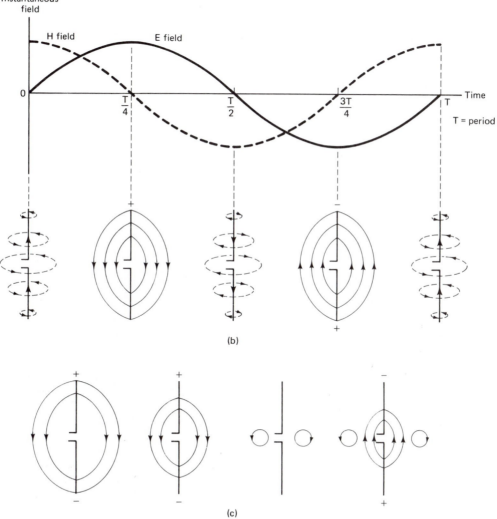

Figure 7-3 Electric and magnetic fields surrounding the half-wave antenna.

collapsing back with the velocity of light. However, since the action is not instanta-neous, the collapse will only be partial, so that closed electric and magnetic loops will be left in space (Figure 7–3c). These loops represent the *radiated* electromagnetic energy which is propagated into free space and travels with the velocity of light. Those flux lines that collapse back into the antenna represent the *induction* field, which is strong only in the immediate vicinity of the antenna.

The fields in space around a half-wave antenna are shown in Figure 7–4. The radiated E and H fields are in time phase, but the two sets of flux lines are 90° apart in space. A vertical Hertz antenna radiates an electric field with vertical flux lines and a magnetic field with horizontal flux lines. In addition, the two sets of flux lines are each at right angles to the direction in which the electromagnetic energy is being propagated. Mathematically, the electric field E (measured in volts per meter), the magnetic field H (measured in amperes per meter), and the transfer of EM energy in a particular direction represent a right-handed system of vectors. The transfer of EM energy is measured by the Poynting vector S, which represents the amount of radiated energy passing through unit area (1 m²) in unit time (1 s). E, H, and S

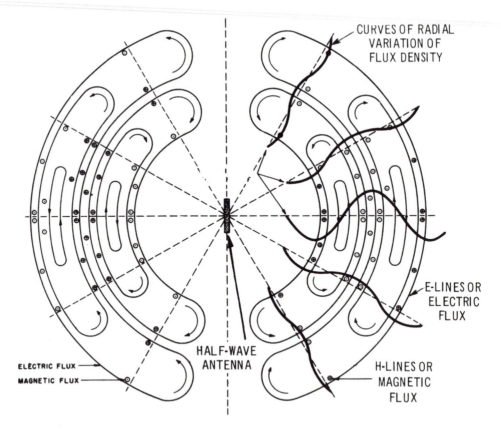

Figure 7–4 Radiated fields surrounding the half-wave antenna.

(in that order) form a right-handed system with S as the vector product of E and H; in terms of units, volts/meter $(E) \times$ amperes/meter $(H) =$ volts $\times$ amperes/ $(\text{meter})^2 =$ watts/$(\text{meter})^2 =$ joules per square meter per second (S). If the instantaneous E direction is rotated through 90° to lie along the instantaneous H direction, the direction of S is that of a right-handed screw which is subjected to the same rotation (Figure 7–5).

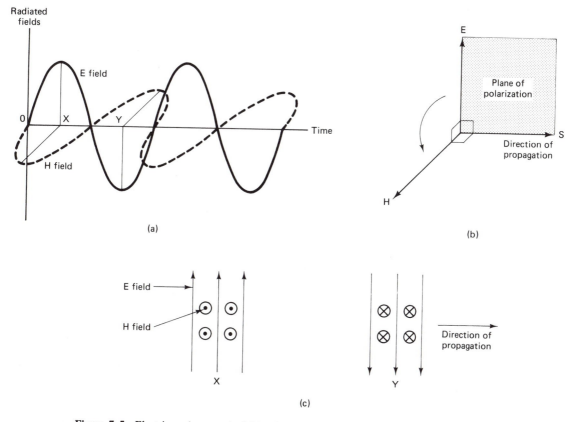

(a)

(b)

(c)

Figure 7–5   Electric and magnetic fields of a vertically polarized EM wave. Part (b) shows the vector diagram of a vertically polarized wave.

Plane of Polarization

The plane containing E and S is referred to as the plane of polarization. A vertical Hertz antenna will radiate a vertically polarized wave with a vertical E field and a horizontal H field. Similarly, a horizontally polarized wave has a horizontal E field, a vertical H field, and is associated with a horizontal antenna. As practical examples, AM broadcast systems have antenna systems which radiate vertically polarized waves while TV broadcast stations and many microwave systems use horizontal polarization.

Summarizing, the Hertz dipole is a half-wave antenna which is equivalent to a 70-Ω resistive load when the antenna is center-fed (current-fed) and is at resonance. The radio wave emanating from the antenna consists of electric and magnetic fields which are in time phase. The two sets of flux lines are 90° apart in space and each set is at right angles to the direction of propagation. The plane of polarization is defined by the directions of the E field and the Poynting vector S, so that, for example, a vertical dipole radiates a vertically-polarized EM wave in which the electric field is vertical and the magnetic field is horizontal.

Example 7–4   A radio wave traveling in free space has an electric field intensity $\mathscr{E}$ of 500 mV/m. The corresponding value of the wave's magnetic field intensity H is 1.326 mA/m. Find (a) the $\mathscr{E}:H$ ratio, and (b) the Poynting vector S.

Solution   (a) The value of the ratio

$$\mathscr{E}:H = \frac{500 \text{ mV/m}}{1.326 \text{ mA/m}} = 377 \text{ }\Omega$$

This value is known as the intrinsic impedance of free space (Section 1–2).

(b) The value of the Poynting vector

$$S = 500 \text{ mV/m} \times 1.326 \text{ mA/m}$$
$$= 663\mu\text{W/m}^2$$

7–4 RADIATION PATTERNS

The radiation pattern surrounding an antenna may either represent the field intensity, $\mathscr{E}$ (volts per meter) or the power density (watts per square meter) distribution. The two patterns are similar in appearance and we will confine ourselves mainly to the field intensity, which directly measures the strength of the electric field component in the radio wave. Theoretically, the value of $\mathscr{E}$ would be equal to the voltage induced in a conductor 1 m long and positioned parallel to the electric flux lines. Although the basic unit of the electric intensity is the volt per meter, more practical units are millivolts per meter and microvolts per meter.

The radiation pattern is a plot of the electric field intensity *at a fixed distance* from the (transmitting) antenna versus an angle measured in the particular plane for which the pattern applies. In the case of the half-wave dipole, it is customary to consider two planes. The $\mathscr{E}$ plane contains the antenna itself and the electric flux lines while the magnetic flux lines lie in the H plane, which is at right angles to the antenna at its center point. Combining the results for the two perpendicular planes produces the antenna's complete three-dimensional radiation pattern.

The radiation patterns for a *vertical* dipole are shown in Figures 7–6a and b. The maximum field strength will be associated with the center of the antenna, where the highest current distribution exists; practically no radiation will occur from the ends, where the current is minimal. The vertical radiation pattern associated with the plane consists of two "oval" shapes (not circles) and is shown in Figure 7–6a.

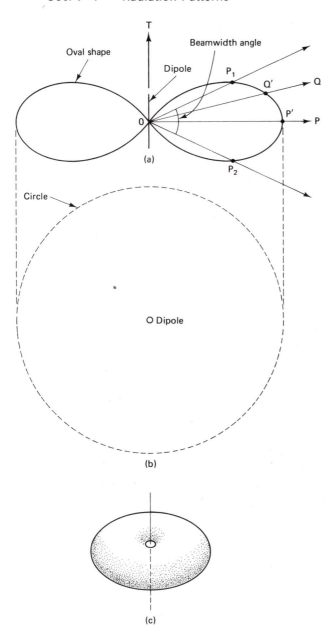

Figure 7-6 Radiation patterns surrounding a vertical dipole: (a) vertical radiation; (b) horizontal radiation; (c) three-dimensional radiation.

In the direction OP the strength of the electric field intensity is at its maximum and is represented by the length OP', while OQ' (which is shorter than OP') is a measure of the field intensity in the direction OQ. For the direction OT, the length intercepted by the radiation plot and consequently the $\mathscr{E}$ value are both zero. Points P_1 and P_2 are those where the electric field intensity is 0.707 of its maximum value

(OP'); the beamwidth is then the angle P_1OP_2 between the half-power points and is about 50° for the half-wave dipole.

The vertical dipole radiates equally well in all horizontal directions so that its pattern in the magnetic plane is a circle (Figure 7–6b). The complete three-dimensional pattern will then resemble a "toroidal" shape or "doughnut" surface with no hole in the center (Figure 7–6c). The same pattern will apply to a vertical dipole used for reception purposes; in any particular direction the length intercepted by the pattern will be a measure of the received signal strength.

Factors Affecting the Field Strength

Antenna power gain and field gain. The value of the vertical field strength $\mathscr{E}$ in the horizontal plane surrounding a thin center-fed vertical dipole is

$$\mathscr{E}=\frac{60I_A}{d}\qquad \text{V/m} \tag{7–5}$$

where

I_A = antenna current at the feedpoint (A)
d = distance from the transmitting dipole (m)

The field strength is therefore directly proportional to the antenna current and is inversely proportional to the distance from the transmitting antenna. This relationship is true only for the radiated field; for the induction field the field strength is inversely proportional to the *square* of the distance and therefore the induction field is important only in the immediate vicinity of the antenna.

Since the transmitter's operating power is directly proportional to the square of the antenna current, the radiated field strength will be directly proportional to the square root of the transmitter power. For example, a thin center-fed dipole has a radiation resistance of 73.2 Ω, so that if the RF power delivered to the antenna is 1 kW, the antenna current is

$$I_A = \sqrt{\frac{P}{R_A}} = \sqrt{\frac{1000}{73.2}} = 3.696 \text{ A}$$

Therefore,

$$P = I_A^2 R_A \qquad \text{and} \qquad R_A = \frac{P}{I_A^2} \tag{7–6}$$

where

P = operating power by the direct method (W)
I_A = antenna current at the antenna feedpoint (A)
R_A = antenna resistance at the antenna feedpoint (Ω)

The field strength at a position 1 mile (= 1609.3 m) from the antenna is

$$\mathscr{E} = \frac{60 \times I_A}{d} = \frac{60 \times 3.696}{1609.3}$$
$$= 0.1378 \text{ V/m}$$
$$= 137.8 \text{ mV/m}$$

This value is used as the FCC (Federal Communications Commission) standard to determine the gain of more sophisticated antennas. For example, when parasitic elements such as reflectors and directors are added to a dipole (Section 7–5), the radiated RF power is concentrated in particular directions so that the antenna gain is increased. This gain may either be considered in terms of field strength or power.

From measurements taken at a distance of 1 mile from the antenna, the field gain is the ratio of the field strength produced by the complex antenna system to the field strength (137.8 mV/m) created by a simple ideal dipole (assuming that this dipole would be capable of directly replacing the complex antenna); in both cases the antenna power is 1 kW.

Since the square of the field intensity is directly proportional to the radiated power, the power gain of the antenna is equal to (field gain)² and is expressed either as a ratio or in decibels. The effective radiated power (ERP) is the power propagated along the axis of the principal radiation lobe associated with a directional antenna. The equations are:

ERP = (RF carrier power delivered to the antenna) × (antenna power gain ratio)
= (RF carrier power delivered to the antenna) × (antenna field gain ratio)²

Antenna efficiency. So far we have only considered the load of a resonant dipole to be of its radiation resistance R_r. However, apart from the electromagnetic energy which is usefully radiated, RF power may also be lost for the following reasons:

1. Heat dissipated in the antenna as a result of its skin effect
2. Eddy currents induced in any metal structures which are near the antenna
3. Corona discharge from the antenna
4. Hysteresis loss in any dielectrics which are used in supporting the antenna

All these effects can be summarized in terms of the antenna's loss resistance R_l, so that if I_A is the effective current at the resonant dipole's center feedpoint,

$$\text{Radiated RF power} = I_A^2 R_r \quad \text{W} \tag{7-8}$$

$$\text{Dissipated RF power} = I_A^2 R_l \quad \text{W} \tag{7-9}$$

$$\text{Total load of the dipole} = R_r + R_l \quad \Omega \tag{7-10}$$

$$\text{Antenna efficiency} = \frac{R_r}{R_r + R_l} \times 100\% \tag{7-11}$$

For VHF and UHF dipoles, efficiencies of the order of 80 to 90% are commonplace.

Notice that since the power gain factor is involved with the amount of RF power *delivered to* an antenna, the losses of that antenna are automatically taken into account when calculating the effective radiated power.

Example 7–5   The field strength is 850 μV/m at a position which is 2.8 miles from the antenna of a transmitter. If the antenna current is now reduced from 5.4 A to 3.7 A, what is the distance of the new 500-μV/m contour from the antenna?

Solution   The new distance from the antenna of the 500-μV/m contour is

$$2.8 \text{ mi} \times \frac{3.7 \text{ A}}{5.4 \text{ A}} \times \frac{850 \text{ } \mu\text{V/m}}{500 \text{ } \mu\text{V/m}} = \textbf{3.26 mi} \tag{7–5}$$

Example 7–6   The input carrier power to an antenna system is 25 W. If the antenna field gain is 15, calculate the value of the effective radiated power.

Solution   The effective radiated power

$$\begin{aligned} \text{ERP} &= 25 \times (15)^2 \\ &= 5625 \text{ W} \\ &= \textbf{5.625 kW} \end{aligned} \tag{7–7}$$

Example 7–7   The input carrier power to an antenna system is 80 W. If the antenna power gain is 7.5 dB, calculate the value of the effective radiated power.

Solution   A gain of 7.5 dB is equivalent to a power ratio of antilog 0.75 = 5.62. The effective radiated power is 80 × 5.62 = **450 W** [equation (7–7)].

7–5 PARASITIC ELEMENTS: FOLDED DIPOLE AND YAGI ANTENNAS

A vertical Hertz dipole has a circular (omnidirectional) radiation pattern in the horizontal plane. The directional properties may be modified by adding parasitic elements to the antenna system. These parasitic elements are metal structures (e.g., rods) which are not electrically connected (nondriven) and are placed in the vicinity of the driven dipole. The radiated field from the dipole induces a voltage and current distribution in the parasitic element which then reradiates the signal. At any position in space the total field strength will be the phasor resultant of the field radiated from the dipole and the reradiated field from the parasitic element.

As a simple example, let us consider a vertical $\lambda/2$ dipole which is separated by a quarter-wavelength from a parasitic element (rod) which is also $\lambda/2$ long (Figure 7–7a). Due to this separation, the field ($\mathscr{E}_{DP}$) reaching the parasite from the dipole will lag the dipole's field ($\mathscr{E}_D$) by 90°. Since the parasitic element is resonant, it behaves resistively and its current I_P will be in phase with $\mathscr{E}_{DP}$ (Figure 7–7b). The reradiated field $\mathscr{E}_P$ lags I_P by 90° and is therefore 180° out of phase with $\mathscr{E}_D$. Again due to the $\lambda/4$ separation, the reradiated field reaching the dipole ($\mathscr{E}_{PD}$) lags $\mathscr{E}_P$ by 90°. In the direction X the resultant field $\mathscr{E}_X$ is the phasor combination of $\mathscr{E}_D$ and $\mathscr{E}_{PD}$; the magnitude of $\mathscr{E}_X$ is the same as that of $\mathscr{E}_Y$, which is the resultant of $\mathscr{E}_{DP}$

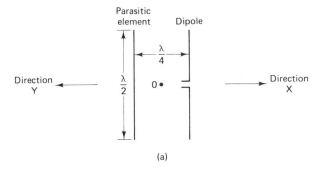

(a)

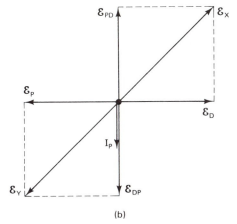

(b)

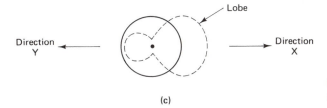

(c)

Figure 7–7 Principle of the parasitic element.

and $\mathscr{E}_P$ in the direction of Y. However, in the two directions which are perpendicular to the plane of the paper and passing through the point 0, $\mathscr{E}_D$ and $\mathscr{E}_P$ are 180° out of phase and the two fields will tend to cancel. The result is to change the shape of the horizontal radiation pattern from a circle (omnidirectional) into two "ovals" or lobes (bidirectional), such that the area of the two ovals is equal to the area of the original circle (Figure 7–7c). In fact, this treatment is oversimplified, since it ignores the reduction in field strength due to the separation between the dipole and the parasitic element. This is taken account of in the field pattern.

Let the parasitic element now be increased in length so that it is more than $\lambda/2$ long and therefore behaves inductively (Figure 7–8a). The parasitic current I_P

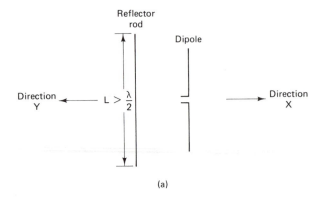

(a)

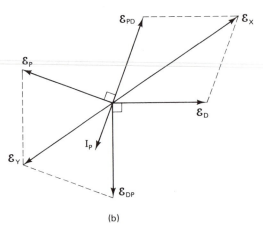

(b)

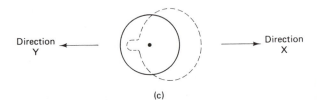

(c)

Figure 7–8 Principle of the reflector element.

lags $\mathscr{E}_{DP}$, so that the various phase relationships are modified as in Figure 7–8b. The magnitude of $\mathscr{E}_X$ is now greater than $\mathscr{E}_Y$, so that the major radiation lobe is pointed toward the X direction while the minor lobe is in the Y direction (Figure 7–8c). The parasitic element is therefore behaving as a reflector since its effect is to increase the radiation in the forward direction (from the parasitic rod toward the dipole) at the expense of the radiation in the backward direction. By comparing the field strengths along the axes of the two lobes we can calculate the "front-to-back" ratio as defined by

$$\frac{\text{forward field strength}}{\text{backward field strength}}$$

For a single reflector rod the front-to-back ratio may be adjusted up to a value of 3:1; to take into account the reduction in field strength between the dipole and the parasitic elements, the separation distance and the length of the reflector rod are typically optimized to 0.1λ and 0.55λ, respectively.

If the length of the parasitic rod is reduced to less than half of the physical wavelength (Figure 7-9a), the rod behaves capacitively, I_p leads $\mathscr{E}_{DP}$, and $\mathscr{E}_X$ is

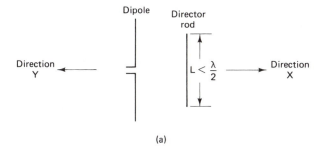

(a)

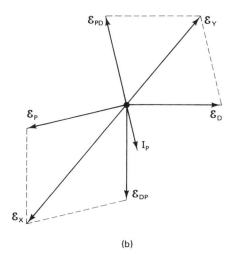

(b)

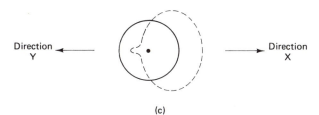

(c)

Figure 7-9 Principle of the director element.

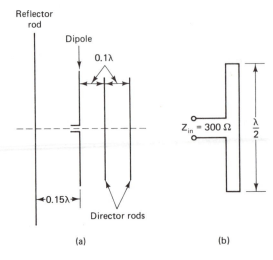

Reflector rod

Dipole

0.1λ

0.15λ

Director rods

(a)

$Z_{in} = 300\ \Omega$

$\frac{\lambda}{2}$

(b)

Figure 7–10 (a) Yagi antenna; (b) folded dipole.

greater than $\mathscr{E}_Y$ (Figure 7–9b): The parasitic element now behaves as a director since it "directs" the radiation toward the major lobe (Figure 7–9c). Typically, the length of the director rod is 0.45 of the physical wavelength and its separation from the driven dipole is 0.1λ.

A combination of a driven dipole, a reflector rod, and a director rod forms a Yagi array (Figure 7–10a), which has a power gain of 5 to 7 dB. The power gain may be increased further by adding more director rods. With the major lobe increased at the expense of the minor lobe, the Yagi system is a unidirectional antenna.

It is common practice to use a folded dipole (Figure 7–10b), which has a lower Q than a thin rod and may therefore be operated effectively over a wider range of frequencies. The input resistance of a resonant folded dipole is approximately 300 Ω and therefore it may be conveniently attached to a 300-Ω twin line. However,

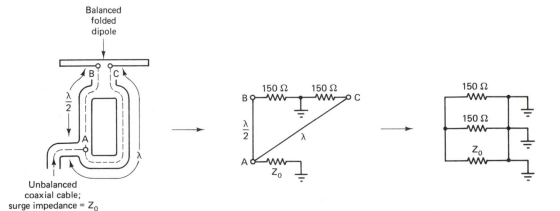

Balanced folded dipole

$\frac{\lambda}{2}$

Unbalanced coaxial cable; surge impedance = Z_0

Figure 7–11 Example of a balun.

if the balanced dipole is fed by a coaxial cable with its outer conductor grounded, it will be necessary to change from the balanced (dipole) to the unbalanced (coaxial cable) condition. This is achieved by means of a "balun" (*bal*anced, *un*balanced), one example of which is the phase-matching transformer (Figure 7–11). The $\lambda/2$ and λ sections provide the necessary 180° phase difference so that the two lines feeding the dipole are balanced with respect to ground. The equivalent circuit shows that for an impedance match, one-quarter of the dipole load must equal the surge impedance of the coaxial cable. In the case of a folded dipole, the coaxial cable's surge impedance must equal $300/4 = 75$ Ω.

7–6 DRIVEN ARRAYS

We have seen that a unidirectional antenna system can be formed by using a driven dipole together with one or more parasitic elements. Even greater directivity and power gain may be obtained from an array of driven antennas whose fields tend to reinforce in the required direction and to cancel in other directions. This effect is achieved by optimizing the spacing of the antennas and the phase relationships between their currents.

The two basic arrangements of driven antennas are called "broadside" and "collinear." In the simplest broadside example two vertical dipoles are spaced $\lambda/2$ apart horizontally and are then fed in phase. Maximum radiation will then occur in the two horizontal directions X and Y (Figure 7–12a), where the distances to the centers of the dipoles are equal. By contrast, there would ideally be no resultant radiation in the horizontal directions P and Q since the two distances would differ by a half-wavelength; this would introduce a 180° phase difference between the two dipole fields at any position along the P and Q directions, so that the fields would

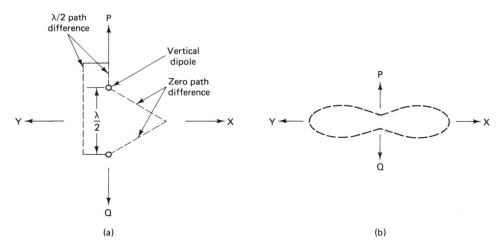

Figure 7–12 Two dipoles in a broadside array.

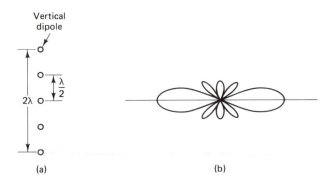

(a)   (b)

Figure 7-13 Five dipoles in a broadside array.

tend to cancel; this is illustrated in the horizontal radiation pattern of Figure 7-12b. Relative to the antennas the horizontal radiation pattern is rotated through a right angle if the dipoles are fed 180° out of phase.

Increased directivity is obtained by using a greater number of broadside antennas. For example, if five vertical dipoles are arranged along a line with λ/2 spacing between adjacent dipoles (Figure 7-13a), the total length of the array is 2λ and its radiation pattern is illustrated in Figure 7-13b. Although the major lobes are more pronounced, there are also a number of parasitic or minor lobes.

In a simple collinear arrangement two horizontal dipoles are placed side by side with their centers separated by about a wavelength. Along certain directions in the horizontal plane the distances to the center of the dipole will differ by half a wavelength, so that cancellation will occur. If the collinear array is increased to four dipoles which are spaced one half-wavelength apart, the horizontal pattern consists of two major and four parasitic lobes (Figure 7-14).

A combination of broadside and collinear arrays is shown in Figure 7-15. This horizontal array consists of four driven dipoles which are delta fed in phase. These dipoles have a vertical spacing of half a wavelength and are mounted about a wavelength apart horizontally. Parasitic reflectors reduce the back radiation so that the single major lobe is concentrated along the forward X direction. For such directional arrays we can illustrate their field patterns in one plane by using either a rectangular or a polar representation (Figure 7-16).

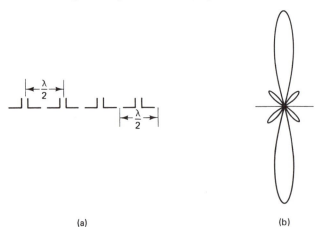

(a)   (b)

Figure 7-14 Four dipoles in a collinear array.

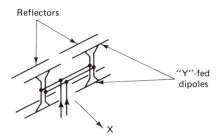

Figure 7–15 Combination of broadside and collinear arrays.

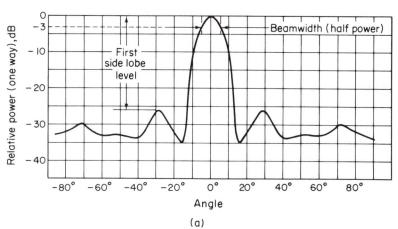

(a)

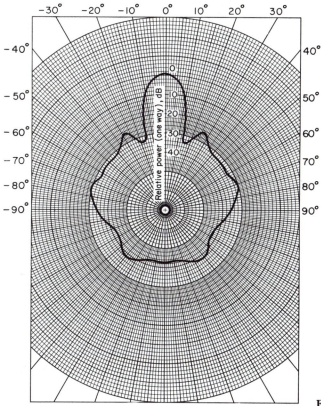

(b)

Figure 7–16 (a) Rectangular and (b) polar field patterns of a directional antenna.

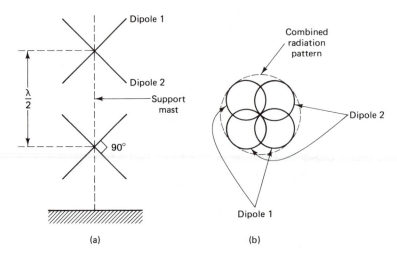

Figure 7–17   Turnstile antenna.

Turnstile Antennas

This antenna name is derived from its appearance. In its basic form it consists of two horizontal dipoles which cross at right angles and are fed 90° out of phase (Figure 7–17a). The horizontal radiation pattern for each dipole consists of two ovals, but when these are combined, the result is roughly circular (Figure 7–17b). This is due to feeding the antennas 90° out of phase so that the resultant field strength must be found by using the Pythagorean equation. To increase the power gain, a number of turnstile antennas may be mounted on a mast with a vertical spacing of half a wavelength between adjacent antennas; this will provide more directivity in the vertical plane. Modified turnstile antennas are commonly used for TV transmission.

7–7 MICROWAVE ANTENNAS

In discussing the dipole systems of the previous sections, we referred to such factors as antenna gain, impedance, radiation resistance, plane of polarization, bandwidth, beamwidth, and the nature of the field pattern. These are the main parameters of an antenna system, but for microwave antennas we must also take into account the following:

1. For such applications as radar and microwave links (Section 7–8) there is a requirement for highly directional antennas with narrow beamwidths of the order of 2° or less. There is virtually no need for omnidirectional antennas.

2. At the higher microwave frequencies the size of the components and the active devices is automatically reduced. This lowers their power-handling capacity,

so that the associated antenna systems must have a very high gain to achieve the required level of effective radiated power.

3. Compared with the lower frequencies there is more receiver noise in the microwave region. Consequently, the receiver signal should be as large as possible in order to achieve an adequate level of signal-to-noise ratio.

4. Since, at microwave frequencies, the wavelengths are only of the order of a few centimeters or less, an antenna system whose dimensions involve a number of wavelengths need not be physically large. Consequently, there are various microwave antenna arrangements which would not be feasible at lower frequencies.

Reflector-Type Microwave Antennas

We have already discussed in Section 1–8 the use of a paraboloidal reflector in providing a high antenna gain and a main lobe whose beamwidth is normally of the order of a few degrees. The parabola itself is a mathematical curve which is defined in relation to a line AOB (the directrix) and a point F (the focus). At point P (Figure 7–18) on the parabola, $DP = PF$ and the parabola is then the result of joining together all such points. As illustrated, the parabola extends mathematically toward infinity, but in practice it must be constructed so that the aperture, or diameter, d is not less than 10 wavelengths long. The source of microwave energy is placed at the focus F. Any ray of electromagnetic radiation that strikes the parabola will be reflected back along the direction parallel to the axis. The theoretical result is a undirectional beam of constant intensity. However, in practice, there are edge effects at the ends of the limited parabola, and no microwave source radiates equally well in all directions. Furthermore, there will be a certain amount of direct radiation which does not strike the reflector.

The most common type of reflector is the paraboloid which is formed by revolving the parabola about its axis. When a microwave source is placed at the focus (Figure 7–19a), it is normally arranged that the power density at the edge of the paraboloid is 10 dB down on the power density at the center C. The practical three-dimensional

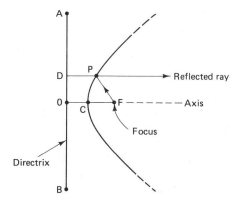

Figure 7–18 Mathematical curve of a paraboloidal reflector.

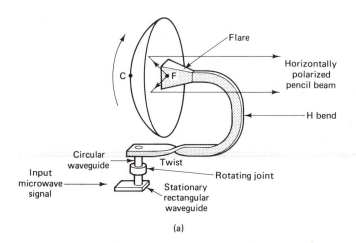

(a)

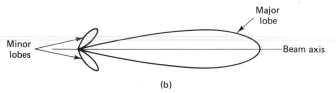

(b)

Figure 7–19 (a) Rotating paraboloidal reflector; (b) lobes.

pattern is a pencil beam consisting of a narrow major lobe, shaped like a cigar, and with equal beamwidths in the vertical and horizontal planes. In addition, there are a number of weak minor side lobes (Figure 7–19b). Recalling equations (1–12) and (1–13), we have

$$\text{beamwidth of major lobe} = 70\frac{\lambda}{d} \quad \text{deg} \tag{7–12}$$

and

$$\text{antenna power gain} = 6\left(\frac{d}{\lambda}\right)^2 \tag{7–13}$$

If the paraboloidal dish is uniformly radiated,

$$\text{beamwidth} = \frac{51\lambda}{d} \quad \text{deg} \tag{7–14}$$

At the lowest microwave frequency of 1 GHz, the wavelength λ is 30 cm and a practical Hertz dipole would be slightly less than 6 in. long. A number of such dipoles may be placed side by side and used with a parabolic cylinder, which is the surface formed by projecting the parabola into the surface of the paper (Figure 7–20a). Such an antenna system produces a "fan beam," whose three-dimensional radiation pattern has an elliptical rather than a circular cross section. In other words, the beamwidths are not the same in the vertical and horizontal planes.

If the ends of the parabolic cylinder are closed, the result is the "cheese" or

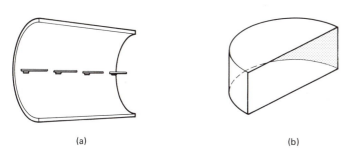

(a)   (b)   **Figure 7–20** Types of parabolic reflector.

"pill box" reflector (Figure 7–20b), which has the advantage of a much smaller size than the full dish. However, although this can be used to create a narrow beam in the horizontal plane, the beam in the vertical plane is much less restricted. An example of a case where this is not a disadvantage is a shipborne surface detection radar system (after all, one ship cannot be positioned *above* another!).

In some applications it is required that the beam be specially fanned out in the vertical plane. For example, we shall assume that a radar set, mounted on an aircraft which is flying at a height h, is required to provide equal microwave illuminations on ships that are located at different ranges d (Figure 7–21). Owing to the spreading out of the spherical wavefront from the transmitter, the electrical field intensity falls off *inversely* as the ship's range increases. From Figure 7–21,

$$\frac{h}{d} = \sin \theta \qquad \text{or} \qquad d = r \csc \theta \tag{7–15}$$

It follows that the field strength diagram of the antenna system must obey the "csc θ" relationship. This result may be achieved in three ways (Figure 7–22). In Figure 7–22a, the mirror is specially shaped to provide some degree of concentration for the downward rays and its cross section is comparable with that of the parabola. An alternative solution is to split the reflector into two parabolic halves and move the upper section a little in front of the lower. Although the focus is not quite the same for each half, the difference may be neglected and a mean focus F is assumed (Figure 7–22b).

Let us consider two rays FAB and FCD, which make equal angles with the axis. The ray FAB arrives at point B in the aperture plane before ray FCD arrives at point D. If we regard points B and D as radiating microwave sources, the source at B is leading in phase on the source at D. The radiation from the two sources will combine in the downward directions BH and DK, provided that their shift in distance BK equals the path difference between FAB and FCD. Mathematically,

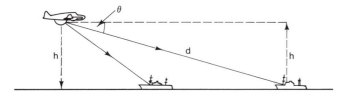

Figure 7–21 Radar detection of ships by an aircraft.

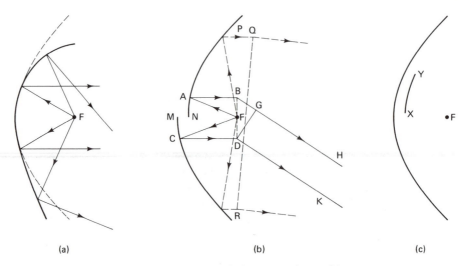

Figure 7–22   Reflectors which produce the csc θ beam.

this path difference is approximately equal to $2MN$, where MN is the fixed displacement between the two halves and the reflector.

The downward deflection of the beam is more pronounced for pairs of rays near the center of the mirror than for those near the edge. This is illustrated by the dashed-line rays in Figure 7–22b. In order that their distance shift PQ should equal $2MN$, the outgoing rays do not have to be deviated very far downward since these rays start from a wide baseline. In practice, we see that only the center portion is important in modifying the radiation pattern. Consequently, instead of splitting the paraboloid, we can include an additional metal plate XY (Figure 7–22c). This plate is attached by adjustable screws (not shown in the figure) and its distance from the main mirror is varied until the required radiation pattern is obtained. The field gain of the Csc θ reflector is only about half that of the full paraboloid, but its radiation pattern should theoretically have no zeros and no minor lobes.

Primary feeds for the paraboloidal reflectors.   As we have already mentioned, the practical paraboloidal reflector cannot be uniformly illuminated with microwave energy, owing to the peculiar field strength distribution that would be required from the "primary" source placed at the focus. Referring to Figure 7–23, the rays FA and FC have been drawn to enclose an angle of 5° and result in the illumination of region BD in the aperture plane. Another pair of rays, FH and FK, also enclose an angle of 5° and illuminate region GN. We observe that GL is about $1.5 \times BD$. Consequently if the aperture is to be equally illuminated, less power must be directed from the primary source in directions FA and FC than in directions FH and FK. It can be shown that the primary source must have a distribution such that the field strength at a given distance in the direction toward the edge of the mirror is twice that (at the same distance) in the direction of the center of the reflector.

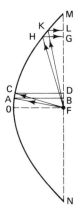

Figure 7–23 Illumination of a paraboloidal reflector by a primary source.

Most practical primary sources provide less field strength at the edge of the aperture than at the center. Although this results in a reduction of the side lobes, it also broadens the main lobe so that the beamwidth is greater. Moreover, the primary source may provide some direct radiation which is not reflected from the paraboloid. Such radiation could spread out in all directions and further spoil the shape of the field pattern.

One example of a primary feed is illustrated in Figure 7–24a. A coaxial cable passes through the center of the reflector and ends in a half-wave dipole with its own parasitic reflector, so that the EM energy is directed toward the paraboloid. However, the field strength diagram of the primary source is different in the E and H planes. In the E plane more energy is concentrated toward the center of the paraboloid, so that the emerging vertically polarized beam is broader in this plane than in the H plane.

Unless a balun device (Section 7–5) is included, RF currents flow in the outer

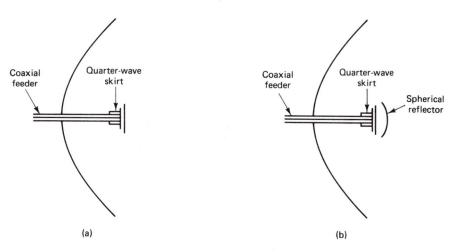

Figure 7–24 Primary sources using coaxial feeders.

conductor, which is directly connected to one-half of the dipole. The quarter-wave skirt inserts an infinite impedance between the outer conductor and the open end of the skirt. This decouples the rest of the outer conductor from the dipole's feedpoint.

Instead of a parasitic reflector, a small metal sheet reflector is sometimes used; this additional reflector is commonly in the form of a spherical shell (Figure 7–24b). If the coaxial cable is replaced by a waveguide (Figure 7–25), a metal tongue can protrude into the guide; the antenna system can then be mounted on both sides of the tongue. When the guide is operated in the dominant TE_{10} mode, the electric field is split into two parts by the tongue, and therefore both sides of the antenna system are excited as the EM wave emerges from the guide. The final radiated beam is again vertically polarized.

In yet another version, which is referred to as a Cutler feed (Figure 7–26), the guide is terminated by a small cavity in which are two slots, each sealed by a dielectric material. These slots act as the radiating elements of the primary source.

All three methods described so far have been examples of rear feed, which has the advantage of reducing the required length for the coaxial cable or the waveguide. However, it is possible to bring the guide around the edge of the reflector and then the end of the guide acts as the primary source; this is illustrated in the horizontally polarized antenna system of Figure 7–27.

To obtain an impedance match with free space, the end of the guide is flared and can be terminated by a dielectric baffle. Although direct radiation from the source is largely avoided, it is difficult to determine its field strength distribution. In general, the final field strength diagram is wide in the plane parallel to the guide's b dimension and narrow in the plane parallel to the a dimension.

We have already referred to the effect created by the direct radiation from the primary source. To reduce this problem there are a number of methods, of which the most popular is the Cassegrain feed (Figure 7–28), appropriately named after

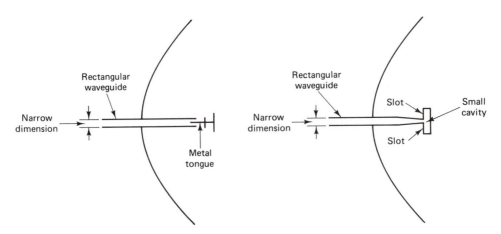

Figure 7–25 Primary source using a rectangular waveguide.

Figure 7–26 Cutler rear feed.

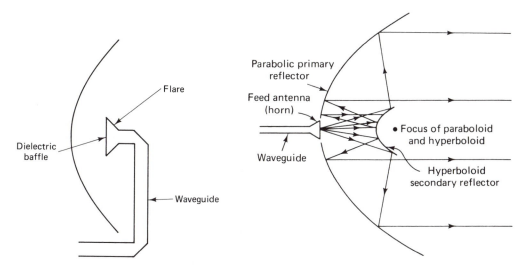

Figure 7-27 Front feed.   **Figure 7-28** Cassegrain feed.

one of the early astronomers. This system incorporates a second hyperboloid reflector one of whose foci, F, coincides with the focus of the paraboloid. The geometry of the arrangement then makes it appear that all rays come from point F before striking the paraboloid, so that the radiation finally emerges as a parallel beam. In addition to reducing the direct radiation from the primary source, the Cassegrain method has the further advantage of shortening the required length for the waveguide feed. However, the secondary reflector obstructs part of the radiation from the paraboloid.

Most of this disadvantage can be overcome by using a large paraboloid and placing the flare of the waveguide as close as possible to the secondary reflector, whose required size is correspondingly reduced. This is an attractive solution when operating with a high level of microwave power. An alternative to the Cassegrain method is the Gregorian system, which uses an ellipsoid rather than an hyperboloid. However, the ellipsoid must be placed farther away from the primary source than the corresponding hyperboloid, and this has the disadvantage of allowing more direct radiation.

Feeding the cheese reflector.   The cheese reflector or mirror consists of a metal parabolic cylinder with flat metal top and bottom plates (Figure 7-29a). The aperture plane usually passes through the focus. The aim is to fill the aperture forming the mouth of the cheese with radiation and thereby obtain a set of microwave sources distributed over this area. The aperture then acts like a complete broadside array of rectangular shape and gives an approximate field strength diagram which is narrow in the plane corresponding to the wide dimension and wide in the other plane. The primary feeding arrangements vary according to the polarization of the wave in the guide.

When the polarization is such that the electric field is parallel to the top and

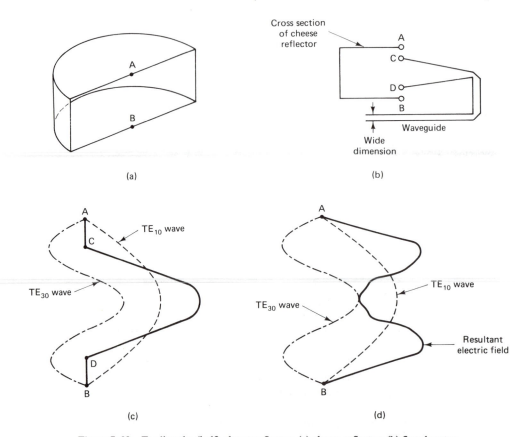

Figure 7–29 Feeding the (half)-cheese reflector: (a) cheese reflector; (b) flared rectangular waveguide; (c) distribution of the electric field at the mouth of the flared waveguide; (d) distribution of the electric field across the narrow dimension of the cheese reflector's aperture.

bottom plates of the cheese, a flared rectangular waveguide is commonly used in the TE_{10} mode. This is illustrated in Figure 7–29b and provides a horizontally polarized output. The flaring takes place gradually in the wide dimension of the guide so that the TE_{10} mode is still present at the mouth of the flare. The width of the mouth is about two-thirds of the height of the cheese. The distribution of the electric field in the TE_{10} mode across the wide dimension is sinusoidal in amplitude, as shown by the full curve between points C and D in Figure 7–29c. The mouth of the guide is CD and the aperture of the mirror is AB, which is the distance between the top and bottom plates. The sinusoidal curve between C and D can be analyzed approximately into two sinusoidal waves fitted into the portion AB. We may then regard the cheese mirror as a type of waveguide in which there are two waves present; these are (1) the dashed curve, representing the amplitude distribution for a TE_{10} wave, and (2) the dotted-dashed curve, which corresponds to a TE_{30} wave. The algebraic sum of these two waves gives the original TE_{10} wave emerging from the mouth *CD* of the flared guide. From the known properties of waveguides (Section 3-4) these waves must travel at different speeds. The depth of the cheese and its height are so adjusted that the two waves, after reflection at the curved surface of the cheese, arrive back in the plane of the aperture with an opposite phase relationship. The resultant amplitude distribution is therefore given by the difference between the dashed and dotted-dashed curves, as shown in Figure 7-29d by the solid-curve. It is noticeable that the field is now distributed right across the aperture, but there is an undesirable dip in the center. However, probably owing to the higher modes, the dip is much less pronounced in practice and a reasonably uniform distribution of amplitude is obtained across the opening of the cheese mirror in the short dimension.

The distribution of the amplitude across the long dimension depends on the field strength diagram of the waveguide opening in this plane. We have already seen (Section 7-7) that for uniform distribution across the opening, more radiation must be sent out from the guide in directions toward the edges of the cheese mirror than toward the center. This is unlikely to be achieved in practice, so that one can expect a tendency toward a tapered distribution of power across the wide dimension of the aperture with a corresponding decrease in gain, broadening of the beam, and reduction of side lobes in comparison with the effects of uniform distribution across the aperture.

When the polarization is such that the electric field is perpendicular to the upper and lower plates of the mirror, the method of flaring the waveguide does not apply. The difficulty is now to obtain a good distribution of power across the long dimension of the aperture since the waveguide is wide in this plane and directs the beam considerably into the center of the cheese. Dielectric lenses, fastened to the waveguide mouth and designed to diverge the beam, have been used successfully.

The quarter cheese of Figure 7-30 allows a more efficient feed from the flare which extends between the top and bottom plates. The flare is placed at the focus of the parabolic cylinder, and the resulting illumination of the reflecting surface is better than that of the half-cheese. Consequently, the quarter-cheese's radiation pattern is superior but its total available EM radiation is less.

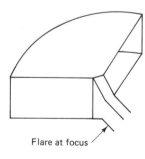

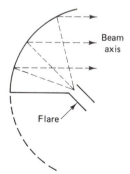

Figure 7-30   Quarter-cheese reflector.

Antenna reflector problems.   The theoretical radiation pattern from an infinitely wide paraboloid is a pencil beam provided that the reflecting surface is uniformly illuminated. The departure from this ideal situation is due to the following difficulties:

1. The practical paraboloid is limited in size with an aperture whose diameter normally exceeds 10 wavelengths. At the edge of the paraboloid, the radiated beam will be diffracted so that a number of side lobes are created.

2. The surface of the reflector must deviate to some degree from that of an ideal paraboloid. The amount of this deviation should not normally exceed one-sixteenth of a wavelength, which at a frequency of 18.75 GHz, would correspond to a distance of 1 mm. With large dishes surface distortion can occur as the result of uneven wind pressure, but this effect may be reduced by using a wide mesh rather than a smooth continuous surface.

3. Ideally, the primary feed would behave as a point source which is exactly located at the focus. Since this cannot be achieved in practice, the result is to increase the beamwidth and strengthen the side lobes. This particular problem can be partially overcome by using a larger aperture together with the Cassegrain or Gregorian methods of providing the primary feed.

When a dipole is used as the primary antenna, we have another problem in the sense that the three-dimensional radiation pattern is doughnut-shaped rather than spherical; this will tend to flatten the major lobe from the reflector. Moreover, if the dipole is replaced by a circular horn, the entire surface of the paraboloid is still not uniformly illuminated; this causes (as we have discussed previously) a broadening of the main lobe but a reduction in the side lobes. In addition, there is a loss of field strength between the primary feed and the reflector.

The problems we have presented are compensated by the narrow beamwidths and high gains available from the antenna system, which employ some form of parabolic reflector. Although we have referred only to the paraboloid and cheese types, there are other specialized forms, such as the hourglass, the truncated paraboloid, the "orange-peel" paraboloid, and the corner reflector (Figure 7–31).

In the case of the hourglass antenna, the reflector is stationary and only the primary source needs to be rotated; however, less of the reflector is illuminated and the beam is less well defined (Figure 7–31a).

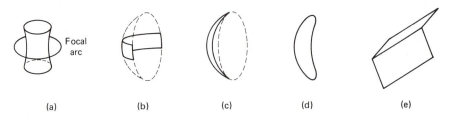

  (a)   (b)   (c)   (d)   (e)

Figure 7–31 Specialized types of reflector.

A truncated paraboloid (Figure 7–31b) is parabolic in the horizontal plane so that the energy is focused into a narrow beam. With the reflector truncated, or cut, so that it is shortened vertically, the beam spreads out in that plane instead of being focused. Such a fan-shaped beam is used to determine the azimuth accurately. Since the beam is wide vertically, it will detect aircraft at different altitudes without changing the tilt of the paraboloid. Such an antenna is also effective for surface search at sea by overcoming the effects of the pitch and roll of the ship.

The truncated paraboloid reflector may be used in height-finding systems if the reflector is rotated through 90° (Figure 7–31c). Since the reflector is now parabolic in the vertical plane, the energy is focused into a narrow beam vertically. With the reflector truncated so that it is shortened horizontally, the beam spreads out in that plane instead of being focused. Such a fan-shaped beam can be used to determine the height of an aircraft accurately.

Figure 7–31d illustrates a paraboloidal section, often called an orange-peel reflector because of its shape. Since such a reflector is short horizontally but long vertically, the emerging beam is wide in the horizontal plane and narrow in the vertical plane; its shape therefore resembles a flat "beaver tail." The RF energy is directed toward the parabolic reflector by a radiating flare connected to the waveguide. Its main advantage is that almost all of the microwave energy illuminates the reflector, with very little escaping from the sides. This type of antenna system is generally used in height-finding equipment.

The corner reflector antenna consists of two flat conducting sheets which meet at an angle to form a corner, as shown in Figure 7–31e. This type of reflector is normally driven by a half-wave radiator located on a line that bisects the angle formed by the sheet reflectors.

For some applications reflectors are too large and expensive, so we now turn our attention to a simple form of microwave antenna, such as the radiating horn.

Horn Antennas

We have seen that when a waveguide is terminated by an appropriate flare (Section 4–7), we can obtain an approximate impedance match between the waveguide and free space. However, the flare or horn can take a variety of shapes (Figure 7–32) and each will possess its own beam shape, beamwidth, and field gain. For example, a sectoral horn flares out in one plane only and may be compared with the cheese reflector previously discussed. By contrast, the pyramidal horn flares out in two perpendicular planes. In all cases, the radiating beam tends to have a spherical rather than a plane wavefront. The performance of a sectoral horn depends on the values of the flare angle θ and the ratio of the horn's length l to the free-space wavelength λ. In the case of the illustrated H-plane sectoral horn, the beam is vertically polarized and is comparatively sharp in the horizontal plane. However, there are conflicting factors which affect the value chosen for the flare angle. If the value of θ is too small, the area at the end of the horn will be restricted; as a result, the amount of diffraction at the edges will be increased and the degree of directivity will be reduced.

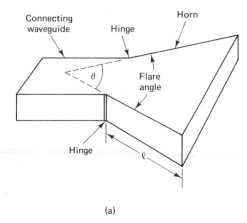

(a)

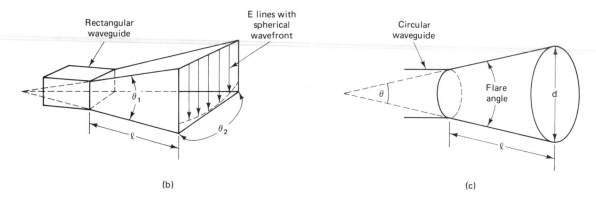

(b)   (c)

Figure 7–32   Types of horn antenna: (a) sectoral; (b) pyramidal; (c) circular.

On the other hand, if θ is too large, there will be a wide spherical wavefront so that the directivity is adversely affected. These results are summarized in Table 7–1.

TABLE 7–1   PARAMETER VALUES FOR TWO SECTORAL HORNS

Flare angle θ (deg)	Ratio l/λ	Beamwidth (deg)	Power gain (dB)
40	6.0	65	16
15	50.0	25	20

     In some cases the side walls of the sectoral horn are hinged so as to adjust the flare angle for optimum results. Maximum radiation ordinarily occurs for radiation angles between 40 and 60°. In addition, the characteristics of the horn may be changed further by using exponentially curved sides as opposed to flat planes.

For the pyramidal horn the radiation pattern depends on the two flare angles, θ_1 and θ_2. Wider vertical patterns are available by increasing θ_1, while broader horizontal patterns are obtained by raising θ_2. In the special case when θ_1 and θ_2 are equal, the radiation pattern is symmetrical in the vertical and horizontal planes. This is also true for the circular horn, which is fed from a cylindrical waveguide; the longer such a horn is in relation to the wavelength, the greater is its gain for which the optimum flare angle is approximately 50°.

In general, horn antennas have a simple mechanical construction and adequate directivity for some applications and a bandwidth of approximately 10°. The directivity may be improved by using a horn antenna as the primary source for some form of parabolic reflector system. Two important examples of such arrangements are the hoghorn and Cass-horn antennas.

Hoghorn antennas.   One version of the hoghorn antenna (Figure 7–33) uses a modified horn into which is built a small parabolic reflector. This serves to direct the microwave energy horizontally into the main cheese reflector so that the vertical beamwidth is reduced. A second version of the same antenna is illustrated in Figure 7–34a. It consists of a paraboloidal reflector which is attached to the end of a pyramidal horn. The diagram of Figure 7–34b shows that the rays converge at a point, F, which is both the focus of the paraboloid and the starting point of the horn. Then as the antenna revolves about its axis of rotation, point F remains stationary; this is an advantage when a low-noise antenna is required for satellite communication.

Another low-noise microwave antenna is the Cass-horn, whose principle is illustrated in Figure 7–35. There are two paraboloidal surfaces, with the upper surface having the greater degree of directivity, and a focal point F, which lies at the center of the lower surface. With the primary conical horn positioned at F, the geometry of the system allows sample rays to be reflected from the top surface to the bottom

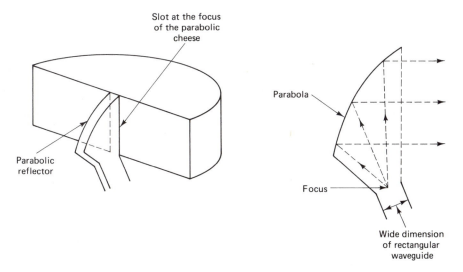

Figure 7–33 Hoghorn antenna with a cheese reflector.

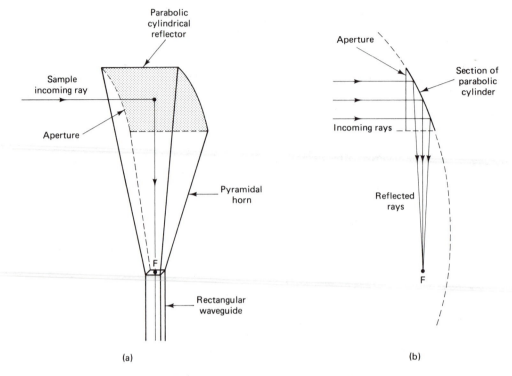

(a)

(b)

Figure 7–34 Hoghorn antenna with pyramidal horn.

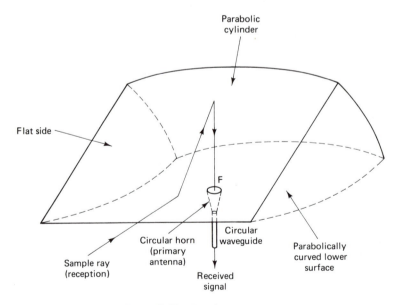

Figure 7–35 Cass-horn antenna.

surface. A second reflection then allows the radiation to emerge from the secondary horn as a collimated beam of parallel rays. For reception purposes the direction of the rays is reversed. Compared with the Cassegrain feed method discussed previously, the Cass-horn has the advantage that the primary source does not prevent any obstruction to an emerging beam.

Lens Antennas

We have already seen that the radiated beam from a pyramidal horn has a spherical rather than a plane wavefront. As a spherical wavefront is propagated through space, it spreads out so that the radiation pattern is not too sharp or directive. By contrast, a plane or constant phase wavefront does not spread out because all of the wavefront moves forward in the same direction. The conversion from a spherical to a plane wavefront can be accomplished by a parabolic reflector, but an alternative method is to use a dielectric or metal-plate lens, which is placed in front of the horn. The principle involves the optical properties that microwaves possess.

The operation of a dielectric lens is shown in Figure 7–36. Such a lens is commonly made of polystyrene, whose relativity permittivity is 2.56. The phase velocity in this dielectric is therefore only $(1/\sqrt{2.56}) \times 100 = 62.5\%$ of the free-space velocity. By using a plano-convex lens, the portion, CC', toward the center of the advancing spherical wavefront will pass through the thickest part of the lens and will therefore suffer the greatest amount of "slowing down." Near the top and bottom edges, EE', of the lens, the dielectric material is thin, so that the wavefront is hardly slowed at all. To achieve the desired result, the lens must be correctly shaped and if the microwave source is to be regarded as a point, it must be positioned at the focus of the lens.

The problem with the dielectric lens is its thickness, since if the wavefront velocity is to be appreciably reduced, the width of the lens at its center must measure

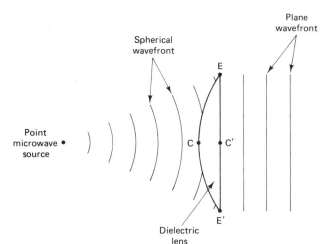

Figure 7–36 Conversion from a spherical wavefront to a plane wavefront by means of a dielectric lens.

several wavelengths. It follows that the lower-frequency limit for a dielectric lens is about 10 GHz (X band) which corresponds to a free-space wavelength of 3 cm. A thick lens also absorbs a considerable amount of microwave power due to dielectric hysteresis, whose effect also imposes an upper frequency limit. Consequently, the dielectric lens is not a practical solution at the lower microwave frequencies and, moreover, would not be recommended for wavefronts with a small radius of curvature. To reduce the required thickness for a dielectric lens, it is possible to *zone* or *step* its shape, as illustrated in Figure 7–37. If the differences in the phase shifts associated with the sample ray paths *SABC*, *SDEF*, *SGHJ*, *SKLM*, and so on, are 360° or multiples of 360°, a plane wave will emerge along line *CFJM*. In terms of the curved wavefront, its center portion is slowed down; the construction of the lens then allows the edges of one or more *preceding* wavefronts to "catch up" and create the final plane wavefront. Although both lenses shown in Figure 7–37 will produce a plane wavefront, the method of zoning the curved surface is more common because the resulting lens has a greater mechanical strength. To correct the wavefront from a flared horn, the mouth of the horn can be directly terminated by a dielectric lens.

Alternatively, the lens may be used separately to replace a parabolic reflector at frequencies in excess of 20 GHz; the main advantage of such an arrangement is the absence of any obstruction due to the primary antenna. Examples of such lenses are illustrated in Figure 7–38.

A special Luneberg lens (Figure 7–39a) is commonly used because of its wide response angle. This lens is in the form of a solid sphere made from a dielectric material whose refractive index varies with the distance from the center of the sphere. Microwave energy that falls on one hemisphere is refracted through the lens and brought to a focus *F* which is located at the center of the surface which forms the opposite hemisphere. Consequently, if a point source is placed at the focus, a plane wavefront will emerge from the lens. Moreover, if a reflecting cap is placed at the focal point, energy will be reflected back in the direction from which it came. If

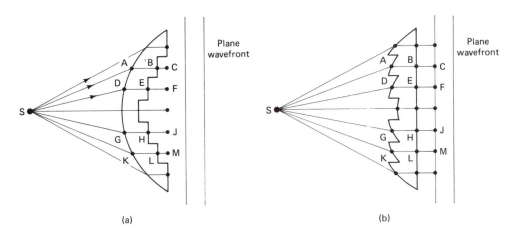

Figure 7–37   Stepped lens.

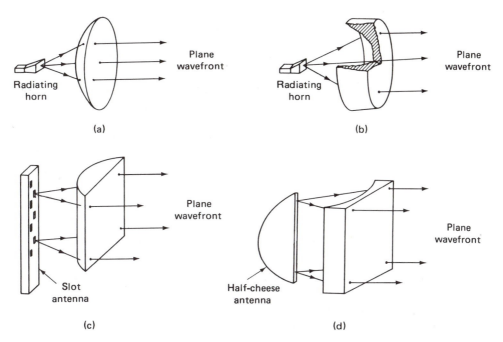

Figure 7–38   Examples of lens antennas: (a) plano-convex lens, relative permittivity > 1; (b) plano-concave lens, made from waveguide sections; (c) plano-convex cylindrical lens, relative permittivity > 1; (d) plano-concave cylindrical lens, made from waveguide sections.

there is a change in the direction of the incident wave, the focal point will shift but if we enlarge the reflecting cap to cover a greater part of the back surface (Figure 7–39b), the target will respond over a wide range of incident angles.

As an alternative to using a dielectric material, a lens may be manufactured from a number of waveguide sections which form a honeycomb.

Remembering that the phase velocity in a waveguide (Section 3–4) is greater than the free-space velocity (compared with the reduction of velocity in a dielectric

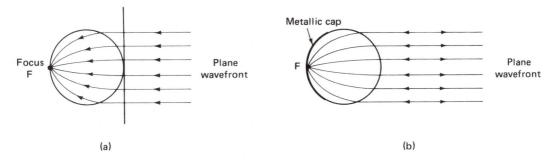

Figure 7–39   Luneberg lens.

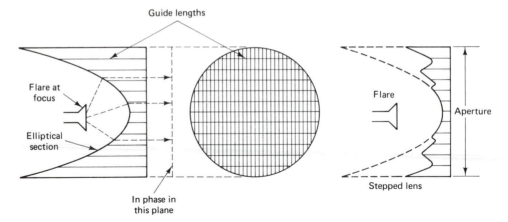

Figure 7–40   Lens formed from waveguide sections.

material), the lens must be plano-*concave* with an elliptical cross section (Figure 7–40); the primary antenna, such as a flared horn, is then placed at the nearer focus. The bulk of the honeycomb is usually reduced by cutting away at the elliptical sections to form another stepped lens. The same principle was applied to our discussion of the zoned dielectric lens.

Slot Antennas

We have already discussed in detail the principle behind radiating and nonradiating slots (Section 3–4). Based on this principle, various kinds of linear arrays have been constructed with waveguides. A simple example is shown in Figure 7–41a, with shunt slots each about a half-wavelength long and spaced half a guide wavelength λ_g along the waveguide. The end of the waveguide is closed with a metal plate which is $\lambda_g/4$ from the center of the last slot. Successive slots are on opposite sides of the centerline to compensate for the reversal of the phase with the standing wave inside the waveguide. In designing this type of array, the displacement of the slots from the centerline is chosen so that, on the whole, the guide is roughly matched to the slots and little power is therefore reflected back.

The type of slotted array in most favor at the moment employs cuts in the narrow side of the guide (Figure 7–41b). These are shunt slots inclined so as to give the required coupling between the slots and the guide. A metal piston is placed at the end so that there is a substantial standing wave inside the guide. Such a system is resonant, so that when the frequency deviates slightly from its correct value, the matching and the radiation pattern deteriorate appreciably. For example, with a 50-element array the overall frequency bandwidth is 1% of the midfrequency, while for 200 elements the bandwidth is only 0.25%. The alternative solution is to use a nonresonant design in which an attempt is made to produce a single traveling wave along the guide. The elimination of the reflected wave is achieved by making

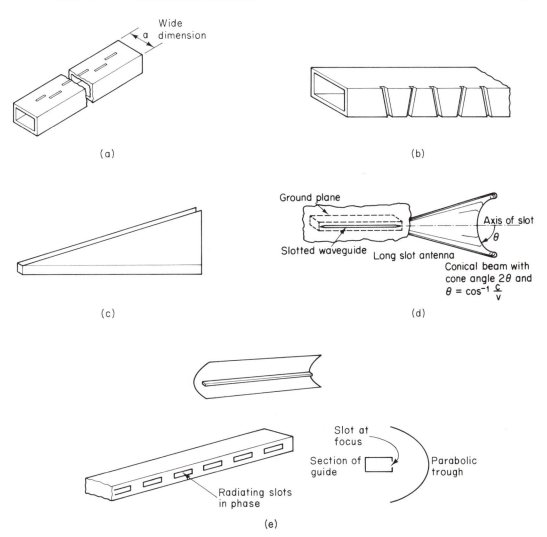

Figure 7–41 Various type of lens antennas: (a) shunt slot array in a waveguide with transverse polarization; (b) shunt slot array in a waveguide with longitudinal polarization; (c) addition of triangular plates to slotted antenna array; (d) end-fire antenna; (e) slotted antenna array feeding a parabolic reflector.

the center-to-center spacing of the slots a little different from $\lambda_g/2$ so that their phase shift is, for example, 200° rather than 180°. Then the waves reflected at the slots do not reinforce one another to create a standing wave but rather tend to cancel one another out. In addition, the coupling of the slots is gradually tightened as the distance along the guide from the input end is increased; the purpose is to enable all the slots to radiate the same amount of power. The small residual amount of power left in the guide after the last slot is often absorbed by a nonreflecting dummy load.

The following points are of interest regarding nonresonant arrays with slots on the narrow side:

1. There is some degree of unwanted polarization in the beam, owing to the inclinations of the slots.

2. The slots have to be cut carefully at the correct angle, and the fact that the angle is different for every slot introduces manufacturing difficulties. The problem is overcome to some extent by allowing the slots to be arranged in a succession of small groups with all the slots in a single group inclined at the same angle.

3. The slots are not driven in phase since they are not spaced at intervals of $\lambda_g/2$. As a result, the beam is swung away from the normal to the array; in fact, the beam forms part of a cone with the array as its axis. This is sometimes undesirable and is eliminated by adding two triangular metal sheets (Figure 7–41c), which confine the waves radiated from the slots to appropriate distances so that they emerge in phase into space. Alternatively, the beam can be used with the radiation from the slots behaving as an end-fire device and the beam being propagated along the antenna's axial direction (Figure 7–41d).

4. The rays emerging from the slotted array of Figure 7–41b form a narrow horizontal beam which radiates approximately at right angles to the array. However, in the vertical plane the beam is very broad. By directing radiation from the waveguide into a parabolic cylinder of about the same length as the array (Figure 7–41e), the original narrow horizontal beam is retained while the vertical beam-width is considerably reduced.

Surface-Wave Antennas

We have just finished discussing the radiation from a nonresonant slot antenna. However, a traveling electromagnetic wave which is guided along a surface containing discontinuities will also radiate power directly from the wave itself. Examples of such discontinuities are a corrugated metal surface, a dielectric rod, a slab of dielectric or ferrite material which may or may not be mounted on a ground plane, and a rod supporting a combination of disks (metallic or dielectric); all of these possibilities are illustrated in Figure 7–42.

The helix antenna is an important example of an end-fire device which uses a traveling wave to produce the required radiation. It should be immediately emphasized that, strictly speaking, this type of antenna is not operated at microwave frequencies but is normally used for satellite communications at an approximate frequency of 140 MHz in the VHF band. The radiation is therefore required to pass through the ionosphere, where the free electrons are set into oscillation. However, due to the presence of the earth's magnetic field, the path of the electrons can be elliptical and the result is to rotate the wave's plane of polarization through an angle whose magnitude is continuously changing. For adequate communication we must use an antenna with a circular polarized wave whose energy can be resolved into vertical and horizon-

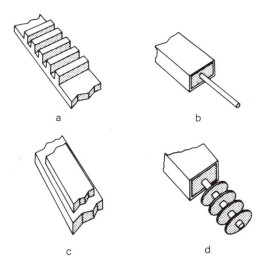

Figure 7-42 Examples of surface wave antennas: (a) corrugated metallic surface; (b) dielectric rod; (c) slab of dielectric on ground plane; (d) disk on rod.

tal components which are 90° out of time phase. The signal can then be intercepted by receiving antennas which are either horizontal or vertical.

The basic construction of the helical antenna is shown in Figure 7-42; the approximate dimensions given correspond to a frequency of 140 MHz. As the traveling wave moves around the turns of the helix, it produces two components of radiation. One component is transverse to the axis of the helix, while the other and more important component has a longitudinal major lobe which is circularly polarized. To create the traveling wave on the helix instead of a standing wave, the last turn is terminated by a resistive load whose value is normally about 150 Ω.

The end-fire beam is fairly broad, with its maximum radiation in the direction of the helix axis. With the dimensions of Figure 7-43 the circumference of each

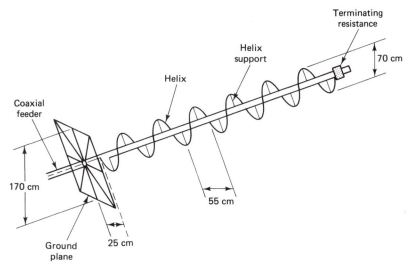

Figure 7-43   Helix antenna.

helix turn is approximately equal to one wavelength; the antenna gain is then of the order of 20 over a frequency range which is 40% of its middle value.

The helical antenna may be used either singly or in a broadside array. Under conditions in which the circumference of the helix is small in comparison with the wavelength, the end-fire radiation may possess an elliptical polarization in which one component predominates over the other. A variety of radiation patterns may also be produced by the use of tapered and/or nonuniform helices.

Phased-Array Antennas

The paraboloidal reflectors we have discussed so far have certain disadvantages when used in radar search or tracking systems. These revolving antennas are mechanically complex and unreliable, have a slow scanning rate, and are limited in the speed at which the beam may be shifted. In many modern systems it is preferable to scan electronically a multielement antenna array which is itself stationary. This is now possible because the required phase shifters are constructed from solid-state integrated circuits and are physically small enough to be mounted on the radiator itself.

The phased-array configuration consists of multiple dipole radiators (Figure 7–44), which are arranged in clusters. Each dipole then receives a signal with a different phase relationship so that the combined radiation from all the dipoles creates a main lobe with a narrow beamwidth. Such an array may require up to several

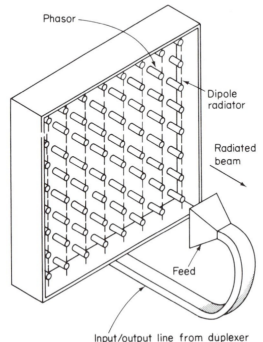

Input/output line from duplexer   **Figure 7–44** Phased-array antenna.

thousand dipole radiators; each radiator is supplied through a variable phase shifter so that the resultant main beam can be shifted, swept, or pulsed over time spans which can be limited to the order of microseconds.

Although the phased array contains a very large number of individual circuit elements, the solid-state circuitry ensures a high degree of reliability.

Effect of the Earth's Surface

In our discussion of microwave antennas, some have produced a vertically polarized beam, while for others the radiation has been horizontally polarized. The question therefore arises: What are the differences, if any, between the two planes of polarization? The answer lies primarily in the effect of the earth's surface.

In general, a considerable part of the radiated output comes into contact with the earth's surface; the majority of this radiation is then reflected to combine with the direct waves. As a result of this interaction, the vertical polar diagram of the antenna can be greatly modified.

In Figure 7–45a the earth is assumed flat for simplicity and an antenna positioned at point O is at a height of h meters above the surface. The radiated energy, which reaches a point P at an elevation θ, is the phasor sum of the direct and reflected waves. The relative phase of these two components depends on:

1. The difference in the path lengths OP and ORP.
2. With horizontal polarization, the wave undergoes a phase inversion when reflected at point R.

It follows that waves will be in phase, giving a maximum resultant signal if θ is such that the path lengths differ by an odd multiple of a half-wavelength. Assuming that $OP = SP$, this difference in the path length is given by QS. The first maximum or lobe of the vertical polar diagram is then given at an elevation θ such that

$$\sin \theta = \frac{QS}{OQ} = \frac{\lambda/2}{2h} = \frac{\lambda}{4h} \tag{7–14}$$

For the second maximum or lobe,

$$\sin \theta = \frac{3\lambda/2}{2h} = \frac{3\lambda}{4h} \tag{7–15}$$

At frequencies in the lower VHF band, such lobes are given at intervals of a few degrees, as shown in Figure 7–45b. For example, if $h = 14\lambda$, lobes are given at approximately 1°, 3°, 5°, and so on, elevations. This means that such frequencies are not suitable for surface warning purposes. However, the lobes enable trained operators to estimate approximate heights of aircraft.

Vertically polarized waves suffer little phase change during surface reflection and if used in our example, lobes would appear between those shown, although the lowest lobe would be seriously attenuated by the earth's surface.

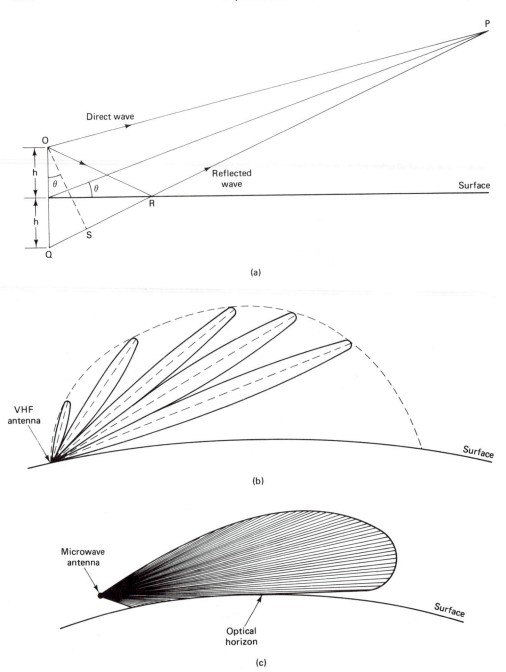

(a)

(b)

(c)

Figure 7–45 (a) Effect of the earth's surface at different frequencies; (b) radiation pattern of **VHF** antenna; (c) radiation pattern of microwave antenna.

At microwave frequencies numerous lobes at intervals of only a few minutes of arc are produced, as illustrated in Figure 7–45c. The effect of separate lobes cannot be considered, although it may be appreciated that they account partially for fluctuations in the echo amplitude. Far more important is the fact that the lowest lobe tends to be given near the surface, making detection of small surface objects possible. When the earth's curvature is taken into account, it will be seen that these lobes do not reach the surface beyond the optical horizon, as seen in Figure 7–45. For this reason it is advisable to place the antenna as high as possible, to increase the range of surface detection. The surface range is slightly increased because of some downward bending of the waves by atmospheric refraction. The approximate maximum range of surface detection is $1.4 \times (\sqrt{h_1} + \sqrt{h_2})$ land miles, where h_1 and h_2 are the antenna and target heights in feet.

The number of lobes and the proximity of the lowest lobe to the surface increases with the radio frequency; this is the reason for the use of 10 GHz for surface detection with navigational radar sets.

Antenna Power Gain Measurements

The power gain of an antenna is usually measured by comparing the antenna to a calibrated "standard." Figure 7–46 illustrates the experimental setup. The output from a microwave source is fed to a radiating antenna which is pointed toward the test antenna. This antenna is moved in azimuth and elevation until the received signal reaches its maximum value (P_1). Then the receiving antenna is replaced by the calibrated standard, usually a horn, and the same detection system is connected to the standard. This antenna is also peaked, and the new maximum received power, P_2, is noted. To simplify calculations, P_1 and P_2 are shown as decibel readings relative to an arbitrary reference level. Then the gain of the antenna under test is

$$G = P_1 - P_2 + G_s \qquad (7\text{-}16)$$

where

$$G = \text{power of the antenna under test (dB)}$$
$$G_s = \text{power of the standard (dB)}$$
$$P_1 - P_2 = \text{difference between the two received signals (dB)}$$

For example, assume that with the test antenna connected, the detector output meter is adjusted for a full-scale reading of zero decibels. Now with the standard horn in place, we find that the output is down 7 dB or reads -7 dB. Then $P_1 - P_2 = 7$ dB. If the standard horn has a gain of 15 dB relative to the basic dipole, the test antenna is 7 dB better and has a power gain of $7 + 15 = 22$ dB.

If the detector does not obey a square-law characteristic, there will be an error in the value of the power gain. This error is small for lower values of $P_1 - P_2$ but can be appreciable if $P_1 - P_2$ exceeds 10 dB. To avoid this difficulty (especially if the detector response is unknown), a calibrated attenuator can be used between the

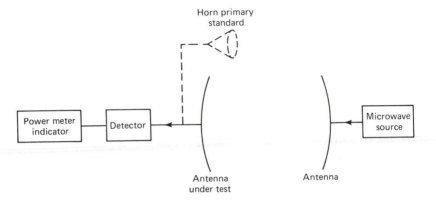

Figure 7–46   Experimental measurement of antenna gain.

higher-gain antenna and the detector. The attenuator can be adjusted to make the two readings identical. The power gain of the test antenna is then

$$G = G_s + A \qquad (7\text{–}19)$$

where A is the setting of the attenuator. It is important that both the test antenna and the standard be well matched since any losses due to reflection will appear as a reduction in the power gain. The detector should also be matched, and the same detector system must be used for both antennas.

The distance between the transmitting site and the receiving site must be far enough so that any slight movements in the positions of the two antennas will not cause appreciable changes in their readings. This requires that the separation between the antennas be at least $2d^2/\lambda$ meters, where d is the antenna's aperture (m) and λ is the free-space wavelength (m).

Primary gain standard.   It is possible to measure the gain of an antenna directly if two identical antennas are available. This method is usually selected to calibrate horns which are to be used as primary standards, but it can also be used for any type of antenna (Figure 7–47). The antennas must be separated by a minimum

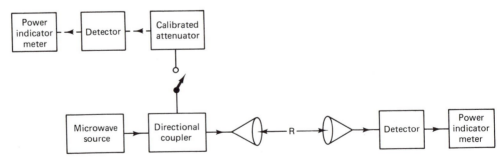

Figure 7–47   Measurement of primary standard's antenna gain.

distance of $2d^2/\lambda$, as indicated previously. Under these circumstances, the gain of an individual antenna, assuming that both antennas are identical, is given by

$$G = \frac{4\pi R}{\lambda} \times \sqrt{\frac{P_r}{P_t}} \qquad (7\text{-}20)$$

where

P_r = power received (W)
P_t = power transmitted (W)
R = separation between the two antennas (m)

The quantities in equation (7-18) can be measured easily and accurately. Although there may be some uncertainty in measuring the separation of the antennas as to what portions of the antenna should be used, the aperture-to-aperture distance is sufficiently accurate provided that R is large. The quantities P_r and P_t are not measured directly; instead, their ratio is determined. First, the detection system is connected to the receiving antenna and a reference level is noted on the output meter. Then the detection system is moved to the transmitter and connected at the output of the calibrated attenuator. The attenuator is adjusted until the output meter reads the same reference level. The ratio of P_t to P_r in decibels is the sum of the coupling of the directional coupler and the loss of the calibrated attenuator.

As an example, assume that the horns which are to be calibrated have an aperture of 18 cm and are measured at 9.6 GHz ($\lambda = 3.12$ cm). Then R must exceed

$$\frac{2d^2}{\lambda} = \frac{2 \times 18^2}{3.12} = 208 \text{ cm} = 2.08 \text{ m}$$

To satisfy this condition, let R equal 3 m. With a 20-dB coupler at the transmitter, assume that 2.5 dB of additional attenuation is needed to bring the power down to the same reference level that was measured at the receiver. This means that P_t and P_r differ by 22.5 dB, which is equivalent to a power ratio of 179. Then, from equation (7-20), the gain of each horn is

$$G = \frac{4 \times \pi \times 300}{3.12} \sqrt{\frac{1}{179}} = 90.3 = 19.5 \text{ dB}$$

Field pattern measurements.   The radiation pattern of an antenna contains information about beamwidth, the sidelobe level, the location of the sidelobes, the locations of the nulls between the lobes, and the level of the back lobe. For example, in Figure 7-48 the 3-dB beamwidth is 20°. One sidelobe, with a peak at about 54°, is 14.6 dB down, and the other major sidelobe, at about 58° on the other side of the main beam, is slightly more than 15 dB down. Other sidelobes are more than 18 dB down, and there is no back lobe. The nulls occur at about 30° on each side of the main lobe.

The radiation pattern of an antenna can be obtained with the experimental

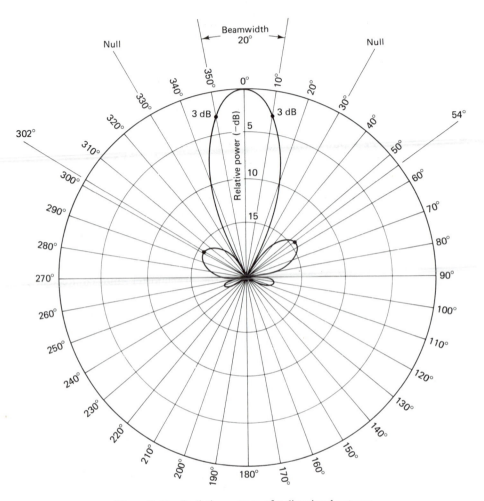

Figure 7–48 Radiation pattern of a directional antenna.

setup shown in Figure 7–49. Again, the microwave source must be separated from the receiver by at least $2d^2/\lambda$. Care must be taken that there are no objects between the two antennas, since such objects might cause reflections which could be picked up in a sidelobe and mistaken for part of the energy in the main beam.

The receiving antenna is mounted on a pedestal which can rotate through 360° in azimuth and can also be tilted in elevation so as to align the two antennas. The motor that runs the pedestal is connected to the turntable of a recorder which duplicates the angle of rotation. The detector output is connected to the pen on the recorder, and then, as the antenna rotates, a plot of its pattern is made automatically. After the antennas are lined up in both elevation and azimuth, so that the power received is a maximum, the gain of the recorder is adjusted so that it reads full scale when the antennas are in line. The antenna is then rotated through 360° and the pattern is recorded.

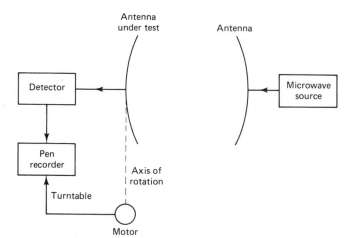

Figure 7–49 Experimental measurement of an antenna's radiation pattern.

For the E-plane pattern of Figure 7–48, the antenna must be rotated in the E plane, so that both the transmitting antenna and the receiving test antenna are horizontally polarized. For an H-plane pattern, both of the antennas are vertically polarized. It is not desirable to obtain either pattern by changing the elevation of the test antenna, since at low elevations the antenna might pick up reflections from the ground.

7–8 MICROWAVE LINKS

As we mentioned in Section 1–8, one of the main applications of microwave antennas is the transmission of TV broadcasts using a number of FM repeater stations. Since the propagation method is *space wave*, we are limited to line-of-sight range, and the typical spacing between two repeaters is approximately 50 km. A simplified block diagram of a microwave repeater is shown in Figure 7–50. The purpose of a repeater is to receive and amplify the FM signal, which is afterward transmitted to the next repeater. In the diagram shown we are receiving the signal from the X direction and transmitting the amplified signal in the Y direction. However, the system is two-way, so that there is also a transmitter which is feeding the X-direction antenna and a receiver which is connected to the Y-direction antenna. To avoid interference effects, the frequencies used for the transmissions in the two opposite directions are typically 200 to 300 MHz apart, while the operating microwave frequency is either 4 or 6 GHz.

Both antennas use parabolic reflectors to achieve the required high gain and the narrow beamwidth. The waveguide from each antenna system is passed to some form of duplexer arrangement, which commonly is a circulator (Section 4–10). As a result, a transmitter is connected to one antenna, which in turn is connected to a receiver, but there is no direct connection between the transmitter and the input circuitry to the receiver. If this were not so, the device in the receiver's mixer stage would immediately be burned out.

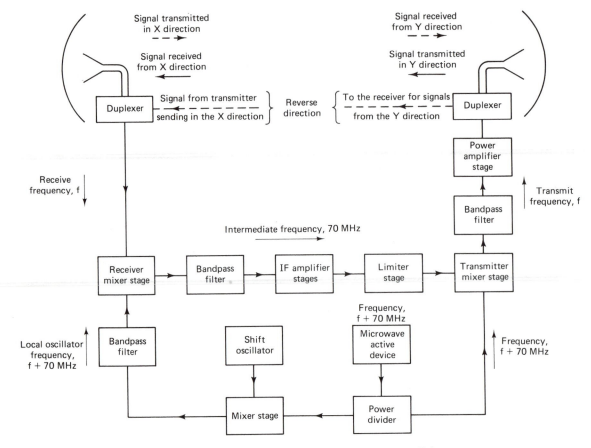

Figure 7–50 Block diagram of a microwave link repeater.

It is important to realize that the repeater does not amplify the FM signal at the microwave frequency. Frequency conversion is used to create a 70-MHz IF signal which is normally amplified with a gain of approximately 70 dB. The purpose of the frequency conversion is noise reduction, since the noise inherent in a 70-MHz amplifier is much less than in an amplifier which is operating in the gigahertz range.

The mixer device is a very low noise Schottky barrier diode, and the output from the mixer stage is fed to a bandpass filter with a center frequency of 70 MHz and a bandwidth of 10 MHz; this filter provides the system with its selectivity. The IF amplifier itself typically consists of several transistor stages which possess the characteristics of good linearity, low noise, and wide bandwidth.

To remove unwanted amplitude modulation, the IF amplifier is followed by a limiter stage. Next comes the transmitter mixer, whose purpose is to recreate the operating microwave frequency. Feeding into this stage is the output from a microwave source whose generated frequency is 70 MHz higher than that of the incoming signal. Such a source could be a Gunn diode or klystron oscillator with an AFC circuit.

However, more modern techniques employ a VHF transistorized crystal oscillator followed by a chain of varactor diode multipliers. The output from this source goes to a divider which sends the majority of the power to the transmitter mixer, while the remainder is fed to the mixer stage of the receiver. However, when the system is passing a signal in the reverse direction, the microwave frequency is a few hundred megahertz higher or lower. This change is accommodated by the shift oscillator, which ensures that the IF remains at 70 MHz.

The bandpass filter which follows the transmitter mixer, eliminates high-order sideband frequencies which may appear as a result of any deviation of the FM signal. The final power amplifier typically raises the signal level to between 2 and 5 W. However, the antenna system with its parabolic reflector has a power gain of several thousand, so that the effective radiated power is several kilowatts with a beamwidth of less than 2°.

In a microwave link the number of carriers is normally between 4 and 12 with up to 2500 channels for each carrier. It should be emphasized that a link only receives, amplifies, and retransmits; it is not involved in the frequency modulation process, which occurs only at the beginning and ending terminals.

Although we have referred to the use of microwave links, their functions can be performed by other means:

1. *Coaxial cables.* Compared with microwave links, such cables are less prone to interference and noise but require a greater number of repeaters for a given distance. As a result, cables introduce a higher degree of amplitude and phase distortion. Such distortion is important in relation to TV signals because of their wide bandwidth.

2. *Optical fibers* (Section 3–7). Fiber optics is a relatively new technology which uses a laser to generate infrared light as the electromagnetic signal. At such high frequencies enormous bandwidths are possible, so that about 5000 channels can be transported by a single fiber with a diameter of only 0.5 mm. Cables containing these fibers are therefore replacing the conventional coaxial lines. Optical fibers also possess a high degree of immunity from electromagnetic interference and have attenuation factors as low as 1 dB/km.

3. *Circular waveguides.* At frequencies above 10 GHz, circular waveguides have very low attenuation when operated in the TE_{01} mode. Consequently, it is possible to use such waveguides for broadband communications over long distances.

7–9 SATELLITE COMMUNICATIONS

Another application of highly-directional microwave antennas occurs in satellite communications. The number of satellite communication systems has increased enormously since 1965 when Early Bird [INTELSAT 1 (INTernational TELecommunications SATellite consortium)] was the first communications satellite to be launched. That satellite provided only 66 telephone circuits, which communicated with five land stations. Now satellite communications relay most of the transoceanic telephone traffic and have consequently superseded the use of submarine cables. This situation may change with the advent of fiber-optic technology, since it is likely that submarine cables containing optical fibers will eventually prove to have more capability and be more economical than satellites. However, at the present time, satellites are used to relay TV signals across the oceans and to provide national television cables directly to the home; this is in addition to handling the telephone traffic and acting as data information links.

A communications satellite is based on the same principle as the microwave links discussed in Section 7–8. The communication is beamed up from a land station to the satellite, which not only amplifies the received microwave signal but also provides frequency conversion. As a result, the downlink frequency is some 2 GHz lower than the uplink frequency. This frequency conversion is essential to avoid interference between the uplink and downlink; moreover, the most common frequencies used (about 6 GHz uplink and 4 GHz downlink) are in the C-band and are already allotted to terrestrial microwave links, so that land stations' receiving antennas have to be carefully located to minimize any reception of the terrestrial communications.

For the particular frequencies chosen in the microwave region, the effect of the ionosphere is negligible and there is little absorption by atmospheric gases and water vapor. The power for the satellite is derived from batteries which are charged from panels of solar cells. The whole satellite unit is designed to be extremely reliable and to have a life expectancy of 5 to 10 years. Maintenance is difficult, to say the least, but may be possible with the aid of the space shuttle.

To maintain their heights exactly, satellites are provided with gas jets. Once this gas is exhausted, the satellite will drift out of its orbit and will be useless; alternatively, the satellite may lose its power if the solar panels fail.

If a satellite is placed in a low elliptical orbit around the earth, the period of this orbit can be less than 2 hours. By contrast, the journey of the moon around the earth is completed in approximately 28 days. Between these extremes there must exist a circular orbit which lasts precisely 24 hours. Such an orbit would allow a satellite's position to be fixed relative to the earth, although the satellite's velocity in space is approximately 11,000 km/h. The satellite is then geostationary or geosynchronous and its orbit is approximately 35,800 km above the earth's surface. The main advantage of such a satellite is the use of fixed antenna systems at the land station since no tracking is required (Figure 7–51). However, the extreme height of the satellite demands that the land station has transmitter power of the order of kilowatts, large directional antenna systems, and a highly sensitive low-noise receiver.

At the end of 1984 there were a total of more than 40 active satellites in synchronous orbit. The U.S. standard for orbital spacing is an assigned minimum of 4°, but in 1982 the Federal Communications Commission proposed a reduction to 2°. The 14 INTELSAT satellites are stretched over the Atlantic, Pacific, and Indian Ocean regions and operate in conjunction with more than 500 land stations; each of these satellites, operating in the 6 GHz/4 GHz band, has a capacity of over 12,000 two-way telephone channels and a number of TV channels. To amplify the received

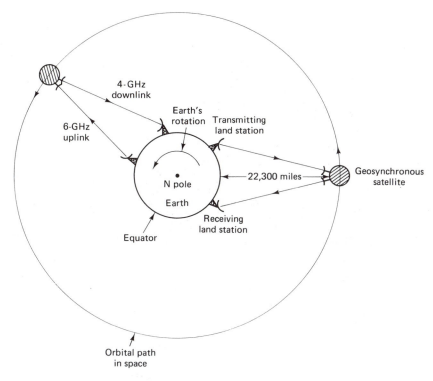

Figure 7–51   Geosynchronous satellite.

signals (approximately 6 GHz) from the uplink and also provide frequency conversion for the transmissions (approximately 4 GHz) on the downlink, a satellite commonly contains 12 transponders, each of which is assigned a 36-MHz bandwidth, while the whole satellite is allowed a bandwidth of 500 MHz. It is possible to double the use of the same frequency range by using 12 transponders which are vertically polarized while another 12 are horizontally polarized. Each of these transponders is linear in the sense that the received signal is not demodulated and processed. In the future, processing transponders will be used for digital satellite communications.

The more modern satellites operate not only in the C band but also in the range 14 GHz/11 GHz. The additional range considerably increases the system's signal capability. The 17 GHz/12 GHz band has been assigned for television broadcasting direct to the home.

Satellite Systems

An exploded view of a 20-ft-high INTELSAT satellite is shown in Figure 7–52, while Figure 7–53 illustrates the block diagram of a particular transponder. This satellite on the uplink receives its frequency-modulated signals within the range 5.93 to 6.42 GHz. These signals pass through a diplexer and are then amplified by a

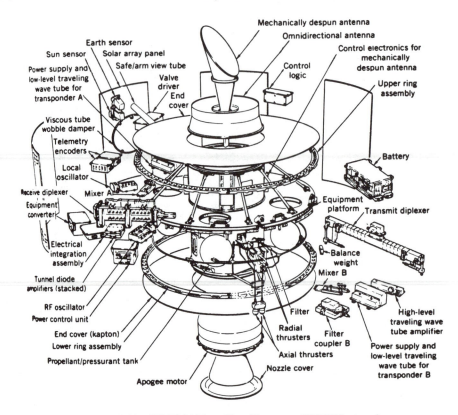

Figure 7-52   INTELSAT satellite. (Courtesy of TRW Systems.)

number of tunnel diodes. After frequency conversion by a low-noise mixer and a local oscillator, the signals in the range 3.705 to 4.195 GHz (frequency difference = 5.93 − 3.705 = 6.42 − 4.195 = 2.225 GHz) are reamplified to drive a traveling-wave tube whose output is passed through another diplexer and then transmitted on the downlink. At the satellite the power level for the downlink transmission is approximately 7 W, while the total power consumption is about 150 W.

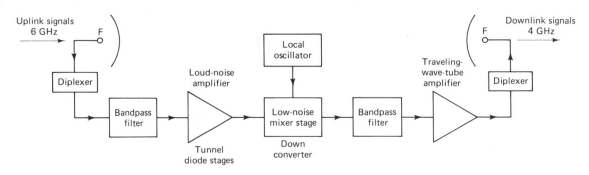

Figure 7-53   Block diagram of a typical satellite transponder.

In more modern satellites a transponder's initial amplification is provided by four silicon bipolar transistor stages. After frequency translation the output TWT circuit is driven by another four-stage bipolar transistor amplifier. In the 14 GHz/ 11 GHz band the uplink frequency is in the range 14.0 to 14.5 GHz, while the downlink transmission lies between 11.45 and 11.70 GHz. Each transponder then has a germanium tunnel diode amplifier to operate a Schottky diode mixer whose output is amplified by a five-stage bipolar transistor amplifier to drive the output TWT stage.

The attitude of a satellite is its position relative to the center of the earth. To stabilize the attitude, the satellite is spun in one direction at 100 revolutions a minute. This means that the antenna must be spun at the same speed in the opposite direction (despun) so that the satellite's antennas are always pointing toward the same positions on the earth.

Transmission of TV Signals

The bandwidth for the composite video and chroma information of a TV broadcast signal is 4.5 MHz, which *frequency* modulates the microwave carrier in the range 5.93 to 6.42 GHz. This bandwidth includes the video and chroma information together with the synchronizing, blanking, and equalizing pulses. The aural (sound) signal has a bandwidth of 50 Hz to 15 kHz and is frequency modulated on to a 6.8-MHz subcarrier which is frequency-division multiplexed with the composite video signal so that it can be relayed by the same transponder.

The maximum frequency deviation of the composite video signal is 10.5 MHz, while the peak deviation of the aural subcarrier is 2 MHz. Consequently, the peak deviation is $10.5 + 2.0 = 12.5$ MHz and the required bandwidth is $2 (12.5 + 6.8) = 36.6$ MHz (Carson's rule). This can be reasonably accommodated by the 36-MHz bandwidth assigned to the transponder. Twelve such transponders would have a total bandwidth of $36 \times 12 = 432$ MHz; allowing for guard bands, the total assigned bandwidth of 500 MHz is adequate.

Transmission of Data and Telephone Signals

A satellite relays data signals by either of two methods:

1. Converting the data to time-division-multiplexed (TDM) signals which then digitally modulate a carrier for transmission purposes
2. Using frequency-division multiplexing (FDM). Modems then convert the data into signals which are compatible with VF (voice frequency, telephone) group bandwidths. These signals then modulate the microwave carrier in an analog manner.

Modulation by telephone signals is also achieved by either of two methods:

1. Using frequency-division multiplexing to convert the VF signal directly into groups

2. Converting the VF signal to a pulse code modulation (PCM) signal, which is then time-division multiplexed and merged with the digital signals.

Multiple access of a particular transponder by a number of uplink and downlink land stations is achieved by either time-division multiplex access (TDMA) or frequency-division multiplex access (FDMA).

Land stations.   The land stations pose considerable problems because the satellite's height will mean considerable signal power losses as a result of the up and down paths. Using some approximate figures, let us assume that the total path loss is 400 dB. The receiving and transmitting antennas of the two land stations (Figure 7–51) have an overall gain of 120 dB, the gain of the transponder is 60 dB, while the satellite's antennas have a combined gain of 60 dB. The total loss is therefore $400 - 120 - 60 - 60 = 160$ dB, so that if the input power to the land station's antenna is 2 kW, the received output signal from the other land station's antenna would be a mere $2 \times 10^3 \times 10^{-16} \times 10^{12}$ pW $= 0.2$ pW. This level of signal would certainly require a low-noise preamplifier, which is located at the antenna. The first stage will be a cryogenically cooled mesa unit or a parametric amplifier whose pump is a varactor multiplier chain or a reflex klystron. The signals are further amplified by a low-noise TWT or a tunnel diode amplifier. In newer receivers the parametric amplifier's pump may be a transistor oscillator whose frequency is stabilized by an AFC circuit incorporating a crystal. The signals are then amplified by a series of cascaded FET stages which can be incorporated in a microwave integrated circuit (MIC).

After preamplification the signals are passed to the receiver, where they are converted to the 70-MHz intermediate frequency and afterward demodulated. Passband filters are used to eliminate the unwanted carriers, while the wanted channels are sent by a terrestrial microwave link or cable to the appropriate international terminal.

The land stations fall into three categories:

1. *Standard A stations.* These stations are allocated a maximum power of 8 kW for the total spectrum of their satellite communications. However, most stations use only a portion of their spectrums, so that the normal power output is about 4 kW. This is usually achieved with two separate high-power output stages, which are either multicavity klystrons or water-cooled traveling-wave tubes. These are driven by low-power TWTs, while all preceding stages employ solid-state devices. The high-gain antennas have parabolic reflectors with 30-m diameters and Cassegrain feeds.

2. *Standard B stations.* These stations are capable of being transported so that their paraboloid antenna systems have diameters of only 11 m. This means that these stations operate only with high-gain satellites.

3. *Standard C stations.* These fixed stations are designed to operate in the range 14 GHz/11 GHz and have paraboloid antennas with diameters of about 16 m.

The antenna systems of land stations have to be automatically steerable to a limited degree because the earth is not a perfect sphere, and consequently the satellites are not totally geostationary but change their positions by as much as 20 km over a period of one day. If the antenna of the land station transmits a single beam, the satellite can return a beam which is wide enough to cover an entire country or even a continent. By contrast, the system may use a number of spot beams, in which case reception may be confined to 20 or less separate areas each of which is only a few hundred miles wide.

PROBLEMS

BASIC PROBLEMS

7-1. What is the electrical length in centimeters of a Hertz antenna designed to resonate at 0.75 GHz?

7-2. A dipole is used to feed a paraboloid reflector whose diameter is 5 m. If the operating frequency is 1.5 GHz, what are the values of the beamwidth and the antenna's power gain?

7-3. A paraboloid antenna is designed to have a power gain of 5000 at a frequency of 10 GHz. What is the diameter of the paraboloid?

7-4. At a particular position the field strength of a transmission's fundamental component is 137 mV/m. The second harmonic of this transmission has a field strength of 93 μV/m at the same position. What is the attenuation of the second harmonic relative to the fundamental component?

7-5. If the current feeding an antenna is reduced by 50%, what is the percentage change in the field intensity at a given point of reception?

7-6. If the physical length of a Hertz antenna is 25 cm, what is its resonant frequency in gigahertz?

7-7. If 15 kW of RF power is delivered to an antenna whose field gain is 3.5, what is the effective radiated power?

7-8. The RF power delivered to an antenna system is 8.5 kW. If the antenna power gain is 7.5 dB, calculate the value of the effective radiated power.

7-9. An antenna has a radiation resistance of 70 Ω and an effective loss resistance of 8 Ω. What is the value of the antenna's efficiency?

7-10. If a microwave transmitter's power is increased from 500 W to 1 kW, what is the percentage increase in the field intensity at a particular distance from the transmitter's antenna? What is this increase expressed in terms of decibels?

ADVANCED PROBLEMS

7-11. At a position which is 2.5 km from a transmitting antenna, the field strength is 550 μV/m. If the antenna current is halved, what are the field strengths at the same position and at a position which is 1.5 km from the transmitting antenna?

7-12. An S-band radar set is operated at a frequency of 3 GHz and provides 50-W pulses to a parabolic reflector with a diameter of 6 m. What is the effective radiated power of each pulse?

7-13. Two identical parabolic antennas are separated by 50 m. The power transmitted by one antenna is 100W while the power received by the second antenna is 0.01W. If the frequency is 15 GHz, what is the power gain of each antenna? Is the separation sufficient for the experimental results to be valid?

7-14. A microwave system feeds 200 W of RF power to a transmitting antenna with a gain of 30 dB. If the operating frequency is 1.5 GHz and the receiving antenna also has a gain of 30 dB, what is the value of the received power if the antennas are separated by 1.5 km?

7-15. The gain of a microwave antenna is checked against a calibrated standard horn with a gain of 10. The received power is registered as -20 dB relative to an arbitrary level. When the standard horn is substituted, the received power level falls to -25 dB. What is the power gain of the antenna under test?

7-16. In a microwave communications system the transmitting antenna has a gain of 15 dB while the receiving antenna's gain is 10 dB. The RF power input to the transmitting antenna is 100 W. If the antennas were replaced with two new antennas, each with a gain of 20 dB, what would have to be the new value of the RF power input to the antenna (assuming the same level for the received power)?

7-17. What is the effective radiated power if the output of a microwave transmitter is 100 W, the waveguide loss is 2.5 W, and the antenna gain is 35 dB?

7-18. At a position 6 km from the antenna of a 5-kW transmitter, the field strength is 200 mV/m. If the power is now reduced to 3.0 kW, what is the new distance from the antenna of the 500-mV/m contour?

7-19. The irradiation of a paraboloidal dish antenna is tapered down 10 dB at its edges. If the frequency of the primary feed is 10 GHz and the antenna has an aperture of 0.5 m, what is the value of the beamwidth (assuming equal E- and H-field patterns). If the antenna were uniformly irradiated, what would be the new value of the beamwidth?

7-20. For a signal of 7.5 GHz, what is the minimum acceptable diameter of a paraboloidal dish reflector? Calculate the value of the antenna's power gain. If the diameter is increased by a factor of 5, what is the new value of the antenna gain?

CHAPTER 8

MICROWAVE MEASUREMENTS

8-1 INTRODUCTION

In ac circuits which are operated at low frequencies, we find it convenient to measure the voltage, the current, and the true power. From these measurements we can calculate the values of the impedance, the power factor, and the phase angle. However, at microwave frequencies, it is difficult to measure the voltage and the current distributions on a waveguide system; as a result, we will tend to concentrate on various methods of determining the power level.

Microwave elements such as waveguides and cavities are concerned primarily with distributed circuitry as opposed to lumped components. Consequently, it is only necessary to obtain the overall circuit impedance, rather than the magnitudes of the distributed constants. Moreover, the value of this impedance is normally important only in relation to the waveguide impedance. For example, to determine what happens at a waveguide's termination, it is only necessary to know the load's normalized impedance or admittance, rather than the actual value in ohms or siemens.

Low frequencies may readily be measured by using some method which involves an oscilloscope. However, this test instrument is useless for the microwave band and we normally turn instead to an adjustable resonant cavity (Section 1–3) for determining the frequency.

Apart from knowing the power, the impedance, and the frequency, we also need to find such quantities as the phase shift, the attenuation, the value of the SWR, and the noise factor. Descriptions of these microwave measurements are covered in the following topics:

8-2 FREQUENCY MEASUREMENTS

The basic experimental arrangement of Figure 8-1 can be used to illustrate many of our required measurements. This X-band setup consists of the following units:

1. A microwave source whose output is of the order of milliwatts. This may be either some form of signal generator (Figure 8-1b), a Gunn diode oscillator (Section 6-8), a backward wave oscillator (Section 5-7), or a reflex klystron tube (Section 5-4). Provisions are normally made for the signal output to be either CW or square-wave modulated at an audio rate (usually 1 kHz). In addition, most modern microwave test generators contain sweep oscillators, which allow the output frequency to be varied periodically over a required range.

2. A precision attenuator (Figure 8-2), which was described in Section 4-11 and can provide 0 to 50 dB attenuation above its insertion loss.

3. A variable flap attenuator (Figure 8-3), whose calibration can be checked against readings of the precision attenuator.

4. A frequency meter (Figure 8-4) in the form of a cylindrical cavity (Section 1-3), which can be adjusted to resonance and is slot coupled to the waveguide.

5. A slotted line carriage (Figure 8-5), in which an E probe is able to traverse a nonradiating slot (Section 3-4) which is cut along the center of the waveguide's *a* dimension. This unit is required primarily for SWR measurements.

6. A crystal detector (Figure 8-6). See below for a description.

7. The SWR indicator (Figure 8-7), which is basically a sensitive tuned voltmeter. It consists of a high-gain amplifier which operates at a fixed audio frequency (1 kHz) with a very low noise level. The amplifier's output is measured with a square-law-calibrated voltmeter which can provide direct readings of the SWR or its equivalent value in decibels (Figure 8-8). A gain control allows the reading to be adjusted to the required value.

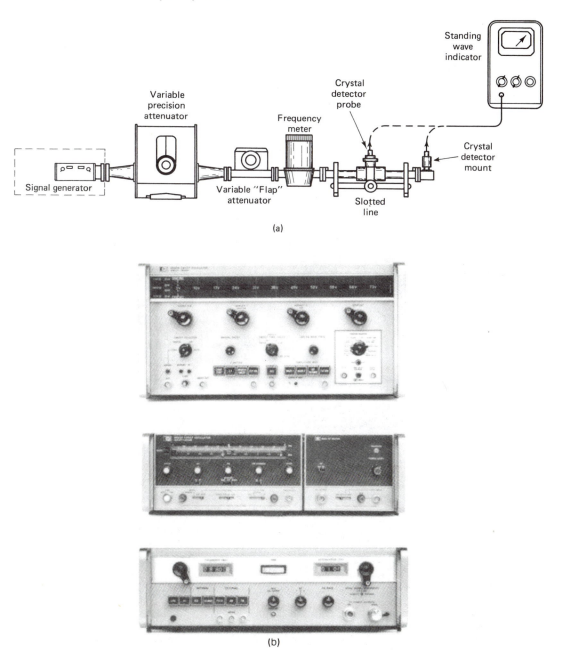

(a)

(b)

Figure 8–1   (a) X-band experimental arrangement; (b) typical signal generators. [(b) Courtesy of Hewlett Packard Company.]

Figure 8–2   Precision attenuator.

Figure 8–3   Variable "flap" attenuator.

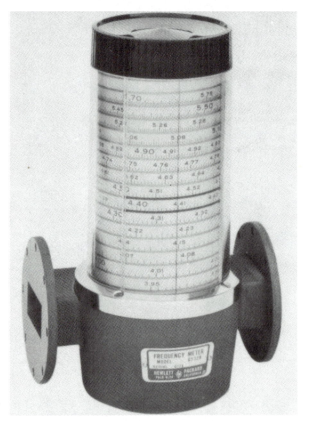

Figure 8–4   Resonant cavity frequency meter.

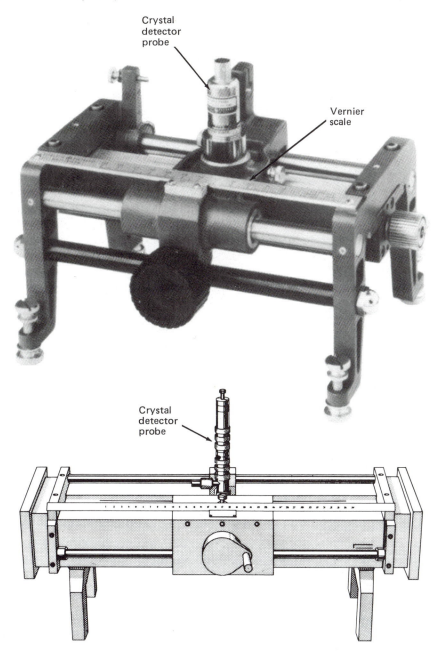

Figure 8–5 Slotted-line carriage. (Courtesy of Hewlett Packard Company.)

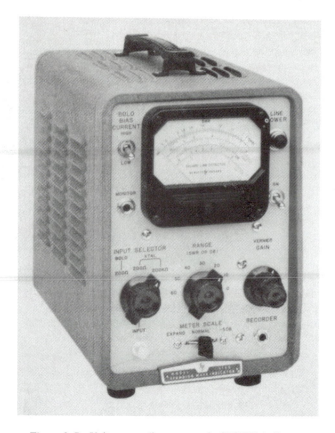

Figure 8–6   Crystal detector in its mount.   **Figure 8–7**   Voltage standing-wave ratio (VSWR) indicator.

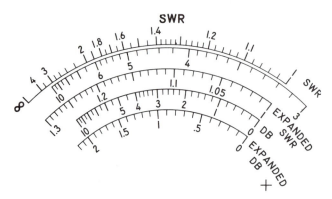

Figure 8–8   Scale of the VSWR indicator.

Crystal Detectors

A crystal detector is inserted in the E probe of the slotted line (Figure 8–9a) and is also contained in the crystal detector mount (Figure 8–9b) at the end of the waveguide run. Each of these crystal detectors is a complete assembly which includes the crystal-rectifier element itself with its holder and the output signal fitting; these crystal detectors are shown in Figures 8–9c and d.

A microwave crystal is a nonreciprocal device which rectifies the input signal. Its essential parts are a semiconducting silicon wafer (about 1.6 mm square) and a pointed tungsten whisker wire (diameter approximately 0.08 mm) which makes a rectifying contact with the wafer. Its forward voltage/current characteristic is approximately parabolic (nonlinear) in shape and is therefore "square law" in the sense that the crystal current is proportional to the (signal voltage)^2, which in turn is directly related to the power. The crystal detector's output is consequently a direct measure of the microwave power level. However, the square-law characteristic holds only for low values of incident power (typically less than 10 W). For higher power levels, the forward voltage/current characteristic becomes increasingly more linear, so that it may be necessary to attenuate the signal before it reaches the detector and to take account of the attenuation when calculating the actual level of the incident power. When reversed biased, the amount of the crystal current falls to an extremely low level.

The dc level of the crystal current can be measured with an ordinary micro-ammeter or milliammeter after the microwave signal has been applied to the detector. It is important to monitor this current continuously since, with a square-law relationship, the current can rise very rapidly if the power is increased. For example, the current of some sensitive crystals can be as high as 20 mA when the incident power is as low as 40 mW.

The mount itself contains matching elements, so that its SWR is typically of the order of 1.3 or less and the microwave power is absorbed and not reflected by the crystal detector. An RF bypass capacitance is also included so that the microwave signal is confined to the crystal detector and is not coupled in any way to the SWR meter. Finally, we must provide a dc path to allow for the flow of direct current through the entire circuit; such a path may take the form of a shorted matching stub (Figure 8–9d). If the microwave signal is amplitude modulated by a square wave, the crystal detector's output can be amplified by the SWR meter and then peaked at the appropriate amplitude and frequency of the square wave.

When using crystal detectors, it is important to prevent the rectifier element from being burned out by an excessive incident power of the order of 50 to 100 mW. Such protection is normally provided by an input attenuator pad. When the crystal detector's output is connected to the SWR meter, it is necessary to match the detector's output impedance to the meter's input impedance. The level of the output impedance is either high (2500 to 10,000 Ω) or low (50 to 200 Ω).

For the E probe used with the slotted line, the depth of the probe's penetration should be limited. If this precaution is not observed, the probe may disturb the configu-

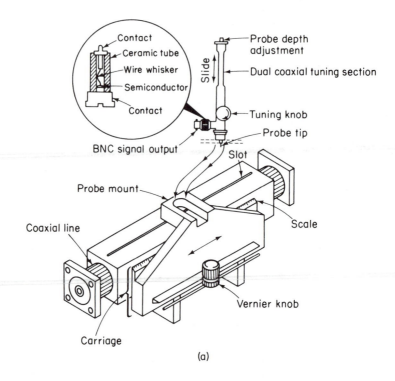

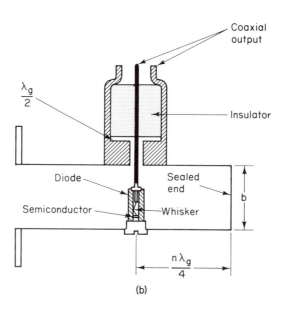

Figure 8–9   Crystal detector and its mounts. [(a) Courtesy of The Narda Microwave Corporation.]

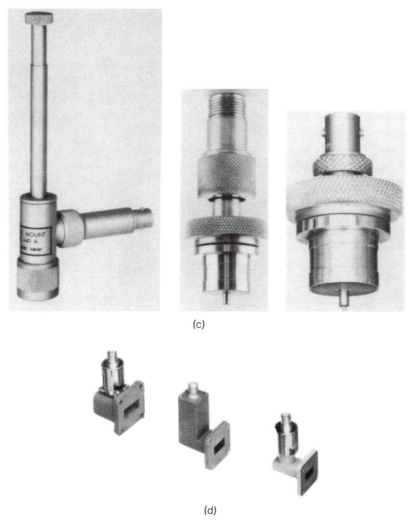

(c)

(d)

Fig. 8–9   (*cont.*)

ration of the electric field and create some degree of mismatch, which will increase the value of the SWR.

The procedure for determining an unknown microwave frequency is as follows:

1. Make sure that the cavity of the frequency meter is not at resonance for the unknown frequency. Adjust both the precision attenuator and the flap attenuator so that each provides a 10-dB loss.

2. Activate the microwave source and switch to square-wave modulation. Vary the amplitude and the frequency (1 kHz) for a peak reading in the SWR meter. If necessary, adjust the meter's gain control and the precision attenuator so

that the needle is on the 0 dB mark on the right-hand side of the scale (Figure 8–8).

3. Adjust the size of the cavity until resonance occurs. The waveguide will then deliver the maximum amount of energy to the cavity. Although this amount is small, the result is an appreciable dip in the reading of the SWR meter. The unknown frequency can then be read directly off the scale of the so-called reaction meter, examples of which are illustrated in Figure 8–10. One advantage of such meters is that even when the cavity is only slightly off resonance, the

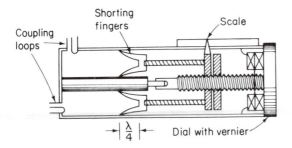

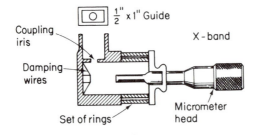

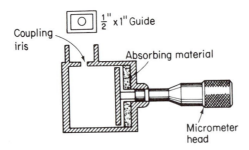

Figure 8–10 Types of resonant cavity wavemeter.

effect of the slot is negligible and therefore the meter may remain in the circuit at all times. A typical Q value for the cavity is about 5000, so that its reaction indicator only covers 2 MHz at a resonant frequency of 10 GHz.

The basic accuracy of the cavity wavemeter is ±1%, but this value is subject to changes in temperature and humidity. These effects can be overcome to some extent by sealing the cavity and providing temperature and humidity calibration curves.

An alternative method of determining the frequency is to use a slotted line. We can replace the crystal mount termination by a short circuit so that a standing-wave pattern will appear. After connecting the output of the E probe to the SWR meter, we can accurately locate the positions of two adjacent nulls by moving the probe along the slotted line and using the vernier scale. However, these two positions are separated by half a *guide* wavelength, $\lambda_g/2$. To derive the free-space wavelength λ, we must use the relationship $\lambda = \lambda_g/\sqrt{1+(\lambda_g/2a)^2}$ [equation (3–6)]. We can then calculate the frequency by using f (GHz) $= 30/\lambda$ (cm) [equation (1–3)]. The accuracy of the result will depend partly on our knowledge of the waveguide's inner wide dimension, a. The overall accuracy obtained is of the order of ±1%.

The methods of measuring frequencies described so far depend on mechanical devices such as the cavity and the slotted line. However, there are also electronic devices such as frequency counters and heterodyne systems which compare the unknown microwave frequency with a high-order harmonic of an accurately determined lower frequency. Although the electronic devices are much more accurate, they are inevitably more complex and expensive.

Example 8–1   Using the slotted-line technique to determine the frequency, the distance between two adjacent nulls is found to be 1.85 cm. If the guide's internal dimensions are 2.86 cm and 1.39 cm, calculate the value of the frequency.

Solution   The guide wavelength

$$\lambda_g = 2 \times 1.85 = 3.7 \text{ cm}$$

The free-space wavelength,

$$\lambda = \frac{\lambda_g}{\sqrt{1+(\lambda_g/2a)^2}}$$
$$= \frac{3.7}{\sqrt{1+\left(\dfrac{3.7}{2 \times 2.86}\right)^2}} \qquad (3\text{–}6)$$
$$= 3.107 \text{ cm}$$

The frequency

$$f = \frac{30}{\lambda}$$
$$= \frac{30}{3.107} \qquad (1\text{–}3)$$
$$= \mathbf{9.66 \ GHz}$$

8-3 POWER MEASUREMENTS

In waveguide systems voltage and current measurements are not only difficult to obtain but the values change according to the position at which the measurements are carried out. By contrast, the power level is independent of the monitoring position (assuming that the losses in the waveguide system are negligible). The various methods used in measuring microwave power depend on the power levels involved:

Low Power

In the range 0.01 to 10 mW it is normal practice to use some form of bolometer, which is any device whose resistance changes as it absorbs microwave power. The two most common types of bolometer are the barretter and the thermistor. The barretter consists basically of a short piece of thin metal wire which has a *positive* temperature coefficient of resistance (a piece of fuse wire behaves in this manner and was used as one of the early forms of barretter). By contrast with the barretter, the thermistor has a negative temperature coefficient of resistance and is a small bead, rod, or disk of semiconductor material which is positioned inside a suitable mount (Figure 8–11).

Both the barretter and the thermistor have square-law characteristics and may therefore be used for the same purpose as crystal detectors. However, we must realize that bolometers only change their resistance and do not by themselves generate any current. It is therefore necessary to provide the bolometer with a dc bias voltage. The current through the bolometer then depends on its resistance, which in turn is determined by the amount of microwave power absorbed. As with crystal detector measurements, bolometers are normally used with modulated microwave sources and meters (Figure 8–12) which contain audio amplifiers. Such meters have a number of internal bias voltages, which can be applied in turn to a particular bolometer and also have a switching arrangement so that they can operate with either a barretter or a thermistor. The impedances of both these bolometers are typically 100 or 200 Ω.

As with crystal detectors, dc return paths are required for bolometers when they are positioned in their mounts. In addition, such mounts must be carefully matched to the waveguide system so that the resultant SWR is typically 1.2 or less. As far as protection is concerned, baretters are far more delicate than thermistors and typically can withstand microwave powers of only a few milliwatts with a 50% overload. By contrast, a thermistor with a maximum rating of only 25 mW can operate up to 200 mW without serious deterioration of its characteristics.

If a signal from an amplitude-modulated microwave source is applied to a bolometer, the modulation is normally a 1-kHz square wave, although both bolometers are capable of operating at a frequency down to 200 Hz. However, if the frequency is too low, the bolometer's resistance will fluctuate at twice the modulating frequency.

Another microwave sensing device is the thermocouple, which consists of two dissimilar metals such as bismuth and antimony. These two metals form a junction to which microwave power is applied. The heating of the junction creates a thermoelec-

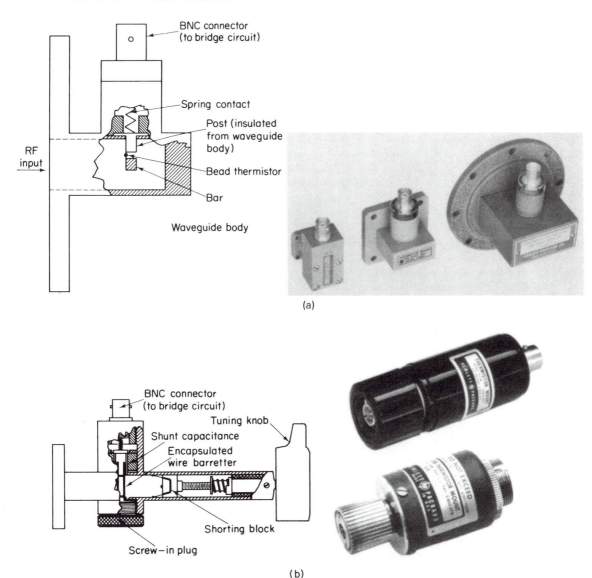

Figure 8–11 Types of bolometers: (a) thermistor and its mounts; (b) barretter and its mounts.

tric electromotive force (EMF), which is then used to operate some form of indicating device. Compared with bolometers, thermocouples have a low sensitivity and are seldom used as detectors; however, thermocouples are more reliable in the measurement of absolute power levels.

When the absolute power to be measured is below 10 mW, we normally use some form of balanced bridge circuit in which one of the arms is the bolometer or

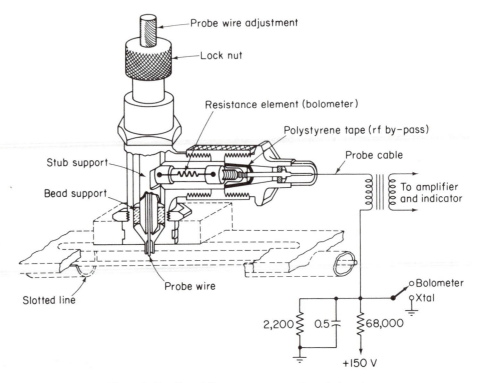

Probe wire adjustment
Lock nut
Resistance element (bolometer)
Polystyrene tape (rf by-pass)
Probe cable
Stub support
Bead support
To amplifier and indicator
Probe wire
Slotted line
Bolometer
Xtal
2,200   0.5   68,000
+150 V

Figure 8–12 Slotted-line measurement using a bolometer.

thermocouple. The microwave power applied to this arm will alter the bolometer's resistance or cause the thermocouple to generate its EMF. In either case the balance is disturbed so that the output from the bridge is no longer zero but can be recorded on a voltmeter which is calibrated to read the level of the input microwave power. To eliminate the problem of calibration inaccuracy, we can alternatively apply to the bridge that equivalent amount of audio power which will create the same degree of imbalance as was produced by the microwave signal. Consequently, the microwave power level must then be the same as the audio power, which is easily measurable. An example of such a bridge arrangement is illustrated in Figure 8–13a. Audio and dc power are applied simultaneously to bring the bolometer to its required operating resistance. When the microwave power is introduced, the action of the circuit causes an equivalent amount of audio power to be removed. This amount of audio power is then amplified and displayed on the meter as a true indication of the microwave power level (Figure 8–13b).

The bridge method may be extended to cover the power range 10 mW to 10 W. This is achieved by inserting a precision attenuator between the meter and the power level to be measured. The recorded power is then multiplied by the power ratio corresponding to the attenuator's decibel loss.

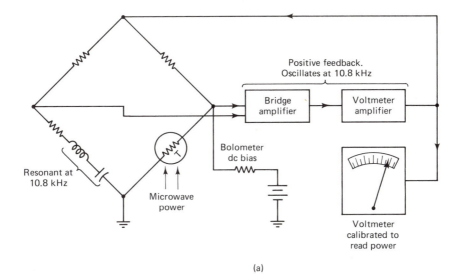

(a)

(b)

Figure 8–13 Low-level microwave power measurements: (a) principle of the bridge circuit for measuring low levels of microwave power; (b) microwave power meter employing the bridge circuit principle.

Medium Power

In the range of 10 mW to 10 W, a typical instrument can measure the average power over the frequency range from dc to the upper limit of the X band. Powers above 10 W may be measured by inserting precision attenuators or directional couplers (Sections 4–9 and 4–11).

Figure 8–14a illustrates a self-balancing bridge which consists of identical tem-

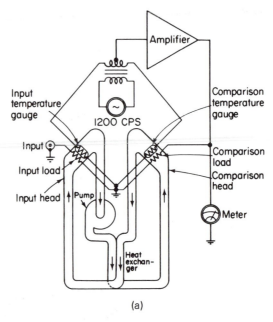

(a)

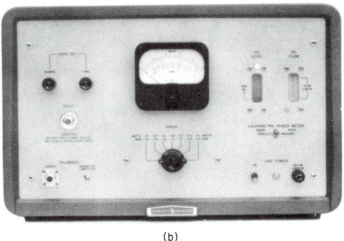

(b)

Figure 8–14   Self-balancing calorimetric wattmeter.

perature-sensitive resistors or gages in two arms, an indicating meter, and two load resistors, of which one senses the unknown input microwave power and the other is associated with the comparison power. The input load power and its associated gage are placed in close thermal proximity so that the heat generated in the input load resistor raises the temperature of the gage and unbalances the bridge. The signal due to the imbalance is amplified and then applied to the comparison load resistor, which is in close thermal proximity to the other gage. Consequently, the heat generated in the comparison load resistor is transferred to its gage and the bridge is virtually rebalanced. The meter measures the amount of power that is supplied to the comparison load in order to rebalance the bridge. The characteristics of the two gages are matched and heat transfer characteristics from each load are the same. Therefore, equal powers are dissipated in the two loads and the meter is calibrated directly in terms of the input microwave power. This is the principle of the calorimetric wattmeter (Figure 8–14b). In order that the bridge can quickly be balanced, an efficient heat transfer from the loads to the gages is achieved by immersing these components in an oil stream; full-scale deflection is then possible in less than 5 s.

The power measurement is normally accurate to within ±5% because the flow rates are the same through the two heads, which have identical characteristics. To ensure a practically constant temperature, the streams are passed through a parallel-flow heat exchanger just before they enter the heads. The accuracy may be improved further by employing techniques to reduce impedance mismatch effects between the meter and the "source" of the microwave power.

High Power

High-power microwave energy (10 W to 50 kW) is usually measured by calorimetric wattmeters which are categorized as either the "dry" type or the "flow" type. A dry-loaded calorimeter usually consists of a coaxial cable which is filled by a dielectric with a high hysteresis loss. The flow type uses circulating water, oil, or any liquid which is a good absorber of microwaves. The power is then directed into a well-matched liquid load whose impedance is normally of the order of 50 Ω.

The fluid flows through the load and its temperature is raised by the microwave energy. The difference between the temperature of a known quantity of liquid before entering the load and the temperature after it emerges is a measure of the power which has been absorbed. If the rate of flow of the liquid is known, the exact value of the power may be calculated. The equation is

$$P = \frac{V \times K \times \rho \times \Delta T}{4.18} \tag{8–1}$$

where

P = measured power (W)
V = rate of flow (cm³/s)
K = specific heat (cal/g)
ρ = specific gravity (g/cm³)
ΔT = difference between the entering and emerging temperatures (°C)

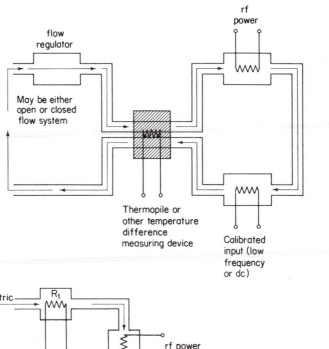

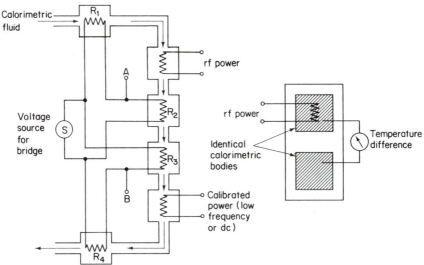

Figure 8–15   Calorimetric microwave wattmeter for measuring power levels above 10 W. (Photos courtesy of The Narda Microwave Corporation, and Weinschel Engineering Co., Inc.)

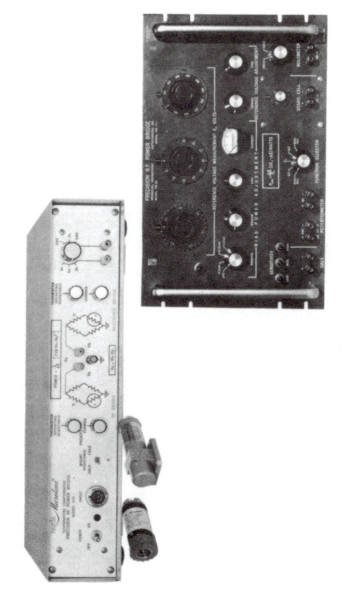

Figure 8–15 (*cont.*)

Water loads are normally connected directly to a tap and a meter is then inserted to indicate the flow rate. By contrast, oil loads are usually divided with a closed circuit and a pump that pushes the oil through at a predetermined rate. Calorimetric meters of this type (Figure 8–15) are connected to an indicator which is calibrated directly in terms of power.

The meters so far discussed measure the incident or forward power. However, there are two-directional wattmeters (Figure 8–16) which measure both forward and reflected power. These instruments are useful when monitoring the efficiency of a microwave system. From a knowledge of the incident and reflected powers, we can determine the values of the reflection coefficient and the SWR.

In many microwave systems (e.g., a navigational radar set) the power is modulated by a rectangular pulse of a short duration t, which is of the order of microseconds (Figure 8–17). These pulses are repeated at a rate whose frequency is normally several hundred hertz or a few kilohertz. The reciprocal of this frequency is the time interval

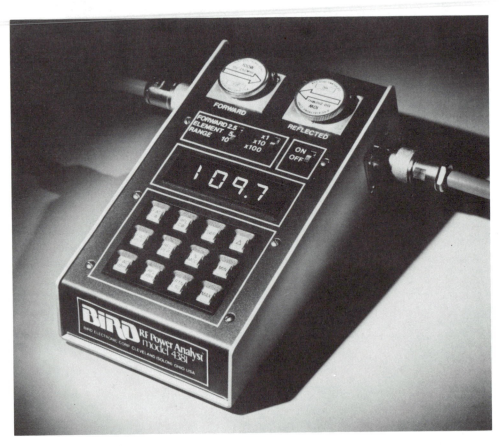

Figure 8–16 Two-directional wattmeter. (Courtesy of Bird Electronic Corp.)

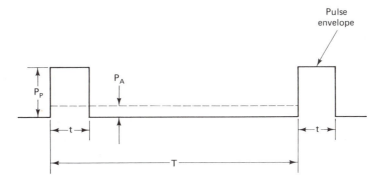

Figure 8–17   Relationship between peak power and average power.

T between two adjacent pulses. Then if the peak power is P_P and the average power is P_A,

$$\frac{P_P}{P_A} = \frac{T}{t} \quad \text{or} \quad P_P = P_A \times \frac{T}{t} \qquad (8\text{–}2)$$

A power meter will record the average power P_A but the peak power can be calculated by using equation (8–2).

Example 8–2   A microwave pulse has an average power of 225 W and a duration of 2.2 μs. If the time interval between pulses is 1750 μs, calculate the value of the peak power.

Solution   The peak power

$$\begin{aligned} P_P &= P_A \times \frac{T}{t} \\ &= 225 \text{ W} \times \frac{1750 \ \mu\text{s}}{2.2 \ \mu\text{s}} \\ &= \mathbf{180 \ kW} \end{aligned} \qquad (8\text{–}2)$$

8–4 ATTENUATION MEASUREMENTS

Some degree of attenuation is provided by various networks of microwave components and devices (Section 4–11). The amount of attenuation may be measured by the ratio of the network's output power P_o to its input power P_i. Since there is a loss involved, $P_o < P_i$ and their ratio P_o/P_i is less than unity; consequently, the attenuation when measured in decibels is a negative quantity. However, we should note that attenuation measurements strictly involve the power which actually enters the network. If there is a mismatch created by the network insertion, there will be a certain amount of reflected power, but this is not part of the attenuation. The total effect of the power reflected and the power attenuated is called the insertion loss.

If there is no mismatch, the insertion loss and the attenuation are the same. In a practical microwave system, the value of the SWR is usually less than 1.2, so that the reflection losses are small. If there is a high degree of attenuation, there is little difference between the insertion loss and the attenuation and the two terms can be used interchangeably. However, to be strictly correct, the insertion loss of a network is related to the difference in the power arriving at the terminating load with and without the network in the circuit.

Attenuation measurements are made primarily by using either the power ratio method or the RF substitution method. In its simplest form the power ratio method involves connecting an unmodulated microwave source to a bolometer mount and an appropriate power meter. After the power level is measured, the attenuating device is inserted between the source and the bolometer and the new (lower) reading is taken. The amount of attenuation may then be calculated from the ratio of the power readings. This method is normally limited to values of attenuation which range up to 20 dB. Attenuation measurements up to 30 or 40 dB can be achieved with a barretter detector mount whose output is fed to a SWR meter. However, the microwave source must be modulated and the RF power must be kept below 200 μW in order to operate over the required square-law detector characteristic. This method also requires that the detector's characteristic law be known over the measurement's complete frequency range and that the reflection effects in the system be negligible both with and without the attenuating device. To reduce the reflections between the source, the attenuator, and the termination, it is desirable to use additional attenuating pads which are well matched to the microwave system.

The absolute power method for determining the attenuation is not essential since we are really concerned with relative power levels. As long as we use a detector with a square-law characteristic, the power ratio can be found without knowing the absolute power levels. This is illustrated in the experimental setup of Figure 8–18.

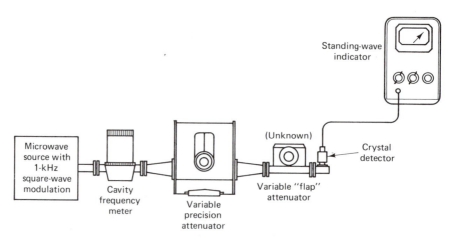

Figure 8–18 Experimental setup for attenuation measurements.

Here we wish to determine the amount of (unknown) loss provided by the flap attenuator. The steps involved with this power ratio method are as follows:

1. Set the precision attenuator to approximately 20 dB and the flap attenuator to 0 dB.

2. Energize the microwave source at the desired frequency and modulate its output with a 1-kHz square wave. Adjust the modulation for a peak reading on the SWR meter.

3. For the crystal detector to operate within its square-law characteristic, the microwave power level at the crystal should be 0.1 mW or less. This is achieved by switching the meter to its 30-dB range (Figure 8–7). The meter's gain control and, if necessary, the precision attenuator's control can then be used to bring the needle to the 0-dB mark.

4. Rotate the control of the flap attenuator to provide the "unknown" loss. The amount of this loss can then be read directly off the decibel scale of the SWR meter. If necessary, the experiment can be repeated for a number of required frequencies.

The main problem with this method lies in the assumption that both power levels (with and without the "unknown" attenuation) are capable of operating the crystal detector within its square-law characteristic. Greater accuracy can be provided by the RF substitution method, which involves the precision attenuator. The steps are as follows:

1. Initially adjust the precision attenuator to approximately 40 dB and the flap attenuator to 0 dB. Repeat the second and third steps of the power ratio method and record the reading of the precision attenuator.

2. Rotate the control of the flap attenuator to provide the "unknown" loss. The needle of the SWR meter will fall back but can be restored to the 0 dB mark by reducing the value of the precision attenuator. Then the difference between the new reading of the precision attenuator and its reading in step 1 is equal to the "unknown" loss.

The advantage of the substitution method is that the power level at the crystal detector is the same when taking both readings of the precision attenuator. However, there are additional errors caused by some degree of mismatch between the various components in a microwave system. For example, if the crystal detector is responsible for an SWR of 1.5 and the unknown attenuating device also creates an SWR of 1.5, the experimental error in determining the attenuation can be as high as 0.5 dB.

Example 8–3   An unmodulated microwave source is connected through an attenuator to a thermistor mount and an appropriate power meter. The reading of the power meter is 7 mW. When the attenuator is removed, the power reading rises to 23 mW. What is the amount of the decibel loss provided by the attenuator?

Solution

$$\text{Attenuation loss} = 10 \log \frac{P_1}{P_2} \quad \text{dB}$$

$$= 10 \log \frac{23 \text{ mW}}{7 \text{ mW}} = \mathbf{5.2 \text{ dB}}$$

Example 8–4   The RF substitution method is to be used in determining the loss provided by an attenuator. Without the insertion of the unknown loss, the reading of the SWR meter is set to the 0-dB mark when the precision attenuator reads 25 dB. After the insertion of the unknown loss, the precision atenuator reads 17 dB when the reading of the SWR meter is reset to the 0-dB mark. What is the value of the unknown loss?

Solution

$$\text{Unknown loss} = 25 - 17 = \mathbf{8 \text{ db}}$$

8–5 SWR MEASUREMENTS

The meaning and importance of a system's SWR value have been fully discussed in Section 2–4. When adjusting a microwave system the SWR measurement is the one most commonly required. Figure 8–19 illustrates a typical experimental setup for obtaining the SWR value. In this arrangement the slotted line is the essential unit. As the E probe is moved along the nonradiating slot to sample the standing-wave voltage, the output of the probe is detected and read on the SWR meter. The ratio

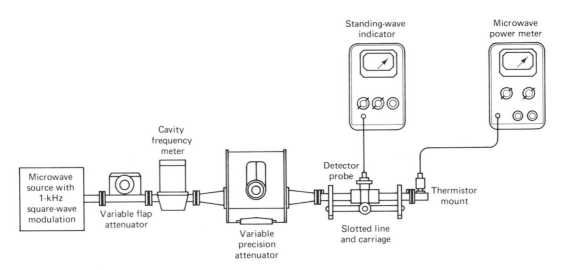

Figure 8–19   Experimental setup for SWR measurements.

of the maximum voltage at the peak to the minimum voltage at the null is the value of the SWR.

The microwave source must present a matched impedance to the system. Otherwise, the reflections from the network being tested will subsequently reflect off the source and cause the peaks and nulls to shift their positions on the standing-wave pattern.

To achieve the required match a pad can be inserted between the source and the rest of the system. Such a pad can be resistive and will absorb some of the reflections; however, it will also attenuate the input signal to the load under test. If the microwave source has too low a power level to permit additional attenuation, the resistive pad cannot be used because the null readings will be lost in the noise level. As an alternative solution, a ferrite isolator is used as a pad. Referring to Section 4–10, the isolator permits signals to pass in one direction with low attenuation, but absorbs most of the power in the reverse direction. The isolator is therefore an ideal pad for SWR measurements as well as for other laboratory work.

The standing-wave ratio as measured will be the SWR at the input to the load under test. As an example, the load could be a thermistor mount which is provided with its correct bias current. Under these conditions we could expect a low value for the SWR measurement.

The microwave source must put out a single frequency. If the signal is frequency modulated or if there are spurious signals present, each frequency will produce a different value of the SWR and will shift the locations of the peak and null positions. This will lead to erroneous readings, and therefore to ensure proper measurements, the source should be modulated only by a square wave (*not* a sine wave).

The procedure for obtaining a low SWR value is as follows:

1. Adjust both the flap attenuator and the precision attenuator to 10 dB.
2. Set the microwave source to the required frequency and modulate its output with a 1-kHz square wave. Supply the thermistor mount with the necessary bias current to obtain an indication on the power meter. Adjust the modulation for a maximum reading on the SWR meter. Slide the E probe along the nonradiating slot until the peak voltage reading is obtained and then vary the meter's gain control so that the needle lies on the SWR = 1 mark (0 dB). In this step it may be necessary to change the probe's depth of penetration in order to keep the meter on either the 30-dB or 40-dB scale.
3. Move the probe to the position of an adjacent voltage null and read the SWR value directly off the scale of the meter. Repeat the experiment for other frequencies as required.

The E probe inserted into the slotted line is a form of discontinuity which may be represented by its normalized conductance. To hold its reflective effect to a minimum, the depth of penetration should be kept as small as possible. At low standing-wave ratios this is no problem, but for higher values the probe must be inserted deeper so that the null value can be recognized above the noise level. The reflection effect of the probe is such that as it moves deeper, it introduces an error which produces a value of SWR lower than that which actually exists in the absence of the probe. This error becomes appreciable for values of SWR above 10, but below 10 the SWR may be read with reasonable accuracy by holding the probe depth to a minimum (Figure 8–20).

To obtain a high SWR value, we can now create a high degree of mismatch by removing all bias current from the thermistor mount. There is now a considerable difference between the peak and the null voltage values, so it would be difficult to remain on the detector's square-law characteristic if we still attempted to use the low-SWR procedure. Instead, we can turn to a method which involves the precision attenuator:

1. Slide the probe along the slotted line until the SWR meter reveals a null value. Adjust the precision attenuator for any convenient reference indication (e.g., 3 dB) on the SWR meter and record the reading of the precision attenuator.

2. Move the probe in either direction until the SWR meter shows the position of an adjacent peak. Readjust the precision attenuator until the SWR meter shows the same reference level as in step 1. Record the attenuator's new reading.

3. The SWR value in decibels is the difference between the two attenuator readings in steps 1 and 2. If this difference is N decibels, the SWR value is antilog $N/20$.

Since the same reference level is used for both attenuation readings, we have eliminated any difficulty associated with the detector's square-law characteristic; however, there remains the problem of the probe's depth of insertion.

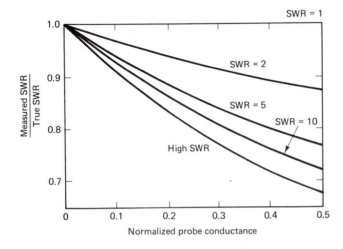

Figure 8–20 Effect of the probe's depth of penetration on the measured value of the SWR.

For high values of SWR, the twice-minimum method should be used:

1. The probe is moved to a null position and is inserted to a depth where the minimum voltage can be read without difficulty.
2. The gain control of the SWR meter is adjusted until the minimum reading is at the 3-dB mark.
3. The probe is now moved to one side and the position is found where the power level is twice the minimum amount; this condition will be indicated by the needle rising to the 0-dB mark. The position is read with vernier accuracy and is recorded as a distance d_1 cm on the scale of the slotted line.
4. The probe is next moved to the twice-power position on the other side of the null. This new position is recorded as the distance d_2 (cm). The value of the SWR is then given by

$$S = \frac{\lambda_g}{\pi(d_1 - d_2)} \qquad (8\text{–}3)$$

where λ_g is the guide wavelength (Section 3–4) in centimeters and is twice the distance on the slotted line between two adjacent null positions.

It should be noted that the probe is moved to the twice-power points. If a standing-wave indicator is used, it is calibrated in conjunction with a square-law detector, and consequently the ratio of the two voltage readings would be 1.414:1, which is equivalent to a change of 3 dB.

Reflectometers

The value of the SWR may be measured without a slotted line by using two directional couplers instead (Section 4–9). Figure 8–21 shows the reflectometer setup, in which the directional couplers need not have identical coupling characteristics.

If the input coupler has 40-dB coupling, the sample incident power from detector 1 will be 40 dB down or one ten-thousandth of the power out from the microwave source. Similarly, if the output coupler has 20-dB coupling, the sample output power from detector 2 will be one-hundredth of the reflected power. Due to the directional properties of the couplers, there is no interaction between the two readings. Comparison of the outputs at the two detectors gives the square of the voltage reflection coefficient P (Section 2–4). We can then determine the value of the SWR from the relationship $S = (1 + P)/(1 - P)$ [equation (2–41)].

The two detector outputs can be combined in a single instrument which will read their ratio directly. This instrument is called a ratiometer and its indicator can be calibrated to read directly in terms of the voltage reflection coefficient and SWR values. If the distance from the reflectometer to the load is accurately measured in terms of the guide wavelength, the normalized load impedance may be found

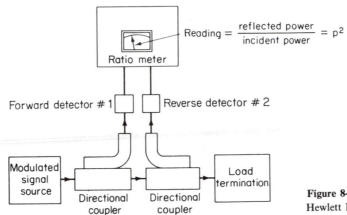

$$\text{Reading} = \frac{\text{reflected power}}{\text{incident power}} = \rho^2$$

Figure 8–21   Reflectometer. (Courtesy of Hewlett Packard Company.)

from a knowledge of the SWR value (Section 2–6, Example 2–11). However, it is not easy, for example, to determine the exact location of a minute thermistor bead in its mount.

Example 8–5   The precision attenuator is used to measure a high SWR value. Under matched conditions the reading of the precision attenuator is 11 dB and the SWR meter at a null position is set to the 3-dB mark. When the mismatch is created, the SWR reading at an adjacent peak is also adjusted to the 3-dB mark by raising the value of the precision attenuator to 29 dB. What is the SWR value?

Solution   The change in the precision attenuator's reading, $N = 29 - 11 = 18$ dB. The SWR value

$$
\begin{aligned}
S &= \text{antilog}\,\frac{N}{20} \\
  &= \text{antilog}\,\frac{18}{20} \\
  &= \text{antilog}\,0.9 \\
  &= \mathbf{7.9}
\end{aligned}
$$

Example 8–6   The twice-minimum power points method is used to determine the SWR value on a waveguide. The separation between two adjacent nulls is 3.14 cm. If the separation between the twice-minimum power points is 2.3 mm, what is the value of the SWR?

Solution   The guide wavelength

$$\lambda_g = 2 \times 3.14 = 6.28 \text{ cm}$$

The SWR value

$$
\begin{aligned}
S &= \frac{\lambda_g}{\pi(d_1 - d_2)} \\
  &= \frac{6.28 \text{ cm}}{\pi \times 0.23 \text{ cm}} \\
  &= \mathbf{8.7}
\end{aligned}
\tag{8–3}
$$

8–6 IMPEDANCE MEASUREMENTS

We have already discussed (Section 2–6, Example 2–12) the method of determining the impedance value of a terminating load by using the slotted-line technique. Briefly summarizing, the load is connected and the value of the SWR is determined. During this measurement the position of the voltage null is located on the slotted line. The load is now disconnected and is replaced by a short circuit; it is then observed that the null shifts a certain distance toward either the load or the source (according to whether the load is capacitive or inductive). The amount of the shift is expressed

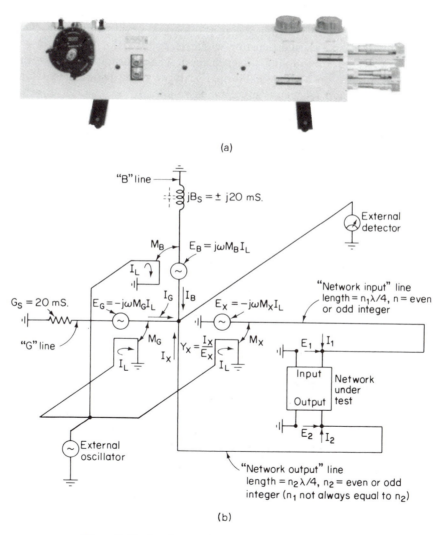

(a)

(b)

Figure 8–22 Immitance bridge. (Courtesy of GenRad, Inc.)

as a fraction of the guide wavelength λ_g (which is twice the distance between two adjacent nulls), and from this information the normalized value of the load impedance may be found. If there is zero shift, the load is entirely resistive with a value which is less than the impedance of the waveguide. By contrast, if the shift is exactly $\lambda_g/4$, the load is again a pure resistance, but its value is greater than the waveguide impedance.

At the low end of the microwave region (1.0 to 1.5 GHz) it is possible to use an immittance (impedance/admittance) bridge (Figure 8–22a). Basically, the circuitry consists of three identical loops which are fed from a common single generator and are magnetically coupled to three lines (Figure 8–22b). The first of these lines contains a resistance/conductance standard, the second includes a reactance/susceptance standard, and the third is connected to the load or device under test. These three lines terminate in an external detector which is adjusted for null reading by setting the correct degrees of coupling for each loop. The information on the three settings is then combined on a single calibrated scale which provides the required resistance/conductance and reactance/susceptance values.

> **Example 8–7**   With a load connected the SWR value on a slotted line is equal to 3.5 and the distance between two adjacent nulls is 2.7 cm. When the load is replaced by a short circuit, the position of a null shifts through a distance of 1.8 cm toward the load. What is the normalized value of the load impedance?
>
> *Solution*   The guide wavelength
>
> $$\lambda_g = 2 \times 2.7 = 5.4 \text{ cm}$$
>
> Then the
>
> $$\text{null shift} = \frac{1.8}{5.4} \times \lambda_g = 0.333\lambda_g$$
>
> Figure 8–23 shows the 3.5 SWR circle and the null shift of $0.333\lambda_g$ toward the load. At point X, the normalized load value is **0.92 + j1.3.**

8–7 PHASE-SHIFT MEASUREMENTS

To determine the phase shift that occurs through a network, we must find the network's electrical length. At the same time we must have an approximate idea of this length, since it is impossible to distinguish between one quarter-wavelength and five quarter-wavelengths as far as phase is concerned (Section 2–6).

The simplest method of measuring the phase shift uses a slotted line and a short circuit. The short is placed at the end of the line and a reference null is noted. Then the unknown network is attached to the slotted line and the short is placed at the output of the network; as a result, the null will shift to a new position. Since this null must be a whole number of half-wavelengths from the short, it is possible to calculate the exact electrical length of the network. For example, suppose that at

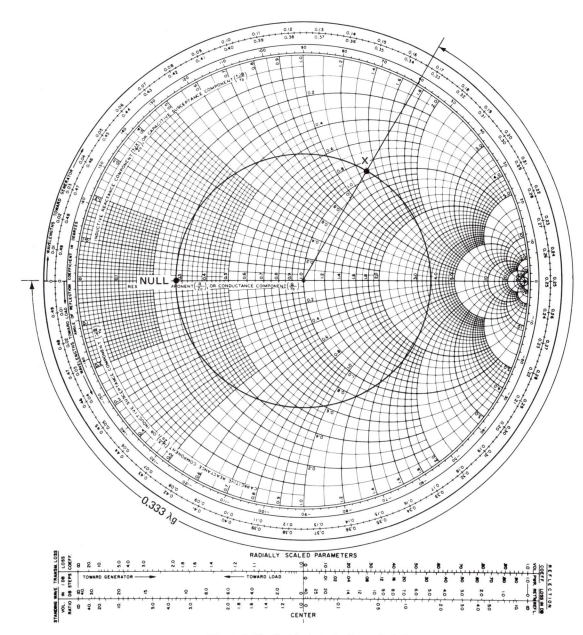

Figure 8–23   Smith chart for Example 8–7.

3.2 GHz a reference null is found at 15.42 cm on the slotted line. The corresponding wavelength at this frequency is $30/3.2 = 9.375$ cm.

If a network that is approximately two wavelengths long, is placed at the end of the line, and with the short across the output of the network the null is found to shift to 14.27 cm, the short has therefore apparently moved $15.42 - 14.27 = 1.15$ cm, which is equivalent to $1.15/9.375 = 0.123\lambda$. Since this apparent movement is in the same direction as the short was actually moved, it is added to the approximate number of half-wavelengths in the network. The total electrical length is then 2.123 wavelengths. This can be multiplied by 2π to give the phase shift in radians or by $360°$ to give the phase shift in degrees. Consequently, the phase shift is $2.123\lambda = 13.34$ rad $= 764.3°$.

Sometimes we are required to find the change in phase shift of a component or network as some parameter is varied. For example, the phase shift changes as a piece of dielectric material is moved from the edge of a waveguide toward its middle. To discover the change in the phase shift, the dielectric's position is varied; the component is attached to the slotted line and is terminated in a short circuit. The dielectric piece is first positioned against one wall of the waveguide, and the position of the null is noted. The dielectric is now moved a small distance toward the center. This in effect increases the electrical length of the component so that the short circuit appears to be farther away. The null's position will follow since it must stay an integral number of half-wavelengths away from the short circuit. The electrical distance that the null moves on the slotted line is the change, d, in the electrical length, which corresponds to the movement of the dielectric. The experiment can be repeated so that we can obtain a graph of the phase shift versus the dielectric's position. Since this is done in small steps, it is possible to calibrate more than $360°$ of phase shift in this way, even though the minimum positions repeat every half-wavelength. This is the principle which governs one of the most common types of phase shifter (Section 4–13, Figure 4–58a).

In the construction of such a phase shifter, a dielectric vane is eccentrically mounted on a shaft and protrudes into the waveguide through a slot. The vane's shape is correctly tapered to introduce minimum reflection effects over the required frequency range. The rotation of the shaft changes the amount of the dielectric in the waveguide and therefore its effective electrical wavelength. As an alternative, the vane may be attached to supporting rods and then moved across the guide's wide dimension.

Some components are unidirectional or nonreciprocal. For example, if it is necessary to measure the phase shift through a ferrite unit (Section 4–10), the short-circuit method cannot be used, since it relies on the signal passing through the unit, being reflected from the short circuit, and then passing back through the unit. For a ferrite component, the phase shift in one direction differs from its value in the reverse direction. Therefore, to determine each of the two phase shifts separately, a different approach is necessary. The setup for measuring a one-way phase shift is shown in Figure 8–24. The microwave power is fed into a splitter which divides the input into two approximately equal signals. Pads are placed in each path to make sure

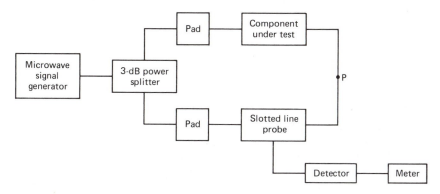

Figure 8–24   Experimental setup for the measurement of a one-way phase shift.

that they are properly matched and to prevent interaction. Nonreciprocal pads are preferable, but ordinary resistive attenuators are satisfactory.

In the slotted line, two approximately equal signals are traveling in opposite directions. Since these are at the same frequency, there is no way to distinguish them from a single signal which suffers total reflection. As a result, there will be deep nulls at half-wavelength intervals. A typical problem might be to determine the variation in the phase shift of a ferrite component as a function of the current through its exciting coil. First, the position of a null is located with zero current. Then as the current is varied in small increments, new null positions are recorded. If the component under test becomes a half-wavelength longer (a shift of 180°), the null will move only a quarter-wavelength toward the component. This is because the other path to the probe has lengthened by a quarter-wavelength while the path through the component is shortened by the same amount. The net change is therefore a half-wavelength, with the result that the phase shift using this setup is twice the shift in the null position. In other words, the shift in centimeters, d, is divided by a half-wavelength instead of by a full wavelength and the answer is then multiplied by 2π or 360° to measure the phase shift in radians or degrees.

An alternative method is to place a calibrated phase shifter at point P in Figure 8–24. Then as the phase through the test piece is varied, the calibrated unit is changed, with the position of the null remaining constant. The required phase shift can then be read directly from the calibrated phase shifter. An example of such a precision phase shifter is described in Section 4–13.

Example 8–8   The phase shift introduced by a waveguide component is measured with the aid of a slotted line and a short circuit. The distance between two adjacent nulls is 2.72 cm and there is a 1.38-cm change in the reference null's position when the waveguide component is inserted. If the amount of the phase shift is known to be of the order of 90°, calculate its value.

Solution   The guide wavelength,

$$\lambda_g = 2 \times 2.72 = 5.44 \text{ cm}$$

Then,

$$\text{phase-shift value} = \frac{360 \times d}{\lambda_g} \quad \text{deg}$$

$$= 360° \times \frac{1.38}{5.44}$$

$$= \mathbf{91.3°}$$

Example 8–9 A slotted line and a short circuit are used to measure the amount of phase shift introduced by a waveguide component. Before inserting the component, the guide wavelength is measured and found to be 6.73 cm; the position of the reference null is located at 13.28 cm (Figure 8–5). When the waveguide component (electrically about one half-wavelength long) is inserted, the reference null moves its position to 11.74 cm. What is the phase shift of the waveguide component?

Solution The shift in the null is in the opposite direction to the movement of the short:

$$\text{phase shift} = 180° - 360° \times \frac{d}{\lambda_g}$$

$$= 180° - 360° \times \frac{13.28 - 11.74}{6.73} \qquad (8\text{--}5)$$

$$= 180° - 82.4° = \mathbf{97.6°}$$

Example 8–10 The setup of Figure 8–24 is used to measure the phase shift of a *ferrite* component. The guide wavelength is 5.7 cm. When the ferrite component is inserted, the reference null shifts position by 0.12 cm. If the phase shift of the ferrite component is less than 90°, determine the amount of the shift.

Solution Owing to its "one-way" property, the phase shift of the ferrite component is

$$720° \times \frac{d}{\lambda_g} = \frac{720° \times 0.12}{5.7} = \mathbf{15.2°}$$

8–8 NOISE-FACTOR MEASUREMENTS

In microwave communications, the weakest signal that can be detected is usually determined by the amount of the noise added by the receiving system. Consequently, any decrease in the amount of noise generated in the receiving system will produce an increase in the output signal-to-noise ratio equivalent to a corresponding increase in the received signal. From a standpoint of performance, an increase in the signal-to-noise ratio by reducing the amount of noise in the receiver is more economical than increasing the received-signal level by raising the power of the transmitter. For example, a decrease of 5 dB in receiver noise is equivalent to increasing the transmitter power by a ratio of 3:1.

The noise at the output of a receiver or an amplifier is the sum of the noise arising from the input source and the noise contributed by the receiver or the amplifier itself. The noise factor is the ratio of the actual output noise power of the device to the noise power that would be available if the device were perfect and merely amplified the thermal noise of the input source without contributing any noise of its own. The noise figure F is the noise factor expressed in decibels; it follows that $F = 10$ log(input signal-to-noise ratio/output signal-to-noise ratio). The noise figure of a receiver may be measured by using a signal-generator input and an output-power meter or square-law detector. However, this method is time consuming and has the added disadvantage that the effective gain–bandwidth characteristics of the device must be determined. Moreover, the available signal power may be difficult to determine accurately at the low levels involved.

Automatic noise-figure measurements utilize standard broadband noise sources; these supply a noise spectrum with a known power which is virtually independent of frequency, so that we can overcome the drawbacks of the signal-generator method. Temperature-limited diodes are suitable as excess noise sources, but at microcomponent frequencies gas discharge tubes which are positioned in suitable waveguide sections provide a greater accuracy and reliability.

Automatic noise-figure measurements depend on the periodic insertion of a known excess noise power at the input of the device under test. Subsequently, detection of the noise power in the later intermediate-frequency stages of the device results in a pulse train of two power levels. The power ratio of these two levels contains the desired noise-figure information. The various contributions of noise power to this output-pulse ratio are shown in Figure 8–25.

The available noise power from the reference load is

$$n = kTB \qquad W \qquad\qquad (8\text{–}6)$$

where

$$k = \text{Boltzmann's constant } (1.37 \times 10^{-23} \text{ W.s/K})$$
$$T = \text{temperature of reference load (K)}$$
$$B = \text{bandwidth of measuring system (Hz)}$$

Excessive noise power added by the noise source is based on the effective fired temperature of the source. An argon gas discharge, for example, is 15.2 dB above the reference temperature power. Then the total noise-power output of the receiver with the noise source "off" is

$$N_1 = GkTB + \text{RCVR} \qquad\qquad (8\text{–}7)$$

where G is the receiver power gain. The total noise-power output of the receiver with the noise source "on" is

$$N_2 = GkTB + \text{RCVR} + \text{excess} \times G \qquad\qquad (8\text{–}8)$$

The noise factor F' as defined above (where T_0 equals 290 K) is

$$F' = \frac{GkT_0B + \text{RCVR}}{GkT_0B} \quad \text{or} \quad \frac{\text{total noise output from device}}{\text{output power if noiselessly amplified}} \qquad (8\text{–}9)$$

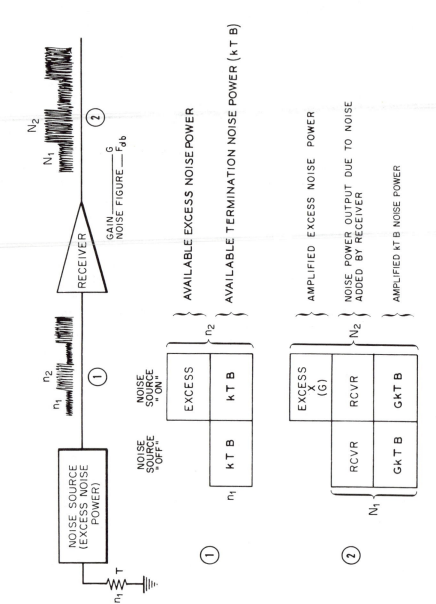

Figure 8-25 Analysis of noise power components.

This yields

$$\text{RCVR} = (F' - 1)GkT_0B \tag{8–10}$$

where RCVR is the noise output power contributed by the receiver. Also, the excess noise power from the gas discharge tube at the point is

$$\text{excess noise power} = \frac{T_2 - T_0}{T_0}kT_0B \tag{8–11}$$

where T_2 is the effective fired temperature of the source. Then the ratio at the output is

$$\frac{N_2}{N_1} = \frac{GkT_0B + (F' - 1)GkT_0B + \dfrac{T_2 - T_0}{T_0}GkT_0B}{GkT_0B + (F' - 1)GkT_0B} \tag{8–12}$$

From equations (8–7), (8–8), (8–10), (8–11), and (8–12),

$$F' = \frac{\dfrac{T_2 - T_0}{T_0}}{\dfrac{N_2 - N_1}{N_1}} \tag{8–13}$$

Note that the gain–bandwidth factor (GB) has disappeared. Finally,

$$\text{Noise figure, } F = 10 \log\left(\frac{T_2}{T_0} - 1\right) - 10 \log\left(\frac{N_2}{N_1} - 1\right) \quad \text{dB} \tag{8–14}$$

The first term is a known quantity, expressed in decibels of excess-noise ratio. For an argon discharge, the excess-noise ratio is 15.2 dB; then

$$F = 15.2 - 10 \log\left(\frac{N_2}{N_1} - 1\right) \quad \text{dB} \tag{8–15}$$

Therefore, the ratio N_2/N_1 contains the desired noise-figure information so that noise-figure meters can measure the noise figure as a function of this ratio. To make noise-figure measurements, the appropriate noise source, and the receiver or amplifier under test, are connected as shown in Figure 8–26. The noise-figure meter's square-wave modulates the noise source at the rate of about 500 Hz and measures the noise figure by comparing the noise output of the device under test when the noise source is "off," to the noise output when the noise source is "on."

The input circuitry consists of a gated-tuned amplifier which operates at either of two frequencies, 30 or 60 MHz, as selected by a front-panel switch. The output from a tuned amplifier is detected, amplified, and alternately applied to two gated integrators. When the noise source is "on," the combined noise power from the noise source and the device under test is amplified by the tuned amplifier, detected, and passed through the AGC integrator. The time constant of the AGC voltage

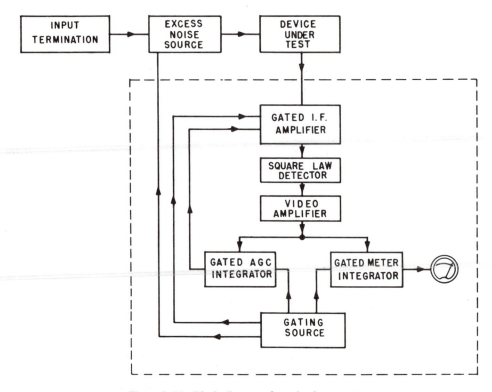

Figure 8–26 Block diagram of a noise-figure meter.

applied to the amplifier is long enough to hold the gain of the amplifier at the same level whether the noise source is "on" or "off."

When the noise source is tuned off, the combined noise power from the source and the device under test is amplified, detected, passed through the meter integrator, and displayed on the meter. Because of the AGC action, the meter deflection is proportional to the ratio of the noise powers (source "on" and source "off"), and since the additional noise from the noise source (excess noise) is accurately known, the meter face is calibrated to read the noise figure directly in decibels.

The AGC action, in addition to establishing a reference against which noise-figure measurements can be made, provides a wide (50-dB) input operating range and also eliminates the necessity for periodic recalibration of the noise-figure meter. AGC voltages appear on a pair of terminals at the rear of the instrument to facilitate measurements which require an indication of the gain of the system in relation to the changes in the noise figure.

Example 8–11 An amplifier has a 5-dB noise figure and an input signal-to-noise ratio of 27:1. What is the value of its output signal-to-noise ratio?

Solution   The noise figure,

$$F = 10 \log \left(\frac{\text{input signal-to-noise ratio}}{\text{output signal-to-noise ratio}} \right) \quad \text{dB}$$

Therefore,

$$5 = 10 \log \left(\frac{27}{\text{output signal-to-noise ratio}} \right)$$

$$\text{output signal-to-noise ratio} = \frac{27}{\text{antilog } (5/10)}$$

$$= \mathbf{8.5}$$

8–9 SWEPT-FREQUENCY MEASUREMENTS

All the measurements described so far have been related to a single microwave frequency. If we needed to determine the values of the SWR, the phase shift, the attenuation, and so on, for a number of different frequencies, we would have to change the frequency of the microwave source and plot the new values for each frequency setting. This is called a point-to-point method of measurement and is carried out whenever accurate quantitative measurements are required. However, for rapid evaluation of a component, qualitative results may be sufficient. For example, if the specification states that the value of the SWR must be under 1.25, it is not necessary to know whether it is 1.10 or 1.20 as long as it meets its specification. In such cases, swept-frequency measurements are much faster than point-to-point measurements and are equally reliable.

The heart of the swept-frequency technique is the sweep oscillator, which electronically changes the microwave frequency at a rapid rate. Sweep oscillators usually cover an octave, so that the ratio of the highest output frequency to the lowest is 2:1, for example, 12 to 6 GHz. They have provisions to sweep any smaller portion of the octave band if the whole octave is not of interest. The sweep rate can usually be varied from one sweep in more than 2 min up to 60 or more sweeps in 1 s.

Since it is not feasible to tune every element in the circuit to each microwave frequency as the oscillator sweeps through its cycle, all components used must have broad-band characteristics. The final display is usually an oscilloscope or an X-Y recorder which displays the output as a function of frequency. In both cases, the horizontal trace is synchronized with the sawtooth voltage which sweeps the microwave oscillator. The signal output is detected in the usual way, fed first into an amplifier and afterward to the vertical motion of the oscilloscope or recorder. The trace will then be a plot of the signal output as a function of the frequency.

The broad-band components used with the sweep oscillator never have a perfectly flat frequency response. For example, the response of a detector is usually optimized at two or three frequencies in the band in order to simulate multiple resonant tuning.

In addition, the swept oscillator itself maintains a minimum power output, but this output varies with frequency. Consequently, we must calibrate the face of the oscilloscope with a known mismatch or a known attenuation.

Figure 8–27 shows a setup for making SWR measurements with a sweep oscillator. The directional coupler is reversed so that only the reflected signal arrives at the detector. Therefore, the oscilloscope presentation will show the amount of reflection as a function of frequency. Before making this test, the oscilloscope must be calibrated by replacing the test piece and its load in Figure 8–27 with certain standard reflections. For example, it might be replaced with a network that has a known SWR of 1.1 : 1 over the entire frequency band. The corresponding trace on the oscilloscope can then be marked with a grease pencil. Similar traces are drawn from mismatches that bracket the expected mismatch of the test component and its SWR specification.

Another method avoids using standard mismatches, which are usually difficult to obtain. Instead, a calibrated attenuator is used to reduce the reflection from a short circuit to the equivalent reflection from the desired SWR. This is done by converting the calibrated SWR values to their equivalent return losses in decibels (return loss $= 20 \log [(S + 1)/(S - 1)]$; Section 2–6). For example, if the SWR, $S = 1.25$ the return loss is

$$20 \log \left(\frac{1.25 + 1}{1.25 - 1}\right) = 20 \log 9 = 19.09 \text{ dB}$$

The experimental setup for the calibrated attenuator method is shown in Figure 8–28. The procedure is:

1. After energizing the circuit, short out the unit under test so that total reflection occurs.

2. Set the precision attenuator at the number of decibels that reduces the SWR from a theoretical value of infinity to one of the desired SWR levels. Record its SWR versus frequency response (Figure 8–29) by using a grease pencil on the face of the oscilloscope and repeat the procedure for each of the SWR values required for calibration purposes.

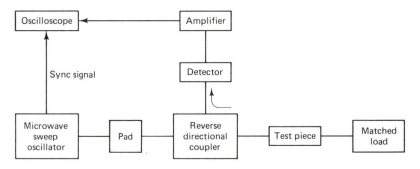

Figure 8–27   SWR measurements with a microwave swept oscillator.

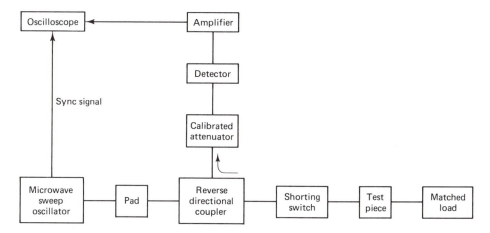

Figure 8–28   SWR measurements with a microwave swept oscillator and a calibrated attenuator.

3. Open the shorting switch and set the calibrator attenuation to zero so that the reflection from the unit under test reaches the detector. Record its SWR/ frequency response.

In our example, calibration response curves are shown for SWR values of 1.1, 1.2, and 1.3. The amplitude variations in each curve is caused by fluctuations that occur in the detector and the outputs from the microwave source and the directional coupler.

The trace for the unknown SWR value after calibration shows a value near

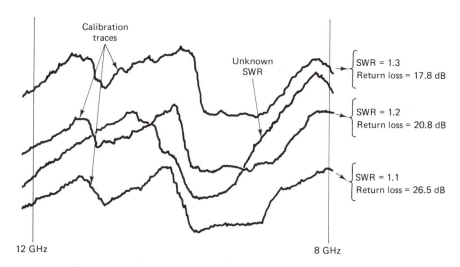

Figure 8–29   CRO SWR display using a microwave swept oscillator.

1.3 at the low-frequency end, dropping to about 1.15 at the center and another peak of about 1.24 near 11 GHz. If the maximum allowed SWR value is 1.3, the component would satisfy its specification. Obviously, the same sort of presentation would appear if a recorder were used instead of an oscilloscope. We should note that the calibration does not have to be repeated, but should hold for subsequent measurements as long as the gain controls and so on are not changed.

If a reflectometer setup, as shown in Figure 8–21, is used with a sweep oscillator, the output of the two detectors can be fed into a ratiometer, the output of which is displayed on the synchronized oscilloscope. Since the output of the ratiometer is the ratio of the reflected power to the incident power, any variations in the power output of the microwave source will cancel out and therefore the calibration lines of the oscilloscope presentation will be almost straight lines.

Attenuation can also be measured by using swept frequency techniques. A simple setup is shown in Figure 8–30. Using a calibrated attenuator, the number of traces are recorded on the oscilloscope; these correspond to certain values of attenuation which will bracket the expected value. For example, if the specification calls for a minimum of 20 dB of attenuation, four traces would be made with values of attenuation such as 15, 20, 25, and 30 dB. The calibrated attenuator is then removed and the unknown attenuation substituted. The new trace that appears on the face of the oscilloscope can be interpreted as a plot of attenuation versus frequency. Strictly speaking, this trace represents insertion loss rather than attenuation, but as discussed previously, the difference between the two is negligible when the attenuation is high. As with SWR measurements, a reflectometer can be used to smooth the lines, but this is not a necessity.

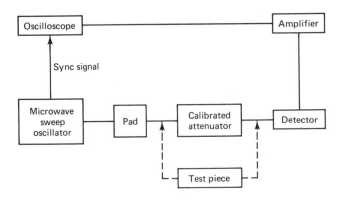

Figure 8–30 Attenuation measurements using a microwave swept oscillator.

PROBLEMS

BASIC PROBLEMS

8-1. A slotted line is used in association with an X-band microwave source. When the line is terminated by a short circuit, adjacent nulls are found at positions which are shown as 9.27 cm and 11.05 cm. What is the value of the guide wavelength?

8–2. In a microwave system the power is modulated by a rectangular pulse whose duration is 1.5 μs. The average power is measured and is equal to 200 W. If the pulse repetition rate is 500 pulses per second, what is the value of the peak power?

8–3. An unmodulated microwave source is connected to a bolometer mount and an appropriate power meter. The microwave power level reads as 25 mW. When an attenuating device is inserted between the source and the bolometer, the power reading falls to 5 mW. What is the amount of attenuation (in decibels) provided by the device?

8–4. An "unknown" attenuation is to be measured with the aid of a precision attenuator. Without the insertion of the "unknown" loss, the needle of the SWR meter is at the 0-dB mark while the precision attenuator reads 38 dB. After the insertion of the unknown loss, the reading of the SWR meter is restored to the 0-dB mark when the precision attenuator reads 25 dB. What is the decibel loss provided by the "unknown" attenuation?

8–5. In measuring the loss provided by an attenuator, the microwave source is modulated by a 1-kHz square wave and the reading of the SWR meter is peaked to 0 dB with the range switch set on 30 dB. When the attenuator is inserted, the range is changed to 40 dB and the needle reads a value of 2 on the SWR scale (Figure 8–8). How much loss is provided by the attenuator?

8–6. A slotted line is used to determine the SWR value on a waveguide. Adjacent null positions are located at 13.31 cm and 15.45 cm. If the separation between the twice-minimum power points is 2 mm, what is the value of the SWR?

8–7. When measuring the SWR value by means of a slotted line, the microwave source is modulated by a 1-kHz square wave and the SWR meter is peaked to the SWR = 1 (0 dB) mark on the 30-dB range. When the probe is moved to an adjacent null, the range is changed to 40 dB and the needle reads 2.2 on the upper SWR scale (Figure 8–8). What is the value of the SWR?

8–8. The phase shift of a network is measured with the aid of a coaxial slotted line and a short circuit. Before the insertion of a network a reference null is located at 17.68 cm. When the network is inserted, the null shifts to 15.82 cm. If the frequency is 3.3 GHz and the network is electrically less than one quarter-wavelength long, what is the phase shift of the network in degrees?

8–9. A slotted line and a short circuit are used to calibrate a dielectric phase shifter. Initially, a reference null is located and the dielectric is then moved by a small amount from the edge of the waveguide toward the center. As a result, the null is shifted through a distance of 0.72 cm. If the distance between two adjacent nulls is 2.15 cm, what is the amount of the phase shift (in degrees) produced by the movement of the slab?

8–10. In the experimental setup of Figure 8–24, a precision phase shifter at point P reads 72° before the component under test is inserted. After insertion, the reference null is restored to its initial position by adjusting the precision phase shifter to 28°. If the component is known to be about one wavelength long electrically, what is its phase shift in degrees?

ADVANCED PROBLEMS

8–11. A slotted line is used to determine the frequency of an X-band microwave source. When the termination is replaced by a short circuit, adjacent nulls are found at positions which are shown as 11.47 cm and 13.78 cm on the slotted line. If the inner wide dimension of the guide is 2.285 cm, what is the frequency of the microwave source?

8–12. A slotted line is used to check the SWR value on a 50-Ω air-dielectric cable. The signal frequency is 2 GHz and the SWR value is high, so that the twice-minimum power method is chosen. If the distance between the positions of twice-minimum power is 0.53 cm, what is the value of the SWR on the line? What is the magnitude of the voltage reflection coefficient?

8–13. Two identical directional couplers are placed in a waveguide to sample the incident and the reflected power. The meter readings show that the power level of the reverse coupler is 10 dB down from the level of the forward coupler. What is the value of the SWR on the waveguide?

8–14. A slotted line is used to determine the value of a load which is terminating a waveguide. The SWR on the line has a value of 3, while the guide wavelength is found to be 3 cm. When the load is replaced with a short, the position of a particular null shifts 0.3 cm toward the generator. What is the normalized value of the load? (*Hint*: Use the Smith chart.)

8–15. An experimental setup is illustrated in Figure 8–31. When the short is put across the waveguide, the reading of the power meter is +5 dB. When the short is removed, the meter reading is −9.5 dB. What are the magnitudes of the reflection coefficient and the SWR?

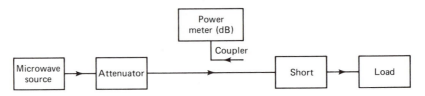

Figure 8–31   Experimental setup for Problem 8–15.

8–16. In Figure 8–32 a load is being tested with the aid of a reflectometer. Determine the values of the return loss, the reflection coefficient, and the SWR.

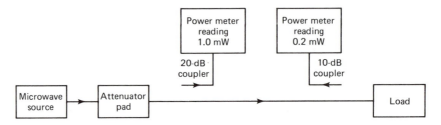

Figure 8–32   Experimental setup for Problem 8–16.

8–17. The experimental setup of Figure 8–24 is used to measure the phase shift created by the insertion of a ferrite component inside a waveguide. When the frequency is 10 GHz, the guide wavelength is 4 cm. If it is found that the null is shifted through a distance of 1.1 cm, what is the phase shift in degrees?

8–18. To measure a high SWR value with the aid of a slotted line, the microwave source is initially modulated by a 1-kHz square wave and the SWR meter reading at a null position is adjusted to the 3-dB mark under matched conditions; the reading of the precision attenuator is then 15 dB. When the mismatch is created, the adjacent peak is located on the slotted line and the precision attenuator is adjusted to 28 dB to bring the SWR reading back to the 3-dB mark. What is the value of the SWR?

8–19. The phase shift of a network is measured with the aid of a coaxial slotted line and a short circuit. Before the insertion of the network a reference null is located at 14.68 cm. When the network is inserted, the null shifts to 15.53 cm. If the frequency is 2.5 GHz and the network is electrically about one wavelength long, what is the phase shift of the network in radians?

8–20. A device has an input signal-to-noise ratio of 7:1. If the device's noise figure is 0.3 dB, what is its output signal-to-noise ratio?

INDEX

426